Principles of Construction Safety

Allan St John Holt
BA, FIOSH, RSP

Foreword by
Sir Frank Lampl

**Blackwell
Science**

© 2001 Mei Wenti Ltd
Blackwell Science Ltd
Editorial Offices:
Osney Mead, Oxford OX2 0EL
25 John Street, London WC1N 2BS
23 Ainslie Place, Edinburgh EH3 6AJ
350 Main Street, Malden
   MA 02148 5018, USA
54 University Street, Carlton
   Victoria 3053, Australia
10, rue Casimir Delavigne
   75006 Paris, France

Other Editorial Offices:

Blackwell Wissenschafts-Verlag GmbH
Kurfürstendamm 57
10707 Berlin, Germany

Blackwell Science KK
MG Kodenmacho Building
7-10 Kodenmacho Nihombashi
Chuo-ku, Tokyo 104, Japan

Iowa State University Press
A Blackwell Science Company
2121 S. State Avenue
Ames, Iowa 50014-8300, USA

First published 2001

Set in 9/14 Trebuchet
by DP Photosetting, Aylesbury, Bucks
Printed and bound in Great Britain by
MPG Books Ltd, Bodmin, Cornwall

The Blackwell Science logo is a trade mark of Blackwell Science
Ltd, registered at the United Kingdom Trade Marks Registry

DISTRIBUTORS

Marston Book Services Ltd
PO Box 269
Abingdon
Oxon OX14 4YN
(Orders: Tel: 01235 465500
                  Fax: 01235 465555)

USA
Blackwell Science, Inc.
Commerce Place
350 Main Street
Malden, MA 02148 5018
(Orders: Tel: 800 759 6102
                       781 388 8250
              Fax: 781 388 8255)

Canada
Login Brothers Book Company
324 Saulteaux Crescent
Winnipeg, Manitoba R3J 3T2
(Orders: Tel: 204 837-2987
              Fax: 204 837-3116)

Australia
Blackwell Science Pty Ltd
54 University Street
Carlton, Victoria 3053
(Orders: Tel: 03 9347 0300
              Fax: 03 9347 5001)

A catalogue record for this title is available
from the British Library

ISBN 0-632-05682-7

Library of Congress
Cataloging-in-Publication Data
is available

The publishers acknowledge with thanks the permission and
co-operation given by IOSH Services Limited in allowing
reproduction of selected material from 'Principles of Health
and Safety at Work'.

For further information on
Blackwell Science, visit our website:
www.blackwell-science.com

Contents

Foreword

Environment, health and safety are already crucial issues in the upper echelons of the construction industry. Sadly, that is the exception rather than the rule, so, if our industry is to prosper in the twenty-first century, everyone at every level needs to understand the importance of these issues and implement the practices that will safeguard both people and our planet.

This book is intended to fill the gap between a technical manual and the many topic guidance notes such as those published by the Health and Safety Executive. Uniquely, the amount of space devoted to construction EH & S management is the same as that given over to techniques and to the law. Experience shows that time spent on pre-construction activity and planning is amply repaid during the construction phase of work, when safety issues can be much more difficult to resolve.

What is needed is to embed these vital factors as values and as a part of all that our industry does, and not to regard them merely as optional extras where time allows.

The author draws upon his lifetime experience in construction-related health and safety to provide the information and the background material important for a full understanding of the issues involved. The book is aimed at every participant in the construction industry needing information and guidance on current and future best practice.

A positive change of attitudes among all players in our industry is vital. I welcome this book as a valuable contribution to that goal. Allan, who has worked with us across the world to further the cause of site safety, is well placed to make it happen.

Sir Frank Lampl
President, Bovis Lend Lease

Abbreviations

ANSI	American National Standards Institute	MHSWR	Management of Health and Safety at Work Regulations 1999
CAWR	Control of Asbestos at Work Regulations 1987	MORT	Management oversight and risk tree
CDM	Construction (Design and Management) Regulations 1994	NADO	Notification of Accidents and Dangerous Occurrences (Regulations)
COSHH	Control of Substances Hazardous to Health Regulations	NEBOSH	National Examination Board in Occupational Safety and Health
CSCS	Construction Skills Certification Scheme		
CTA	Certificate of Training Achievement	PMEP	Project Major Emergency Plan
FMEA	Failure modes and effects analysis	PPE	Personal protective equipment
HSE	Health and Safety Executive	RIDDOR	Reporting of Injuries, Diseases and Dangerous Occurrences Regulations 1995
IOSH	Institution of Occupational Safety and Health		
JSA	Job safety analysis	RPE	Respiratory protective equipment
MEWPS	Mobile elevating work platforms		

Introduction

This book is not intended simply as a 'how to' guide. It aims to be the first in its field to introduce the reader to the principles behind the often-familiar requirements of law, and to inform about standards of international best practice for the control of risks in the construction process. It has been designed for the student and the professional alike, the manager and the managed. Some of its contents have, to the best of my knowledge, received little or no coverage, while others have been written up copiously. Books of this kind typically place much emphasis on the law and particular control measures for common hazards, but comparatively little on best practice in the construction safety management process. Nevertheless, a book entitled 'Principles' can only scratch the surface of a huge subject and some readers may find their personal favourite topics have received only passing attention.

Inevitably, this book contains quite a lot about law and control measures. It assumes no previous knowledge of either on the part of the reader. An attempt has been made to separate legal issues, in an effort to concentrate on the principles involved rather than to produce a legalistic guide to what are inevitably minimum standards. This has not always been possible, and it is hoped that readers will agree that clarity has prevailed.

Many books have been written about construction safety which have painted a black picture of an industry that can easily give the impression that little is cared or done to improve working conditions and reduce the toll of human suffering. But there is much that is good in present construction safety management, and the UK record stands as an example to others in Europe, and, less obviously, in America. Yet the fact remains that in recent years around a hundred people have been killed each year in the construction industry, and asbestos is the largest single cause of work-related fatal disease and ill-health. There is much still to be done.

I wish to acknowledge the help and encouragement given to me by many professional colleagues, especially Jim Allen and Trevor Laws who have put in long hours correcting my mistakes. At the Institution of Occupational Safety and Health, advice has been given generously by Wilson Lambe, Mike Totterdell and Nick Cornwell-Smith. I am grateful to all my colleagues at Bovis Lend Lease Europe for their help, and particularly to environment specialist Kevin Mundy for correcting my errors and suggesting improvements to the text.

As global Director of Environment, Safety and Health for Bovis Lend Lease I have been generously supported in writing this book by the Company President Sir Frank Lampl and Chief Executive Ross Taylor. I am very grateful for the Company's permission to reproduce a selection of checklists and ideas, many of which began their existence and development under the watchful eye of Bryan Toone. This book is a testament to the professional expertise and vigilance of many such people, whose mission is to strive continually to improve the standards of environment, health and safety in the construction industry throughout the world.

Please note that all publications are from HSE Books, unless otherwise stated.

Allan St John Holt
Southampton, England
September 2000

Part 1
Construction Safety Management

1 Fundamentals

What is 'safety'?

We use the word 'safety' so much, often in company with its partner 'health', that it should be easy to find a definition. Yet the dictionaries do not offer much assistance — 'Safety [is] the absence of danger' one says, unhelpfully supplying the entry for 'danger' as 'absence of safety'! Others suggest 'a state of protection' and 'a condition not involving risk'. Perhaps the best we can do is to agree that there is no arbitrary state of 'absolute safety', as there is always a chance — a risk — of something going wrong, however small that chance may be.

In the same way, a little thought about 'health' brings the same conclusion — it is a relative notion, in the sense that in any population there will be those in varying states of wellness. But this does not stop us using the word in an everyday sense to convey the idea that, in the workplace at least, the aim should be that workers do not leave their work less 'healthy' than when they arrived.

The management of workplace health and safety is done together, and in the same way, so that often in speech the word 'safety' is used to mean both. In recent years, it has been recognised that environmental issues also need to be managed, and again often by using the same techniques and practices. So, for reasons of space and ease of understanding, in this book the reader will often find the word 'safety' used alone although the presence of its natural partners 'health' and 'environment' should be understood.

Basic terms

An **accident** is an incident plus its consequences; the end product of a sequence of events or actions resulting in an undesired consequence (injury, property damage, interruption, delay). An accident can be defined more formally as 'an undesired event, which results in physical harm and/or property damage, usually resulting from contact with a source of energy above the ability of the body or structure to withstand it.'.

In normal conversation we use the word 'accident' loosely, and in doing so we often couple in a sense of bad luck on the part of the injured person, and a feeling that it could not have been foreseen. In safety management, we need to be clear that the luck, or the element of chance, is only concerned with the physical outcome of the **incident**, which is 'that sequence of events or actions resulting in the undesired consequences'. For ease of reading, this book uses the word 'accident' to describe injury events, except where an important distinction has to be made between 'accident' and 'incident'.

An **injury** is thus a consequence of an incident — but not the only possible one. It has been shown that hundreds of incidents occur in the construction industry for every one that causes injury or loss. But all have the potential to do so. That is why it is important to look at all incidents as sources of information on what is going wrong. Relying on injury records only allows a review of a minority of incidents — those which happened to result in a serious injury consequence. We can make some reasonable estimates about the likelihood of, say, failure of a lifting appliance. Only chance will decide whether an injury rather than, or as well as, property damage will occur on a particular occasion, and how severe either will be.

Hazard means 'the inherent property or ability of something to cause harm — the potential to interrupt or interfere with a process or person'. Hazards may arise from interacting or influencing components, for example two chemicals interacting to produce a third.

Risk is 'the chance or probability of loss', an evaluation of the potential for failure. It is easy to confuse the terms 'hazard' and 'risk', but a simple way to remember the difference is that 'hazard' describes **potential** for harm, risk is the **likelihood** that harm will result in the particular situation or circumstances, coupled with a measure of the degree of severity of that harm. Comparisons between risks can be made using simple numerical formulae.

What causes accidents?

Accidents are the direct results of **unsafe activities and conditions**, both of which can be controlled by management. Management is responsible for the creation and maintenance of the working environment and tasks, into which workers must fit and inter-react. Control of workers and their behaviour is more difficult. They have to be given information, and the knowledge that accidents are not inevitable but are caused. They need training to develop skills and recognise the need to comply with and develop safe systems of work, and to report and correct unsafe conditions and practices. Their safety awareness and attitudes require constant improement, and the social environment of the workplace — the **safety climate** — must be one which fosters good safety and health practices and conditions, not one which discourages them.

On investigation, and after a little thought, it can be seen that accidents are relatively complex events. A man falls off a ladder. It seems straightforward — the ladder was not tied and witnesses say that it was set at the wrong angle and not secured against slipping. This incident could be put down to carelessness on the part of the man, having failed to appreciate the physical situation. Carelessness, though, is rarely either a good or an adequate explanation of events like accidents.

Unsafe acts and unsafe conditions are often referred to as **immediate** or **primary causes** of accidents, because they are the most obvious causes and because they are usually directly involved or present at the moment the accident happens. **Secondary causes** are also important, although they are usually harder to seek out and identify. They are the failures of the management system to anticipate, and include lack of training, maintenance, adequate job planning and instruction, and not having safe systems of work in place.

Some examples of unsafe acts and conditions are given below.

Unsafe acts

- Working without authority
- Failure to warn others of danger

- Leaving equipment in a dangerous condition
- Using equipment at the wrong speed
- Disconnecting safety devices such as guards
- Using defective equipment
- Using equipment the wrong way or for the wrong tasks
- Failure to use or wear personal protective equipment
- Bad loading of vehicles
- Failure to lift loads correctly
- Being in an unauthorised place
- Unauthorised servicing and maintaining of moving equipment
- Horseplay
- Smoking in areas where this is not allowed
- Drinking alcohol or taking drugs

Some of the reasons why people fail — to behave safely, to conform to policies and procedures, for example — are discussed in Chapter 8.

Unsafe conditions

- Inadequate or missing guards to moving machine parts
- Missing platform guardrails
- Defective tools and equipment
- Inadequate fire warning systems
- Fire hazards
- Ineffective housekeeping
- Hazardous atmospheric conditions
- Excessive noise
- Not enough light to see to do the work

These are all deviations from required safe practice, but they must be seen as the symptoms of more basic underlying **indirect** or **secondary** causes which allow them to exist and persist.

Secondary causes of accidents

- Management system pressures
  - financial restrictions
  - lack of commitment
  - lack of policy
  - lack of standards
  - lack of knowledge and information
  - restricted training and selection for tasks
  - poor quality control systems resulting from the above
- Social pressures
  - group attitudes
  - trade customs

- industry tradition
- society attitudes to risk-taking
- 'acceptable' behaviour in the workplace
- commercial/financial pressures between contractors

The primary causes of accidents in the construction industry have been the target of safety law for many years — specifying details of scaffolding and ladders, for example. Relatively recently, legal requirements in several countries, notably the Member States of the European Union, and Australia, have begun to address the secondary causes as well, forcing attention to be paid to all organisational aspects of safety management.

Techniques of accident prevention

Accident prevention in construction is not just a matter of setting up a list of rules and making safety inspections, although both of these have their place. What is required is a system for managing health and safety which meets the needs of the business and complies with the law. A discussion of the ideas involved in safety management can be found in Chapter 4, and most of the law on construction safety is covered in Part 3 of this book.

There are seven principles to be observed in setting up strategies for control and management of health and safety at work in the construction industry. If they are followed, accident prevention is more likely to be achieved.

1. If possible, avoid a risk altogether by eliminating the hazard

It is always more effective to remove a problem altogether rather than to establish a control strategy, especially one which relies upon people to work in the correct way. This and the next principle demonstrate the fundamental importance of design and planning in safety management.

Examples: *Do not specify fragile roofing materials through which people can fall. This is obviously more effective than specifying solutions designed to minimise the risks from falls through fragile material. Specification of lighter materials, such as blocks and bags, is preferable to arranging ways to handle heavy materials on site. Inevitably, at some stage the latter will result in someone having to lift manually a load heavier than it could have been. Avoid using hazardous substances at all where possible, or substitute those known to be less hazardous.*

2. Tackle risks at source

Design and specification can make a significant difference to site conditions. Design is likely to dictate the way the construction work is carried out on site, and particularly to force contractors to work in potentially unsafe ways. A review at the design stage repays the time spent handsomely because of later savings in time and money, and possibly even lives.

Examples: *Designing floor slabs with fewer voids removes the need to control the risk of falling through them by setting up barriers or covering them. Avoid dust-producing processes: specify off-site finishing.*

3. Adapt work to the individual when designing work areas and selecting methods of work

Ergonomics aims to improve the interface between people and their workspaces, by seeking to adapt the workspace rather than the person. Thought given to layout can improve working conditions and reduce risks.

Example: *Locating equipment such as a saw bench in a corridor could block access for others, may cause lighting difficulties and allow offcuts to pile up, increasing the risks. Asking how much room the carpenter needs and planning a suitable spot for a powered saw bench takes little time.*

4. Use technology to improve conditions

Keeping up to date with new developments can bring a safety benefit when plant is being replaced, or when work operations can be mechanised.

Examples: *Specifying a quieter design of machine when ordering replacement equipment. Use of a wheeled manhole lifter to replace hand hooks or other hand tools brings gains in productivity as well as minimising the chance of a back injury.*

5. Give priority to protection for the whole workplace rather than to individuals

Reliance on personal protective equipment (PPE) as a sole means of risk control is rarely acceptable. This is because no PPE is 100% effective for 100% of the time for 100% of the people who use it (see Chapter 15). One of many reasons for this is that it may not always be possible to identify everyone at risk and issue the PPE to them.

Example: *Extensive work on a flat roof may require the presence of a large number of workers from different employers, in addition to supervisory staff, clients, etc. In these circumstances, the appropriate protection would be provided by perimeter barriers, rather than by giving 'everyone' a safety harness. Installing permanent edge protection during the main construction process instead of at the end gives protection to both constructors and end users.*

6. Ensure everyone understands what they have to do to be safe and healthy at work

Safety training is not just a matter of handing out booklets — it is unwise to assume anything about people's previous experiences in construction work, or even their ability to read and understand instructions and information. Safety awareness is not inherited, and induction training on sites is required to make sure that everyone knows what the hazards and the control measures are.

Example: *Specific site induction must include the local emergency evacuation procedure, and understanding may need to be confirmed by holding regular practice fire drills.*

7. Make sure health and safety management is accepted by everyone, and that it applies to all aspects of the organisation's activities

A single contracts manager, joinery manager, or other member of senior management who believes that the company safety management system does not apply to situations where time is short can destroy the safety climate overnight. If someone is injured as a result, the penalty can be severe for that person, and possibly for the company as well.

Examples: *Failure of a senior manager to wear safety footwear and safety helmet on a site visit gives the impression to the workforce that the rules do not apply to senior management. Failure to carry out risk assessments because of pressure of work could lead to criminal prosecution of an individual.*

Benefits of accident prevention

There are generally said to be five main reasons why accident prevention in construction is worthwhile.

The **cost in human suffering**, physical pain and hardship resulting from death and disability is impossible to quantify — we know that there are hundreds of lives lost each year in construction and related industries, with tens of thousands of serious injuries and countless numbers of more minor injuries. We can only guess at the disruption to lives of workers and their families which these cause, but we know that construction safety is not an impossible dream; it is an achievable goal.

Moral reasons stem from a developing public awareness that something needs to be done to raise the quality of life at work. Attention is focusing on the ability of employers and project managers in the industry to handle a wide variety of issues, previously seen only as marginally relevant to the business. Environmental affairs, pollution, design safety, maintainability and other matters are now commonly discussed. There is a growing belief that it is morally unacceptable to put the safety and health of others (inside or outside the construction site) at risk, for profit or any other reason.

Worker morale is strengthened by active participation in accident prevention programmes, and is weakened by accidents. Adverse publicity affects the fortunes of the organisation both internally in this way and externally, as public confidence may weaken local community ties, market position, market share, shareholder value and reputation generally.

Legal reasons are contained in **statute law**, which details steps to be taken and objectives to be met, and which carries the threat of prosecution or other enforcement action as a consequence of failure to comply. **Civil law** enables injured workers and others to gain compensation either as a result of breach of statutory duties or because a reasonable standard of care was not provided under the particular circumstances. The cost in terms of money and adverse publicity of a prosecution or civil claim can be very high, and there is the potential for a prison sentence in some circumstances. For a discussion of these issues, see Chapter 17.

Financial reasons for accident prevention ensure the continuing financial health of a business and avoid the costs associated with accidents. These include monetary loss to employers, community and society from worker injuries and ill-health, damage to property and production delays. Some, but not all of these costs are insurable and these are known as **direct** costs. They include the cost of compensation (for which insurance is a legal requirement). Increased premiums will be a consequence of claims, so an increase in overheads is predictable following accidents. **Indirect** costs include:

- Uninsured property and material damage
- Delays
- Overtime costs and temporary labour
- Management time spent on investigations
- Decreased output from those replacing the injured worker(s)
- Clerical work
- Fines
- Loss of expertise and experience

A study carried out by the Health and Safety Executive into the costs of accidents showed that for the construction site under review, the direct costs were a small proportion of the total and produced a direct:indirect ratio of 1:11. This ratio is commonly illustrated as an 'iceberg', because of the invisible hidden costs below the waterline. On the site studied over a period of 18 weeks, 120 people were working, and in that time there were 56 minor first-aid injuries and no lost-time injuries. But there were also 3570 non-injury accidents. The results for major, minor and non-injury accidents are often reproduced in the form of an 'accident triangle'

The conclusion to be drawn from this and other accident triangles is that serious injuries are much less frequent than less serious ones, and of course it would be strange if that were not the case. The same amount and quality of information is potentially available for each incident, yet we frequently limit investigations to those incidents where injury or damage is serious — at the tops of the triangles — and thus miss the chance of obtaining a lot more information about what is going wrong. This is why counting and investigating 'near miss' incidents is useful.

Finally, a good safety record and documented safety management system can more than repay the time spent on it because of its value in gaining **new business**. Many clients and project management operations have extensive vetting procedures to identify those contractors and suppliers who are competent in safety matters. The vetting may be a requirement of their quality programme under BS EN 9000 accreditation, but in any event in the UK the competence issue is at the heart of the Construction (Design and Management) Regulations 1994 (CDM).

Conversely, inability to satisfy requirements for competence in safety under CDM can result in loss of significant contracts as well as public reputation. An electrical contractor working for a local authority was successfully prosecuted together with the authority for contravening the Control of Asbestos at Work Regulations when hand-drilling holes in domestic cupboards for wiring. The fine was relatively modest, but the contractor's hard-won local reputation for workmanship and reliability suffered, and the conviction had to be disclosed on every subsequent application to join a tender list.

2 | Where are We Now?

It is widely recognised that health and safety injury statistics offer only the depressing and not especially useful prospect of counting our failures to control injury and ill-health. What is needed is an agreed system of positive measures of how well controls are working — but there is no sign of agreement or even substantive moves towards agreement about the nature of the measures that might provide answers. The prospect of common use of standards such as the British Standards OHSAS 18000 safety management model is brighter, but opinion internationally is divided on whether the objectives of continuous improvement and accreditation should be included in such a standard.

Meanwhile, international statistical comparisons mean little, as there are so many measuring tools and so many different criteria for measurement. There is no general agreement on how to calculate frequency rates, or on what counts as a reportable injury. Even in the UK, the rules change every 5 to 10 years, and invalidate previous data sets. And there is good evidence that many injuries go unreported.

Recent studies by the Health and Safety Executive using the Labour Force Surveys (which send questionnaires to households) indicate that only 55% (1997/98) of all non-fatal reportable injuries in construction are actually reported to the Executive and local authorities. This is actually an improvement — in 1989/90 the figure was as low as 38%, rising to 46% in 1994/95. Commentary on the disparity between reality and what is reported is presented annually in the Health and Safety Commission's Statistical Review. The information can be regarded only as an indication of the position for any given year.

An improved picture can be drawn from a study of fatalities, which are less easy to ignore, although these too may be under-reported because of poor diagnosis of ill-health exposures, for example, and also where a significant time may elapse between injury and consequent death.

On average, a worker is killed in the industry every three working days, and a member of the public is killed every two months by construction activities. There are now about 80 fatalities a year to workers in the construction industry, including the self-employed and trainees. This number has remained roughly the same since the early 1990s (Table 2.1). Taking the long-term view for all industries, the accident rates overall are a quarter of those reported in the early 1960s, and less than a half of those in the early 1970s. This is thought to be due only partly to changes in employment patterns.

Table 2.1: Fatalities to employees in the construction industry, sample years	
Year	Number
1961	272
1971	156
1981	105
1990/91	96
1995/96	62
1996/97	66
1997/98	58
1998/99*	48

* Data for 1998/99 were provisional at the time of writing

Types of injury

The most common source of fatalities in recent years has been the head injury, accounting for almost one-third of the total. There is claimed to be a marked reduction, of about 25%, in the head injuries rate overall, following the introduction of the Construction (Head Protection) Regulations on 1 April 1990. In most years, at least 40% of all construction fatalities have been falls from a height. The figure for falls as a percentage of total fatalities has been remarkably constant over past years, justifying the attention given to fall protection in its own right in the Construction Regulations (Table 2.2).

Table 2.2: Fatal falls to employees in the construction industry over 12 years			
Year	Falls	All employee fatalities	Falls as % of all fatalities
1987/88	47	99	47.48
1988/89	49	101	48.52
1989/90	53	100	53.00
1990/91	45	96	46.88
1991/92	37	83	44.58
1992/93	27	69	39.13
1993/94	35	73	47.95
1994/95	24	56	42.86
1995/96	21	62	33.87
1996/97	33	66	50.00
1997/98	29	58	50.00
1998/99*	22	48	45.83

*Data for 1998/99 were provisional at the time of writing

In 1998/99 the percentage of falls in the 'non-fatal major injury' category was 37%, with an additional 21% as falls from the same level (slips, trips and falls); 35% of reportable, over-3-day injuries in the industry were due to manual handling.

Accident causes

Canadian studies have shown that active involvement in safety management by the most senior levels in a construction company is directly correlated with reductions in numbers of accidents and injuries.

Knowledge of causation patterns provides a starting point for focusing particular preventive measures. Case studies and descriptions of accidents can be used to give information about prevention techniques — the Health and Safety Executive's publication *Blackspot Construction* is still recommended reading, although now out of print. It commented that in a sample studied, 90% of fatalities were found to be preventable, and in 70% of cases positive management action could have saved lives. The three worst task areas found by the study (75% of all deaths) were maintenance (42%), transport and mobile plant (20%) and demolition/dismantling (13%).

Table 2.3 shows the distribution of causes of fatalities in the years 1997/98 and 1998/99, considering all workers in the industry. Some activities, of course, are frequent sources of injury, but rarely result in a fatality — manual handling, for example. Others occur relatively infrequently, but when they do there is a higher than usual chance of not surviving them — becoming trapped by collapse or overturning and electrocution are examples of this. Table 2.4 compares the proportionate outcomes of various causes of injury.

We need this information in order to estimate risk, which is a measure combining the chances of something happening with the potential outcome in terms of injury. Too much reliance should not be placed on the data because of the under-reporting factor mentioned above.

Although the total numbers engaged in some occupations can be expected to fluctuate with time, together with the amount and type of work available in the construction industry, it is hard to believe that there has not been an improvement in health and safety standards in the last 10 years. The improving data are unlikely to be due solely to changes in the pattern of work and numbers employed.

Information on rates is not easy to acquire. The Health and Safety Commission and Executive use comparisons between industries based upon accidents recorded per 100 000 workers, rather than upon hours worked. For construction in 1998/99 the provisional rate is 399.0 for fatalities and major injuries combined per 100 000 workers, and for all reported injuries the rate is given as 1254. The previous year's rates were 388.1 and 1354.4, respectively.

Cross-industry comparisons are of little significance to individual employers unless their business covers several employment sectors. The best information of this kind is to be found by benchmarking against other similar businesses, and comparing internal figures over similar periods.

Occupational health and hygiene

Traditionally the construction industry's high level of injury accidents has received the attention of enforcement, media publicity and management action. Arguably, the size of that problem has led to a neglect of the less tangible consequences of occupational hygiene and health problems, apart from well-publicised topics such as asbestos. There is little general awareness of just how big the occupational health risks are in construction; compare the numbers already discussed for conventional injuries with the fact that mesothelioma, a form of cancer linked specifically to asbestos exposure, kills around 1400 people

Table 2.3: Fatalities in the construction industry by causation								
Categories of accident causation	1997/98				1998/99			
	Employees	Self-employed	Total	% of total	Employees	Self-employed	Total	% of total
Falls: >2 m	29	12	41	51.25	21	14	35	53.03
Falls: <2 m	–	1	1	1.25	–	3	3	4.55
Falls: unknown heights	–	4	4	5.00	1	–	1	1.52
Total falls from heights	29	17	46	57.50	22	17	39	59.09
Contact with moving machinery	3	–	3	3.75	2	–	2	3.03
Struck by moving or falling object	11	1	12	15.00	7	1	8	12.12
Struck against fixed or stationary object	1	–	1	1.25	–	–	–	–
Struck by moving vehicle	5	-	5	6.25	9	–	9	13.64
Lifting, handling, carrying	1	–	1	1.25	–	–	–	–
Trapped by collapse	3	1	4	5.00	3	–	3	4.55
Asphyxiation or drowning	–	–	–	–	1	–	1	1.52
Exposure to harmful substances	–	–	–	–	1	–	1	1.52
Explosion	–	–	–	–	1	–	1	1.52
Electrocution	5	2	7	8.75	2	–	2	3.03
Other	–	1	1	1.25	–	–	–	–
Total	58	22	80	100.00	48	18	66	100.02

each year. A good proportion of these exposures are related to construction work.

A general downgrading of normal health can also occur. Reports suggest that construction workers age prematurely due to hypothermia caused by working in the cold and wet. Respiratory diseases such as bronchitis and asthma are also thought to occur at above average levels in construction workers.

Experience with the implementation of the Control of Substances Hazardous to Health (COSHH) Regulations in construction shows that there is little awareness of the principles of assessment, or significant appreciation of the risks to workers from substances brought onto the site — and especially from those created there. Also, there is said to be a disappointing response from the industry to the noise controls (mostly managerial action and measurement

requirements) imposed by the Noise at Work Regulations 1989.

Dangerous occurrences

'Dangerous occurrences' are sets of circumstances which must be reported to the enforcing authorities if they occur, and which are defined within the current set of regulations containing reporting requirements for injuries. Currently the relevant regulations are the Reporting of Injuries, Diseases and Dangerous Occurrences Regulations 1995 (RIDDOR). They were introduced in approximately their current format in 1980 as the Notification of Accidents and Dangerous Occurrences Regulations, or NADO, which introduced the concept of dangerous occurrences for the first time. Since the list of circumstances which are to be reported has changed over the years, and again because of significant under-reporting, statistical comparisons are

Categories of accident causation	1997/98				1998/99			
	Fatalities	Non-fatal major injuries	>3 day injuries	Total reportable	Fatalities	Non-fatal major injuries	>3 day injuries	Total reportable
Falls: >2 m	41	754	395	1190	35	813	360	1208
Falls: <2 m	1	734	752	1487	3	824	843	1670
Falls: unknown heights	4	163	179	346	1	120	147	268
Total falls from heights	46	1651	1326	3023	39	1757	1350	3146
Contact with moving machinery	3	138	237	378	2	136	283	421
Struck by moving or falling object	12	858	1913	2783	8	122	1745	1875
Struck against fixed or stationary object	1	137	542	680	–	153	474	627
Struck by moving vehicle	5	104	155	264	9	132	142	283
Lifting, handling, carrying	1	370	3633	4004	–	360	3324	3684
Slip, trip or fall, same level	–	790	1694	2484	–	929	1655	2584
Trapped by collapse or overturn	4	55	49	108	3	62	52	117
Asphyxiation or drowning	–	9	12	21	1	14	6	21
Exposure to harmful substances	–	78	218	296	1	86	200	287
Fire	–	13	37	50	–	7	17	24
Explosion	–	4	21	25	1	14	18	33
Electrocution	7	55	96	158	2	75	78	155
Animal injury	–	2	36	38	–	–	27	27
Acts of violence	–	14	39	53	–	15	41	56
Other	1	46	252	299	–	57	187	244
Total	80	4324	10260	14664	66	3919	9599	13584

Table 2.4: Reported injuries to all construction workers, 1997/98 and 1998/99*

*Data for 1998/99 were provisional at the time of writing

likely to be misleading. The intention is to bring to the notice of the enforcing authority those incidents and conditions which are sufficiently serious to be likely to cause an injury, even though such an injury did not necessarily occur.

In 1998/99, 4173 reports were made of dangerous occurrences. Some 25% dealt with failures of lifting machinery of various kinds, 13% advised of substance escape as defined within RIDDOR, and 11% notified pipeline failures. These dominant categories and percentages have remained generally constant over time since 1995. The construction industry is likely to be involved in reporting lifting machinery failures of several kinds, as well as the unintended collapse of buildings.

Acknowledgement: The Health and Safety Executive's assistance with the provision of statistical information is gratefully acknowledged.

Reference

(Health and Safety Commission) *Statistical Review* (published annually). HSE Books, London.

3 Measuring Performance and Recording Information

There are many good reasons for measuring safety performance; some of them are often forgotten. Measuring can enable management to:

- Identify the causal factors involved in injury and loss
- Locate areas where controls are not working adequately
- Have a basis for comparing trends
- Describe the level of safety within the organisation
- Predict future safety problems
- Evaluate the success of the control programme
- Maximise cost-effectiveness of decisions on the allocation of resources
- Assess the costs of injuries and losses
- Benchmark against other similar organisations

Before any useful conclusions can be drawn from any set of data, the information collected must be both reliable and meaningful. Are the right things being recorded? Do the numbers give the whole picture? How many injuries and incidents are unreported? What distortions may be present? And what gives information about 'safety performance'?

Until recently, the only measuring tools available were the counting of failures (lost-time injuries however defined, first-aid cases, property damage incidents and 'near misses'), and attempts at measuring the financial costs of losses resulting from failures to control safety, health and the environment. All of these involve studying the evidence of failures in one form or another, rather than the performance achieved. And there are difficulties in collecting the evidence, for example:

- How severe must an injury be to be counted at all? As mentioned in the previous chapter, this definition varies widely between countries. The UK minimum reporting requirement is set at 3 days' absence from work, although many of the larger construction companies now measure lower severity level indicators such as first-aid treatments and lost workday cases. The US standard is to count 'disabling injuries', which are those that cost a full person-day. The trend is towards lowering the severity thresholds for reporting, but this tends in turn to result in under-reporting.

- How can different levels of risk and exposure variations be allowed for? Using numbers of hours worked as the baseline for exposure measurement does not reflect differences in risk. What can be learned from comparing the injury data per million person-hours worked between, say, carpet fitters and roofers? The roofers' injury rate might be expected to be much higher, because they appear to be at greater risk, but that takes no account of the amount of time spent, the relative exposure to risk, whilst at work. The best that can be done is to compare like with like where possible, and to include severity rates where available.

For several reasons, measuring tools which are purely injury-related do not do a good job of representing the quality of the performance effort or the safety climate in an organisation.

Behaviour-based safety

A set of techniques known generally as 'behaviour-based safety' has introduced ideas and methods from the behavioural sciences into performance measurement and safety management. A full discussion of the techniques used is beyond the scope of this book, but they are based on the claim that measuring the frequency of safe behaviour generates more, and more accurate, predictive data, allows for precise reinforcement and provides positive accountability. The general principle involves sampling, recording and publicising the percentage of safe (versus unsafe) behaviours noted by observers drawn from workforce and management, and specially trained. This gives more data on potential system and individual failures than could be obtained from a study of accident records.

Staff at the University of Manchester Institute of Science and Technology (UMIST) have extensive experience in the application of behaviour-based safety systems to the construction industry. In the United States, the work of Dr Thomas Krause and his colleagues at Behavioral Science Technology Inc, California, is particularly well known. It has shown that a behaviour-based approach to health and safety management can be an effective tool for increasing safety on construction sites and elsewhere, despite some practical problems of implementation.

Employers investing in these techniques say they have found that the involvement of workers in the measuring process generates interest and improved commitment to the employer's safety objectives. The results are said to be significant in that the techniques lead to a reduction in loss-producing incidents as well as to the improved assessment of performance by the positive step of measuring workers' safe actions.

For the purpose of this chapter, it is noted that the attraction is that the system offers a method of measuring the potential for harm, independent of the accident record. Disadvantages may include the need to achieve an altered safety climate for both management and workforce to adopt the techniques, and employee suspicion of hidden motives for the observations.

'No injuries — no problems!'

Because the numbers of recorded incidents and injuries are relatively low in most companies, they produce a limited amount of information about risk and there is a temptation to believe that all is well. The argument often put forward by managers — 'We haven't had any accidents, therefore we must be safe' — takes no account of the potential for injury, or risk, which must be evaluated when deciding on appropriate measures to take.

When things do go wrong, the information which can be obtained from recording and investigating incidents can be both substantial and useful, depending to a large degree on the methods used to collect information on individual incidents and throughout the organisation. The success of any collection method also depends on the commitment of individuals to supplying the information in a timely and appropriate way. This means that organisations need to have a system for reporting and recording injuries and other losses which is seen by all as reasonable under all the circumstances, is sufficiently thorough to capture all sig-

nificant information and is recorded for analysis in a suitable format. Basic training is also required, to ensure that those given the responsibility are fully committed to the goals of investigating and recording all significant incidents, and that the techniques involved are understood.

The way in which injuries, potential injuries ('near-misses'), occupational ill-health exposures and environmental incidents are investigated and the statistics are collected should be written into the organisation's safety policy or equivalent document. This gives the process a mandate from the person in charge, without which no efforts of significance in safety are likely to be successful.

One test of the efficiency of a data collection system is whether 'near-miss' incidents are reported and recorded. Their investigation will provide the same information on causation as 'real' injury incidents.

'All I want to know is the facts!'

In addition to distortions caused by under-reporting and factors introduced by the way the data are collected, whatever data are provided to management may be subject to misinterpretation by anyone who does not fully understand the nature of what is being presented to them. For example, it has been reliably estimated that about 375 000 people were killed by accidents in the United States during World War II. About 408 000 were killed by war action. It has been claimed as a result that it was nearly as dangerous to stay at home as it was to be in the armed services.

A moment's thought should lead to questions about rates rather than numbers. It turns out that the death rate in the US armed forces during World War II was about 12 per thousand men per year, which compares with the overall civilian accidental death rate of about 0.7 per thousand per year.

Other influencing factors include the ages of those exposed to the hazards, and the duration and type of exposure to the hazard under analysis. The raw numbers do not, in fact, tell us much at all about the chances of survival at the time.

Calculating rates

The simple formulae for calculating frequency and severity rates which follow produce the rates used most often in the industry. In most countries, there are no standard or formal requirements for these formulae.

Frequency rate

$$F = \frac{\text{Number of injuries} \times 100\,000}{\text{Total number of hours worked}}$$

In the USA, figures of 1 000 000 or 200 000 often replace the 100 000 in the numerator, depending upon the collecting agency.

Severity rate

This is essentially a weighted frequency rate, allowing the days lost due to temporary total disability to be recorded, and also a notional number of days to be recorded for fatalities and permanently disabled cases. The notional days often used are 6000 (20 working years at 300 days per year) for a fatality and 1800 for loss of an eye. The American National Standards Institute (ANSI) has developed an arbitrary schedule of notional days in relation to particular permanently disabling injuries, of which the foregoing are two examples.

$$S = \frac{\text{Total days lost plus notional days charged} \times 100\,000}{\text{Employee hours of exposure}}$$

The same numerator should be selected as in the frequency rate calculation.

The frequency rate can be improved 'artificially' by controlling temporary total absence cases (often by management policy, and by making jobs available for temporary workers convalescing). Similarly, the severity rate can fluctuate wildly because the schedule of notional days can impose a severe notional time penalty for some injuries.

Other performance measures

Many companies are now introducing positive measures of performance in the field of environment, health and safety. Once a safety management programme has been developed, key points in it can be identified and measured to provide information on whether the system is working properly. For example, it might be decided that one of the critical features in such a programme is the need to ensure that necessary safety information is supplied to subcontractors prior to tendering. Whether this happens or not can be measured by examining project documentation, at perhaps quarterly intervals and a score produced based on the percentage of projects where this is being done. Or, if a company operates a pre-qualification scheme for subcontractors, measures could be developed to show the percentage of subcontractors working on a project that had actually been pre-qualified. The percentage of subcontractors who had submitted a competent method statement prior to starting work could be another.

These internal yardsticks can also be given targets to achieve, so that continuous improvement can be made by raising the targets year on year. The setting of such **key performance indicators** is one of the features that distinguishes those companies interested in establishing themselves as 'world class' rather than being merely 'compliance orientated'.

Accident investigation and recording

The elements of an **accident recording system** consist of:

- Report form(s)
- Investigation reports — format
- Summary analysis forms used by the data collector
- Statistical analysis
- Summary reports for management

In large organisations, or those with widely-spread sites, the system may also use a fax or email notification system giving early warning to senior management that an incident has occurred which requires their attention. Care should be taken to avoid making personal comments in emails, as they are likely to become 'discoverable' in legal proceedings — visible to everybody. A good rule to follow is: 'If you don't want your opinion or comment known to potentially hostile strangers, don't send the email'. Deleting an email from the system usually has no effect on its viability on a server, somewhere.

Standard report format

Use of a suitable standard report form allows the collection of information in a uniform way. The design of the report form is important. As the forms will usually be completed at site level it will assist site staff if they are only asked to give information which is likely to be readily available to them — social security numbers and other personnel information may be restricted, or held elsewhere. The penalty for using a format calling for answers that site staff cannot provide can be delays in the return of forms. Whatever detail is asked for, and whatever the final design of a report form, it will be helpful to require at least those answers to be given which are required by local or national authorities, when notifying them in turn.

Report forms should **not** require the senders to:

- Assign blame for the incident
- Make comments which the senders cannot substantiate, or
- Draw conclusions which may be beyond their level of competence

This is because the form may be required as evidence in legal proceedings, where liability is an issue and statements of this type may be found to constitute an admission of liability where none was intended.

Speed is of the essence — it may be desirable to design an initial report form to be faxed to senior management as soon as practicable, to be followed by a more detailed report when time and circumstances permit. The detail may be completed later, following a deeper investigation and interviews with witnesses.

Investigation report format

The person or team carrying out an investigation into an incident should record their comments in writing. This may or may not involve the completion of an internal investigation report form. Again, care should be taken to avoid making statements or comments which are not factual. Recommendations to prevent a recurrence should be made in the form of a letter attachment rather than on a report form, again for legal reasons.

When completing an investigation report, it is important to be aware that a potential reader may know nothing about circumstances or techniques which the investigator may take for granted. For this reason, no assumptions should be made about the level of knowledge possessed by the report's readership. Sketches, plans, drawings and especially photographs should be included to amplify the written report.

The use of a standard layout in a typed report can assist the investigator, because it can help to clarify thought. A suitable format for a written report is discussed later in this chapter.

Summary analysis forms used by the data collector

In order to make the best use of data supplied, the system of performance measurement needs a way of recording the types of incident and their consequences so that common causal factors can be isolated, future problem areas can be predicted, training can be focused and trends assessed. Depending upon the size of the organisation, it may be necessary to collect data site by site and collate that into a monthly executive summary report.

If benchmarking is carried out against other organisations or against industry records, it will be helpful to design summary forms which use the same categories as the benchmark targets. Otherwise, the format used by the regulatory authorities to present data can be used.

Statistical analysis

It is always tempting to compare the results from the organisation with national or industry data, but the relative sizes of the samples must be remembered when doing so. Also, industry frequency and severity rates, where published, are often based on guesswork on the hours worked, and can take no account of under-reporting of injuries to the authorities. Generally, the best comparison to use is a previous time period within the same business or benchmarking partners.

Summary reports for management

Monthly or quarterly summaries of injury figures should be presented in a format which makes valid comparisons easy, and with a short written account to provide a summary of selected incidents. The use of pie charts and other presentation aids should be considered. It is important that the person designated as the competent person for the purposes of the Management of Health and Safety at Work Regulations 1999 is given a summary report at regular intervals. A clutch of statistics alone rarely provides enough meaningful information for senior management. A written summary of lessons learned, actions taken and the current status of the organisation's rolling safety programme should also be included.

Principles of accident investigation

The hardest lessons to be learned in accident prevention come from the investigation of accidents and incidents which could have caused injury or loss. Facing up to those lessons can be traumatic for all concerned, which is one reason why investigations are often incomplete and simplistic. Nevertheless, the depth required of an investigation must be a function of the value it has for the organisation and other bodies which may make use of the results, such as enforcement agencies. Conducting one can be expensive in time.

Purpose

The number of purposes is large; the amount of detail necessary in the report depends upon the uses to be made of it. Enforcement agencies look for evidence of blame, claims specialists look for evidence of liability, trainers look for enough material for a case study. From the viewpoint of prevention, the purpose of the investigation and report is to establish whether a recurrence can be prevented, or its effects lessened, by the introduction of safeguards, procedures, training and information, or any combination of these.

Procedure

There should be a defined procedure for investigating all accidents, however serious or trivial they may appear to be. The presence of a form and checklist will help to concentrate attention on the important details. The management team of the project where the accident occurred will be involved; for less serious accidents they may be the only people who take part in the investigation and reporting procedure. Workers' representatives may also be involved as part of the investigating team.

Equipment

The following are essential tools in the competent investigation of accidents and damage/loss incidents:

- Report form, possibly a checklist as a routine prompt for basic questions
- Notebook or pad of paper
- Tape recorder for on-site comments or to assist in interviews
- Camera — Polaroid instant-picture cameras are useful (but further reproduction of the results may be difficult and expensive, or they may be of poor quality). Their improved performance and the ability to insert photographs into text now makes the use of digital cameras attractive
- Measuring tape, which should be long enough and robust, like a surveyor's tape
- Special equipment in relation to the particular investigation, e.g. meters, plans, video recorder

Documentation

Information obtained during investigations is given verbally, or provided in writing. Written documentation should be gathered to provide evidence of policy or practice followed on site, and witnesses should be talked to as soon as possible after the accident. The injured person should also be seen promptly and interviewed.

Key points to note about investigations are:

- Events and issues under examination should not be prejudged by the investigator
- Total reliance should not be placed on a single source of evidence
- The value of witness statements is proportional to the amount of time which passes between the events or circumstances described and the date of a statement or written record. (Theorising by witnesses increases as memory decreases)
- The first focus of the investigation should be on when, where, to whom and the outcome of the incident
- The second focus should be on how and why, giving the immediate cause of the injury or loss, and then the secondary or contributory causes
- The amount of detail required from the investigation will depend upon (a) the severity of the outcome and (b) the use to be made of the investigation and report
- The report should be as short as possible, and as long as necessary for its purpose(s).

The report

For all purposes, the report which emerges from the investigation must provide answers to the following questions. Only the amount of detail provided should vary in response to the different needs of the recipients.

- What was the immediate cause of the accident/injury/loss?
- What were the contributory causes?
- What is the necessary corrective action?
- What system changes are either necessary or desirable to prevent a recurrence?
- What reviews are needed of policies and procedures (for example, risk assessments)?

It is not the task of the investigation report to allocate individual blame, although some discussion of this is almost inevitable. Reports are usually 'discoverable'; this means they can be used by the parties to an action for damages or criminal charges. It is a sound policy to assume that accident investigation reports will be seen by solicitors acting on behalf of the injured party. They are entitled to see the factual report, and this will include anything written in it or in connection with it (which might later prove embarrassing), so certainly it should not contain comments on the extent of blame attaching to those concerned, or advice given to management. It is appropriate, necessary and quite proper that professional advice is given, but it should be provided in a covering letter or memorandum suitably marked 'confidential'. Changes in the rules governing the civil liability claims process mean that 'side letters' and other formerly acceptable means of conveying concern 'off the record' have become potentially discoverable.

Whether the report is made on a standard form, or specially written, it should contain the following headings:

- A summary of what happened
- An introductory summary of events prior to the accident
- Information gained during investigation
- Details of witnesses
- Information about injury or loss sustained
- Conclusions
- Recommendations
- Supporting material (photographs, diagrams to clarify) added as appendices
- The date, and it should be signed by the person or persons carrying out the investigation

Sample standard report forms are included at the end of this chapter (Figs 3.1 and 3.2).

Inspections and audits

Audits look at systems and the way they function in practice, **inspections** look at physical conditions. So, while inspections of a site, or particular items of equipment, could (or possibly should) be done formally at least weekly, an audit of the inspection system throughout an organisation would look at whether the required inspections were themselves being carried out, the way they were being recorded, who received copies of the report, whether action was taken promptly as a result, and so on. More information on audits can be found in the next chapter.

Inspections

Inspections should be based on a positive approach, seeking to establish what is good and well done as well as what is not. Too often the 'inspection' process has a negative implication associated with fault-finding. The inspection of sites and premises has three main objectives:

- Identification of hazards, triggering the corrective process
- Improving conditions and reducing risks
- Measuring safety performance

Some common system should be followed for every inspection to make sure that everything relevant has been covered. Checklists can be used, and an adequate reporting system must be present so that a record is made of what needs attention, and management can be advised of the results of the inspection. Ideally, inspections should be measurable so that comparisons can be made with standards elsewhere in the business.

Inspection for health and safety purposes often has a negative implication, associated with fault-finding. A positive approach based on fact-finding will produce better results, and co-operation from all those taking part in the process.

There are a number of types of inspections, for example:

- Statutory — for compliance with health and safety legislation
- External — by enforcement officials, insurers, consultants
- Executive — senior management tours
- Scheduled — planned at appropriate intervals, by supervisors
- Introductory — check on new or reconditioned equipment
- Continuous — by employees, supervisors, which can be formal and preplanned, or informal

For any inspection, knowledge of what is being looked at is required, also knowledge of applicable regulations, standards and Codes of Practice. Some system must be followed to ensure that all relevant matters have been considered, and an adequate reporting system must be in place so that remedial actions necessary can be taken and that the results of the inspection are available to management.

Some experts believe that 'assurance' is a better description of the activity — there is a need to assure that the system is working properly (safely). Inspection measures how good or bad things are, allowing comparison with standards set either locally, corporately or nationally. Corrective action can then be taken.

A special kind of inspection often overlooked is the readiness check, where operations are evaluated for safety before work on them begins. Examples of situations where these inspection checks are appropriate include the setting up of tower cranes and other temporary works such as installation of falsework, preparation for excavations and façade support work. Two checklists at the end of this chapter are intended to give useful examples of readiness inspection checks for specific topics (Figs 3.3 and 3.4), and Fig. 3.5 provides a general site checklist.

Before any inspection, certain basic decisions must have been taken about aspects of it, and the quality of these decisions will be a major influence on the quality of the inspection and on whether it achieves its objectives. The decisions are reached by answering the following questions:

1. What needs inspection? Some form of checklist, specially developed for the inspection, will be helpful. This reminds those carrying out the inspection of important items to check, and serves as a record. By including space for 'action by' dates, comments and signatures, the checklist can serve as a permanent record.
2. What aspect of the items listed needs checking? Parts likely to be hazards when unsafe — because of stress, wear, impact, vibration, heat, corrosion, chemical reaction or misuse — are all candidates for inspection, regardless of the nature of the equipment or where it is used.
3. What conditions need inspection? These should be specified, preferably on the checklist. If there is no standard set for adequacy, then descriptive words give clues to what to look for — items which are exposed, broken, jagged, frayed, leaking, rusted, corroded, missing, loose, slipping, vibrating.
4. How often should the inspection be carried out? In the absence of statutory requirements, or guidance from standards and Codes of Practice, this will depend upon the potential severity of the failure if the item fails in some way, and the potential for injury. It also depends upon how quickly the item can become unsafe. A history of failures and their results may give assistance.

5. Who carries out the inspection? Everyone has a responsibility to carry out informal inspections as they move around a site. Managers, supervisors and foremen should plan to make general inspections. Workers' representatives may also have rights of inspection, and their presence should be encouraged where possible. Management inspections should be made periodically; formal compliance inspections should take place in their presence.

Techniques of inspection

Inspections should be carried out by people who are competent to do so. In this context, 'competent' means knowledgeable — the person needs to know what safe conditions look like, have experience of a range of potentially unsafe conditions and know how to convert the latter into the former. This does not mean that the physical ability to make the change has to be possessed by the person — it is not necessary to be or have worked or trained as a scaffolder in order to inspect and comment on the safety of a scaffold structure, for example.

The following observations have been of assistance in improving inspection skills:

1. Those carrying out inspections must be properly equipped to do so, having necessary knowledge and experience, and knowledge of acceptable performance standards and statutory requirements. They must also comply fully with local site rules, including the wearing or use of personal protective equipment, as appropriate, so as to set an example as well as to protect themselves.
2. Develop and use checklists, as above. They serve to focus attention and record results, but must be relevant to the inspection.
3. The memory should not be relied upon. Interruptions will occur, and memory will fade, so notes must be taken and entered onto the checklist, even if a formal report is to be prepared later.
4. It is desirable to read the previous findings before starting a new inspection. This will enable checks to be made to ensure that previous comments have been actioned as required.
5. Questions should be asked, and the inspection should not rely upon visual information only. The 'what if?' question is the hardest to answer. Site workers are often undervalued as a source of information about actual operating procedures and of opinions about

possible corrections. Also, systems and procedures are difficult to inspect visually, and carrying out an appropriate inspection depends upon the asking of the right questions of those involved.

6. Items found to be missing or defective should be followed up and questioned, not merely recorded on the form. Otherwise, there is a danger of inspecting a series of symptoms of a problem without ever querying the nature of the underlying disease.

7. All dangerous situations encountered should be corrected immediately without waiting for the written report, if their existence constitutes a serious risk of personal injury or significant damage to plant and equipment.

8. Where appropriate, measurements of conditions should be taken. These will serve as baselines for subsequent inspections. What cannot be measured cannot be managed.

9. Any unsafe behaviour seen during the inspection should be noted and corrected, such as removal of machine guards, failure to use personal protective equipment as required, or smoking in unauthorised areas.

10. Risk assessments, method statements, safety plans and the like should be checked as part of the inspection process. This is because it is important to be able to demonstrate not only good site conditions, but also a trail of action, through CDM, to the assessment process and the employer's definite commitment to taking the necessary actions to ensure safety.

Reference

Petersen, D.C. (1998) *Techniques of Safety Management*, 3rd edn. McGraw-Hill, Kogakusha, USA.

| Figure 3.1: Initial accident report form |

YOURCO   INITIAL ACCIDENT REPORT — SUMMARY

| Project Name: | Report distribution (the day after accident): Project Manager Area/Senior Manager Safety Manager |

1. Name of the injured person

2. Date and time of the accident

3. Where on site did the accident happen?

4. Injured person's nationality

5. Employer of the injured person (state if self-employed)

6. Trade or occupation

7. Details of what happened and the injury as known at this time:

8. Witnesses: [give their names, location and employers' details].

9. What action has been taken to investigate and prevent a recurrence, and by whom?

10. Have local procedures been complied with? [All necessary authorities, insurers and the like have been informed]

11. Name and status of the person sending this Report

12. Signature

13. Date of this Report:

	Day	Month	Year

FOR SAFETY OFFICE USE ONLY:

DATE REPORT RECEIVED:	PROJECT MANAGER CONTACTED:	FURTHER ACTION			

Figure 3.2: Detailed accident report form

YOURCO   DETAILED ACCIDENT REPORT

(Sheet 1 of 2)

Accident Report No: Date Received:	Project Name:	Distribution: Project Manager Senior Manager Safety Manager

INJURED PERSON

1. Name:

2. Nationality:

3. Address:

4. Employer:

5. Age: 6.   Married   Single

ACCIDENT

	Day	Month	Year	Time
7. Date and time of accident:				

8. Exact location on site:

9. What time did the injured person start work the day of the accident?

10. When did they stop work as a result of the accident?	Day	Month	Year	Time

11. What was the injured person doing at the time of the accident?

12. On what level was the injured person working?

Excavation	Basement	Ground	Roof	Upper Level	Other	*

13. Was this the injured person's authorised work?	YES	NO	*

14. Nature and extent of injury: (state exact parts of body injured)

15. If injured person fell, or an object fell on them, state the height of the fall:

* — Delete as appropriate

(Sheet 2 of 2)

16. Was first-aid given to the injured person on site? | YES | NO | *

17. If yes, by whom?

18. To whom on site was the accident first reported?

19. What action has been taken to prevent a recurrence of this accident?

24. What were the causes of the accident?
(attach drawings/sketches if necessary)

25. Was scaffold involved? | YES | NO | *

26. Was machinery involved | YES | NO | *

[If YES, supply name of machine and details of the part causing the injury]

GENERAL

27. Was the injured person taken to hospital?

28. If YES, state name and address of hospital:

29. How long is the injured person likely to be off work?

30. Were there any witnesses to the accident? | YES | NO

31. If YES, are names/addresses/statements attached? | YES | NO

32. Was the injured person working under supervision at the time of the incident? | If YES, give the supervisor's name |

33. Was the task/activity covered by a risk assessment? | YES | NO

34. Was a method statement prepared for the task/activity? | YES | NO

35. Was the method statement being followed? | YES | NO

36. Report author and signature

Figure 3.3: Sample checklist — set-up of temporary works — tower cranes				
PROJECT NAME: CRANE SUPPLIER: CRANE BASE DESIGNER:		IDENTIFY CRANE(S):		
NO.		YES	NO	COMMENTS
1	Design			
a)	Are the calculations available on site?			
b)	Are sketches/drawings available on site?			
c)	Is the base design still applicable to the actual crane being supplied, as the available crane type may change?			
2	Design Approval			
a)	Has the design been checked by an engineer?			(who/date?)
b)	Has a formal design certificate been issued?			(date?)
3	Clearances			
a)	Have oversail and tailswing clearances been checked?			
b)	Have adjacent property owners been notified of overswing?			
c)	Have local Statutory Authorities/Utilities etc. been notified?			
4	Public Utilities			
a)	Has identification of buried services or services exposed during the works been carried out and recorded?			
5	Concrete Base			
a)	Has concrete base bearing pressure been agreed? (Gravity Bases)			
b)	Do pile loads specified for a piled base provide a minimum factor of safety of 2.0?			
c)	Is it confirmed that the pile design can resist the tensile loads imposed?			
d)	Have the clearances between adjacent piles been approved by the engineer?			
6	Steel Grillage Base			
a)	Has foundation grillage been weld tested after fabrication?			
b)	By whom?			(name)
c)	Tests applied? List:			
d)	Have foundation grillage bolts been tightened to correct torque?			

Contd

NO.	Figure 3.3: *Contd*	YES	NO	COMMENTS
7	**Crane Erection and Dismantling**			
a)	Has the crane erection and dismantling method statement been made available?			
8	**Inspection and Maintenance**			
a)	Has the 3 monthly mast and grillage bolt torque checking programme been agreed?			
b)	Has the inspection and maintenance programme been agreed?			
c)	Who will undertake statutory inspections?			(name/agency)
9	**Safety Legislation**			
a)	What are the requirements of Regulations? List:			

The checklist has been completed in respect of the equipment identified.

Signed: Position:

   Date:

Figure 3.4: Sample checklist — set-up of temporary works — hoists

PROJECT:

HOIST SUPPLIER:

HOIST LOCATION:

DESIGNER:

NO.	SITUATION	YES	NO	REMARKS
1	Type of hoist:			
	Goods only:			
	Goods and passenger?			
2	State the following hoist particulars:			
	Manufacturer:			
	Model No:			
	Safe Load Carrying Capacity:			
3	Design			
a)	Are calculations available on site:			
b)	Are sketches/drawings available on site?			
4	Design Approval			
a)	Has design been checked by an engineer?			(who/date?)
b)	Has a formal design certificate been issued?			(date?)
5	Mast Base			
a)	Is the base founded upon the ground or a ground-bearing slab and has this been correctly prepared for hoist loads?			
b)	Is the base founded upon a suspended floor slab or deck and have structural props been provided below the hoist mast?			
c)	Is the base supported by a scaffold or structural steel gantry and is the layout designed to withstand impact buffer loads?			
6	Mast Ties			
a)	Are all the mast ties installed in accordance with the layout shown upon the design sketches/drawings? (Note any installation modifications)			
b)	Are the fixings to the main structure installed with the correct fixing bolts (particularly when fixing into brickwork) and in accordance with the design details?			
c)	Have fixings into brickwork been tested?			
d)	Are all nuts and bolts within the tie-arm assemblies tightened to the correct torque and of a self-locking type?			
e)	Where tie-arms are fixed to main structural beams, are the bottom flanges of these beams adequately restrained?			
f)	Where tie-arms are of a length in excess of 4 metres, is a support against the self-weight of the tie-arm required and if so, has this been fitted?			

Figure 3.5: Example basic checklist — general: site physical hazards and arrangements				
PROJECT NAME/ADDRESS: INSPECTION BY: DATE:				
NO.		**YES**	**NO**	**COMMENTS**
1	**Public protection and information**			
a)	Are the working areas fenced off, or is there other protection for the public?			
b)	Is access to the site restricted to authorised visitors?			
c)	Is there a notice displayed at the entrance(s) requiring all visitors to check in before proceeding onto the site?			
d)	Is a copy of Form 10 displayed, to provide details of the principal contractor and planning supervisor?			
e)	When work finishes, are the following checked?			
	■ Site entrances are secured			
	■ Perimeter fencing is intact and functional			
	■ Flooded areas, excavations, openings are covered or protected by barriers			
	■ All site plant and remaining vehicles are immobilised			
	■ Flammable materials and COSHH substances are removed from work areas and locked away			
	■ LPG and other gas supplies are locked off and keys are removed			
	■ Access to scaffolding is removed or blocked			
	■ All piled and stacked material is safe			
2	**Fall prevention/protection**			
a)	Have all exposures to falls been assessed for possible methods of prevention?			
b)	Are all open edges and holes appropriately protected to guard against falls of people and materials?			
c)	Are appropriate guardrails and toeboards (kickboards) present at every edge where falls of more than 2 m could occur?			
d)	Where platforms cannot be used as first choice for working places, has an adequate evaluation been made of the preventive measures being used?			
3	**Safe access**			
a)	Is there safe access to all places of work, with all access equipment and gangways, and other walking surfaces clear and unobstructed?			

Contd

Figure 3.5: *Contd*				
NO.		**YES**	**NO**	**COMMENTS**
b)	Are all structures being worked on stable, safe and not overloaded?			
c)	Is the site tidy and meeting an acceptable standard of housekeeping?			
d)	Has the collection and disposal of waste been adequately organised?			
e)	Is there adequate lighting in main and common areas, in addition to task lighting?			
4	**Scaffolds**			
a)	Are checks made to ensure that scaffolds are only erected and altered by competent persons?			
b)	Is safe access available to platform levels in all cases?			
c)	Are there effective means of warning people not to use incomplete or unsound scaffolds?			
d)	Have all scaffolds been designed and erected appropriately to meet the needs for which they are constructed?			
e)	Are all scaffolds inspected at least weekly by a competent person and the results recorded, and always following substantial alteration or damage?			
f)	Are all working platforms fully and properly boarded out?			
g)	Are all scaffold bracing and support members in place?			
h)	Are all scaffolds appropriately braced or secured to structures appropriately to prevent collapse?			
i)	Have all requirements for guardrails, toeboards and platform width been met? [Guardrails at least 910 mm above work area, no unprotected gap of more than 470 mm between toeboards and guardrails, toeboards at least 150 mm high, minimum platform width of 600 mm]			
j)	Are stored materials evenly distributed on the scaffold platforms and not excessive?			
5	**Ladders and stepladders**			
a)	Are ladders used only as means of access, except for short duration work?			
b)	Are ladders checked for defects before use and in good condition?			
c)	Are ladders secured before use to prevent slipping?			
d)	Is there always a handhold available at landing places?			

Contd

NO.		YES	NO	COMMENTS
Figure 3.5: *Contd*				
e)	Are steps always used fully extended, with adequate ropes or stays to prevent misuse?			
f)	Are there firm site rules to prevent workers from standing or working on the top third of stepladders?			
6	**Powered access equipment, including MEWPs**			
a)	Is the equipment properly erected by a competent person to comply with suppliers' instructions?			
b)	Is fixed equipment firmly braced and supported, and connected to an appropriate structure?			
c)	Are working platforms in good condition, fitted with edge protection to prevent falls of persons and materials?			
d)	Are precautions in place to protect other people from movement of the equipment and falling materials?			
e)	Are the operators appropriately trained and competent?			
f)	Are the procedures for isolation at the conclusion of work adequate and appropriate?			
g)	Is there a safe system of work in place for users to follow, including the wearing of safety harnesses in high risk situations?			
7	**Cranes and lifting appliances**			
a)	Are crane erection and dismantling method statements available for all cranes and lifting appliances?			
b)	Are the safe working loads and other constraints known to operators?			
c)	Are all operators trained and competent?			
d)	Are safe load indicators fitted where appropriate to all lifting appliances?			
e)	Have slingers or banksmen been appointed and identified to operators as the only persons entitled to give signals?			
f)	Have all slingers or banksmen been adequately trained in signalling and slinging?			
g)	Have all slingers or banksmen been specifically instructed in the identification of weight and centre of gravity before lifting a load?			
h)	Are all lifting appliances inspected at least weekly by a competent person with the inspection results recorded?			
i)	Do all lifting appliances have appropriate in-date test certificates issued by a competent authority?			

Contd

Figure 3.5: *Contd*				
NO.		**YES**	**NO**	**COMMENTS**
j)	Are drivers of visiting mobile cranes and lifting appliances required to produce inspection documents and evidence of competence?			
k)	Are all lifting appliances required to operate outriggers only with additional timber beneath to spread the load?			
8	**Plant and equipment**			
a)	Is all plant and equipment being used appropriately — the right equipment for the job?			
b)	Are all dangerous parts guarded? [Shafts, gears, pulleys, drive belts for example]			
c)	Is all plant and equipment in good repair?			
d)	Is a maintenance log maintained to record any defects and the date of their repair as well as preventive maintenance?			
e)	Where appropriate, do all operators possess the appropriate certification or authority to operate equipment?			
9	**Hoists**			
a)	Are all hoists inspected at least weekly by a competent person and the results recorded?			
b)	Do all hoists have appropriate in-date test certificates issued by a competent person?			
c)	Are the gates kept shut other than when the platform or cage is at the landing?			
d)	Are effective gates and barriers in place at all landings, including ground level?			
e)	Are the controls arranged so that operation is from one position only?			
f)	Are all operators trained and competent?			
g)	Are materials hoists marked to prevent people riding on them?			
h)	Is the safe working load clearly marked on all hoists?			
i)	Are all hoists protected to prevent anyone being struck by any moving part, or materials falling down the hoistways?			
10	**Site traffic and vehicles**			
a)	Is there a site traffic plan in place, with details given to drivers where necessary?			
b)	Does the plan identify separate pedestrian and vehicle routes where possible?			

Contd

	Figure 3.5: *Contd*			
NO.		**YES**	**NO**	**COMMENTS**
c)	Has the need for reversing been minimised?			
d)	Where reversing is necessary, are such movements controlled by banksmen?			
e)	Are regular physical checks made on the condition of vehicles, including the functioning of warning devices?			
f)	Are drivers trained and in possession of appropriate certification?			
g)	Are vehicles properly loaded, with protection against falling loads where appropriate?			
h)	Are passengers provided with secure riding positions and prevented from riding in dangerous positions?			
i)	Are speed limits in force and enforced?			
11	**Excavations**			
a)	Is an adequate supply of appropriate supporting material present before work starts? [Sheets, timber, trench boxes, props, etc.]			
b)	Has a safe method of work been agreed, which will prevent people from the need to work within unsupported areas?			
c)	If sloping sides are selected as the appropriate protection method, is the angle of batter sufficient to prevent collapse?			
d)	Has safe access by ladders been provided to all excavations?			
e)	Are all excavations protected where necessary to prevent people, materials or vehicles falling in or causing them to collapse?			
f)	Are stop blocks provided to prevent tipping vehicles falling in?			
g)	Is the stability of any nearby structure likely to be affected by an excavation?			
h)	Are all excavations inspected by a competent person before each shift and after any event likely to compromise stability, with the results recorded?			
i)	Is there 1 m of clear space at the edges of all excavations, between the edges and spoil heaps or material stacks?			
12	**Fire and other emergencies**			
a)	Is there a site emergency plan, detailing steps required to evacuate the site in case of fire, and for confined space rescue where appropriate?			

Contd

Figure 3.5: *Contd*				
NO.		YES	NO	COMMENTS
b)	Are those on site made aware of the emergency plan and procedures?			
c)	Is there an appropriate means of raising the alarm, which is known to be clearly audible at all points on the site?			
d)	Are appropriate numbers of fire extinguishers present and marked?			
e)	Are arrangements made to remove waste regularly, and for storing it in bins prior to removal?			
f)	Are LPG and other gas cylinders correctly stored externally?			
g)	Is there an effective smoking ban where flammable materials are used, stored or installed?			
h)	Is the quantity of flammable material on site kept to a minimum?			
13	Electricity			
a)	Has identification of buried services or services exposed during the works been carried out and recorded?			
b)	Are appropriate precautions in place to guard against striking underground services during the work?			
c)	Have all overhead lines been identified and appropriate steps taken to remove, divert or mark the lines to prevent contact?			
d)	Is the supply voltage for tools and equipment the lowest practicable?			
e)	Have all temporary circuits been fitted with residual current devices?			
f)	Are all cables and leads protected from damage?			
g)	Are all connections properly made with suitable plugs and connectors?			
h)	Is all site lighting at reduced voltage, with bulbs and lamps protected against damage?			
i)	Are all halogen lamps raised up above head height and fitted with shields?			
j)	Is there an appropriate system in place to ensure that the temporary electrical supply system is checked by a competent person at appropriate intervals?			
k)	Are all electrical tools and equipment subject to a documented examination and testing regime?			

Contd

Figure 3.5: *Contd*				
NO.		**YES**	**NO**	**COMMENTS**
14	**Hazardous substances**			
a)	Have all hazardous substances been identified and their risks assessed?			
b)	Are appropriate precautions in place, including signs, limitations on access and use of protective clothing and equipment?			
15	**Manual handling**			
a)	Have mechanical handling solutions been identified and their use maximised?			
b)	Have positive steps been taken to identify materials likely to be supplied in bulk, in unacceptably large size units? [e.g. blocks, dry goods, cement]			
c)	Have steps been taken to minimise risks to those who have to handle materials manually?			
16	**Welfare**			
a)	Are all welfare facilities reasonably accessible to all workers on site?			
b)	Is there accommodation available for sitting, heating water and preparing food?			
c)	Are there changing, drying and clothes-storing facilities?			
d)	Are there adequate numbers of toilets, and wash basins with warm water, cleaners and towels?			
e)	Is there a sufficient supply of drinking water?			
f)	Has an adequate assessment of first aid requirements been made and appropriate facilities provided?			
17	**Noise**			
a)	Has quieter equipment been specified where appropriate?			
b)	Has the work been planned to minimise noise exposure on and off the site?			
c)	Has noisy work been segregated to reduce exposure to those not involved?			
d)	Has suitable hearing protection been provided for workers in noisy areas?			
e)	Are boundary noise measurements taken to establish baseline data?			

Contd

Figure 3.5: *Contd*				
NO.		YES	NO	COMMENTS
18	**Personal protective equipment**			
a)	Have appropriate personal protective equipment requirements been defined by risk assessment for all site workers?			
b)	Have the appropriate items of personal protective equipment been issued to users and signed for? [e.g. helmet, eye protection, hearing protection, respiratory protective equipment, gloves, safety footwear, protective outdoor clothing for adverse environments including dusty, wet, or dirty conditions]			
c)	Is all personal protective equipment in good condition and properly maintained?			
d)	Have arrangements for replacement of worn or lost equipment been made, and explained to users?			
e)	Have arrangements been made for maintenance of the equipment where necessary and storage when not in use?			

4 Techniques of Construction Safety Management

The term 'safety management' is used for convenience and for brevity, and wherever it is used it should be taken to refer to the management of occupational health and the environment as well as safety. Safety management is concerned with, and achieved by, all the techniques which promote the subject. Some have been described already, others will be covered in later chapters.

Safety management is also concerned with influencing human behaviour, and with limiting the opportunities for mistakes to be made which would result in harm or loss. To do this, safety management must take into account the ways in which people fail (fail to do what is expected of them and/or what is safe), and Chapter 11 contains an outline of those ways. Generally, safety management techniques are aimed at the recognition and elimination of hazards, and the assessment and control of those risks which remain. Many risks cannot be confined to the construction process — there are overlaps with clients, other contractors and third parties.

Objectives

The practical objectives of safety management are:

- Gaining support from all concerned for the health and safety effort
- Motivation, education and training so that all may recognise and correct hazards
- Achieving hazard and risk control by design and purchasing policies
- Operation of a suitable inspection programme to provide feedback
- To ensure that hazard control principles form part of supervisory training
- Devising and introducing controls based on risk assessments
- Compliance with regulations and standards.

To co-ordinate and achieve these objectives, the keystone is a safety policy statement. The design and other considerations for this policy are discussed in the next chapter.

Benefits

As discussed in Chapter 1, successful safety management can lead to substantial cost savings, as well as a good accident record. Inadequate safety management can lead to financial ruin, especially for smaller businesses.

Some companies have become well known for the success of their safety management system. UK leaders in this field include Bovis Lend Lease, AMEC and Taylor Woodrow, and many smaller companies have devoted substantial time and money to the development of sophisticated management systems.

Although neither a construction operation nor UK-based, the work of Du Pont is noteworthy. This company claims that several of its plants with more than 1000 employees have run for more than 10 years without recording a lost-time injury accident. Du Pont uses 10 principles of safety management which are worthy of study:

1. All injuries and occupational illnesses are preventable
2. Management is directly responsible for doing this, with each level accountable to the one above and responsible for the level below
3. Safety is a condition of employment, and is as important to the company as production, quality or cost control
4. Training is required in order to sustain safety knowledge, and includes establishing procedures and safety performance standards for each job
5. Safety audits and inspections must be carried out
6. Deficiencies must be corrected promptly, by modifications, changing procedures, improved training and/or consistent and constructive disciplining

7. All unsafe practices, incidents and injury accidents will be investigated
8. Safety away from work is as important as safety at work
9. Accident prevention is cost-effective; the highest cost is human suffering
10. People are the most critical element in the health and safety programme. Employees must be actively involved, and complement management responsibility by making suggestions for improvements

It is important to appreciate that these basic ideas are just as applicable to the construction industry as chemical manufacturing — or indeed any other industry. Backing for these principles can be found in British Standards and in Health and Safety Executive (HSE) publications. Critics complain that it is much harder to apply the principles to the construction industry, and this is probably correct. Construction differs from manufacturing industry in many respects, but from an organisational viewpoint it has a transient, highly mobile workforce carrying out work that is constantly changing and where the risks are generally higher. The workforce is correspondingly more difficult to train, motivate and involve in corporate safety efforts. Therein lies the challenge.

Key elements

The key elements of successful health and safety management are:

- Policy
- Organising
- Planning and implementing
- Measuring performance
- Reviewing performance and auditing

The reader interested in a complete treatment of this topic is referred to the HSE publication HSG65, *Successful Health and Safety Management*. The central idea is viewed as a process which depends upon continual feedback, certainly from reviews and audits, but also during the earlier stages, so that there is a continual, dynamic system in place. This model, with some variations, also lies at the heart of other management systems such as BS8800, ISO 18001, and the somewhat older principles of quality assurance and management which began as BS5750 and became the ISO 9000 series. In the latter case, as with ISO 14000, the importance of documenting the system is stressed, as this provides an audit tool that is specific to the system.

Policy

Successful safety management demands comprehensive health and safety policies which are effectively implemented and which are considered in all business practice and decision-making. Since April 1975, the law has required written safety policy statements to be created by all employers, except for the smallest organisations. This is simply a reflection in the law of what has been known for many years — written policies are the centrepieces of good health and safety management. They insist, persuade, explain and assign responsibilities. An essential requirement for management involvement at all levels is to define health and safety responsibility in detail within the written document, and then to check at intervals that the responsibility has been adequately discharged. This process leads to ownership of the health and safety programme, and it is based on the principle of accountability.

Organising

To make the health and safety policy effective, both management and employees must be actively involved and committed. Organisations which achieve high standards in safety create and sustain a **culture** which motivates and involves all members of the organisation in the control of risks. They establish, operate and maintain structures and systems which are intended to:

- Secure **control** — by ensuring managers lead by example
- Encourage **co-operation** — both of employees and those representing them as their trade union safety representatives or in other ways
- Secure effective **communication** — by providing information about hazards, risks and preventive measures
- Achieve **co-ordination** of their activities both internally between projects, sites, departments and other operating areas, and with other organisations which interface with them
- Ensure **competence** — by assessing the skills needed to carry out all tasks safely, and then by providing the means to ensure that all employees (including temporary ones) are adequately instructed and trained

Recognising the fundamental importance of this, the Management of Health and Safety at Work Regulations 1999 encourage by requiring employers to recruit, select, place, transfer and train on the basis of assessments and capabilities, and to ensure that appropriate channels are open for access to information and specialist advice when required.

Planning and implementing

Planning ensures that health and safety efforts really work. Success in safety management relies on the establishment, operation and maintenance of planning systems which:

- Identify objectives and targets which are attainable and relevant
- Set performance standards for management, and for the control of risks which are based on hazard identification and risk assessment, and which take legal requirements as the accepted minimum standard
- Consider and control risks both to employees and to others who may be affected by the organisation's activities, the structures they construct and complete and the services they provide
- Ensure documentation of all performance standards

Organisations which plan and control in this way can expect fewer injuries and claims resulting from them, reduced insurance costs, less absenteeism, higher productivity, improved quality and lower operating costs.

Monitoring

Just like finance, the safety management system has to be monitored to establish the degree of success of the operation. For this to happen, two types of monitoring system need to be operated. These are:

- **Active** monitoring systems. They are intended to measure the achievement of objectives and specified standards before things go wrong. This involves regular inspection and checking to ensure that standards are being implemented and that management controls are working properly. Examples are regular inspections by site management, and technical inspections of equipment at specified intervals.
- **Reactive** monitoring systems. They are intended to collect and analyse information about failures in health and safety performance, when things do go wrong. This involves learning from mistakes, whether they result in accidents, ill health, property damage incidents or 'near misses'. Examples are investigation reports, and reviewing of risk assessments and method statements following incidents.

Information from both active and reactive monitoring systems should be used to identify situations that create risks and enable something to be done about them. Priority should be given to the greatest risks. The information should then be referred to people within the organisation who have the authority to take any necessary remedial action, and also to make any organisational and policy changes which may be necessary.

Reviewing and auditing performance

Auditing enables management to ensure that their policy is being carried out and that it is having the desired effect. Auditing complements the monitoring programme. Economic auditing of a company is well established as a tool to ensure economic stability and it has been shown that similar systematic evaluation of safety performance has equal benefits.

As mentioned in the previous chapter, an audit is not the same thing as an inspection. Essentially, the **audit** assesses the organisation's ability to meet its own (defined) standards on a wide front, rather than providing a 'snapshot' of a particular site or premises. Each has its uses, but inspections are part of the active monitoring system (see above) and audits are an element of a coherent ongoing management system. Inspections are often made as a part of the audit process.

The two main objectives of an audit are:

- To ensure that standards achieved conform as closely as possible to the objectives set out in the organisation's safety policy, and
- To provide information to justify carrying on with the same strategy, or a change of course

The best health and safety audit systems are capable of identifying deviations from agreed standards, analysing events leading to these deviations and highlighting good practice. They look especially at the 'software' elements of health and safety such as systems of work, management practices, instruction, training and supervision as well as the more traditional 'hardware' elements which include scaffolding, machinery guarding and the use of personal protective equipment.

To carry out a simple audit, it is only necessary to take company documentation and turn its requirements into questions. For example, if part of the written arrangements (in the safety policy document) require a particular person to be responsible for making risk assessments, the audit paperwork could ask 'Have all necessary risk assessments been made' and 'Who made them?' It would be usual to verify the answers by interview and by seeing the documented evidence backing up the answers.

Another way of reviewing performance is by benchmarking. This has become popular amongst larger companies in recent years. Benchmarking is simply a process of comparison with other similar organisations in terms of methods and systems used, and also their cost savings and other results from using the systems. It offers an opportunity of evaluating comparisons of the best use of scarce resources in the interests of safety. Cynics may point out that it also gives employers the ability to judge what (minimum) steps are 'reasonably practicable' by reference to an industry norm.

And if companies do not do all this?

It is, of course, quite possible to remain ignorant of all the foregoing, and yet to run a construction business for some time without injuring anybody. Luck is a substantial ally, but a fickle friend. The risks remain. The aim of the management process is to identify and reduce the risks to as low a point as can be achieved — but even then there may still be some which have not been identified or which are considered to be negligible or tolerable. Where the risks have not been reduced, the danger is greater. Often, businesses that ignore safety to a greater or lesser extent say that they rely upon people 'to take care of themselves', and on foremen and other supervisors to control conditions on site. In some cases they are even unwise enough to put these philosophies into writing within their safety policy statements.

In all control measures, reliance is placed upon human behaviour to carry out the solutions, so a major task of any safety management system is to assure safe behaviour by motivation, education, training and the creation of work patterns and structures which enable safe behaviour to be practised. Discussion of some aspects of human failure can be found in Chapter 11.

In a major study carried out on construction sites in England in 1977, supervisors in the construction industry were asked to explain why deviance from good safety practice had occurred on sites in their charge. It had been widely believed up to that time that the main factor in achieving a good standard of safety on site was to make sure that the foreman or supervisor was appropriately trained. The results of the study showed that the main reasons put forward for the presence of defects were organisational and social, rather than any perceived lack of knowledge on the part of the supervisors. It may, of course, be felt that erring supervisors were always unlikely to admit their own incompetence! They gave a wide variety of reasons for their inactivity on safety matters: in order of frequency, the most common responses are tabulated in the list on page 39.

This study showed that reliance upon supervisors alone to achieve safe conditions on site may not be a successful strategy. Supervisors are part of the overall safety management system which must operate within a favourable environment, with clear responsibilities given and accountability practised, together with necessary training in the complex nature of the accident phenomenon and in solutions to health and safety problems. Supervisors' and workers' attitudes to safety generally reflect their perception of the attitudes of the employer.

Attempts to motivate the individual meet with greatest success when persuasion rather than compulsion is used to achieve agreed safety objectives. Consultation with workers, through representatives such as trade unions, and locally through the formation of safety committees, is generally a successful strategy, provided that an adequate role is given to those being consulted. The importance of consulting employees on matters affecting their health and safety has been recognised in UK law.

Benefits of safety committees include the involvement of the workforce, encouragement to accept safety standards and rules, help in arriving at practicable solutions to problems and recognition of hazards which may not be apparent to management.

The contribution of behaviour-based safety management as a technique that fosters employee involvement has been mentioned in the previous chapter. The technique is not without its critics, the more sensible of whom have pointed out the danger that physical hazards may be overlooked amid the enthusiasm to 'spot workers doing something wrong'.

Is it all just a pile of paperwork?

At this point the reader may reasonably be asking whether all the inevitable paperwork is really necessary. The fact is

The most common responses of supervisors to questions on safety practice

1.	Resource limitations	'There are not enough staff on site to do the job properly and my attention has to go to production'
2.	Safety tasks seen as outside the boundaries of their duties	'It's not my job to spot other people's mistakes'
3.	Acceptance of hazards as inevitable	'Construction work is dangerous, so people have to look out for themselves'
4.	Influences of the social climate on site	'I don't want to become unpopular by going on about safety — I'd always be complaining and we wouldn't get the job done'
5.	Industry tradition	'We've always done it that way though I know it's wrong'
6.	Lack of technical competence	'I don't know what the safe way is to do that'
7.	Incompatible demands upon their time	'I don't have enough time to do my job properly'
8.	Reliance upon the worker to take care	It's up to the men to look after themselves, not my job to nurse them'
9.	Lack of authority	'I can't stop them doing that, because the progress of work would suffer'
10.	Lack of information	'I thought it was dangerous, but I didn't know for sure'

that many organisations in the construction industry can point to 'a good safety record', but one which can be shattered in a moment. The experience of workers, coupled with the enthusiasm and concern of site management, can and does count for a lot. But when the magnifying glass is turned on the management process itself, there is often no evidence that the desires of the Board of Directors are translated through a measurable chain reaction down to site level. That is what UK law requires, and it is also what is needed in order to provide some guarantee that the safety record is a matter of planning, not luck.

Nobody would expect to run the financial management of a business without keeping accounts, without basic rules on controlling costs. Strangely, though, many people expect to be successful in controlling site safety, the health of their employees and minimising environmental impacts without having any system in place to handle it.

The emphasis is now shifting away from the 'employer responsible' towards the 'individual responsible'. One reason for the introduction of the offence of 'corporate killing' is to allow individual directors to be more clearly visible to the regulators and the courts. The trend is towards increasing personal accountability of directors, and increasing responsibility in corporate governance and sustainability. The control systems to meet those challenges are not difficult to manage, but they need to be measured, recorded and audited. In construction, the emphasis is towards pre-planning. Studies by the European Union have shown that 65% of fatalities studied in the industry have their causal origins in the pre-construction phase — design, work planning, scheduling and the like. The solution that will allow us to eliminate construction hazards lies in recognising the sense in getting the paperwork right; enough to do the job, but not too much so that the spirit of the business is stifled by it.

World best practice

Benchmarking against the world's most successful construction companies is not for the faint-hearted. It is plain that corporate minimum requirements must be set in order that all parts of the business can comply, where the construction work is spread around the world. Nevertheless, even the global minimum may be relatively high. In the

United Kingdom, compliance with the CDM Regulations allows those who do to demonstrate nearly all aspects of world best practice.

Ten major elements can be identified that are common to all the world's major contractors. There are many different ways used to measure the extent to which each element is complied with, but, in businesses where they are all found, directors can be sure that they are meeting the challenge of world best practice.

- All work is done according to a managed design that has taken safety, health and environmental issues into account, not only as they affect the end user, but also those constructing and maintaining the structure, and the population of the surrounding area
- All work has been assessed and steps have been taken to identify and control significant hazards and their risks
- All work is managed by staff with an appropriate knowledge of the safety, health and environmental issues involved in the work
- All work is carried out by contractors and their workers who are competent in safety and health matters as well as in their particular skills, who have been verified as competent and who have been given a job-specific induction to the work
- The work is done by contractors who have made appropriate allowance in their tenders for necessary health and safety measures required by the demands of the contract
- Workers have been given necessary information and training about the hazards and risks, and the control measures used to remove or minimise them
- There is a system for ensuring work is co-ordinated between groups of workers and different contractors, and that work safety issues are discussed and solutions agreed before the work begins
- The work is carried out in compliance with national or local standards where these exist, and in accordance with international good practice where they do not
- There is a safety plan specific to the work, which includes the details of the control methods applied to the hazards and risks, and a comprehensive fire, emergency and environmental plan, which is in place before the work to which it relates begins
- There is a system of reward for safe behaviour and compliance with the safety management system, and unsafe behaviour is penalised or otherwise discouraged.

Perhaps the most interesting feature of the elements is that most of them are organisational 'software' rather than consisting of 'hardware' items relating to the actual construction process. Many companies do have such 'software' requirements, but their prominence and technical details vary. World best practice lies in the organisation, coupled with in-company standards.

The role of the client

The client or owner is a key figure in the construction safety management process, as the instigator of the work and the source of funds. Curiously, with the exception of the European Union, Australia and New Zealand, the client is exempted worldwide from responsibility for any part of the safe conduct of the work he has commissioned. Yet, by withholding or providing adequate funds to programme (design) and carry out the work with due regard to safety, health and the environment, the client has a major influence in what happens on site. Gradually, this role is being recognised by national legislation. Meanwhile, some companies are being driven by stakeholder concern into adopting policies that call for refusal to work on some projects, or withdrawal from contracts where the client is not willing to pay for the work to be done safely.

In the Construction (Design and Management) Regulations 1994 (CDM), the client's main duties are contained in Regulations 6, 10, 11 and 12. A client is defined as 'any person for whom a project is carried out, whether carried out by another person or in-house'. Where projects are carried out for domestic (non-commercial) clients, most of the CDM requirements apply neither to the contractor nor to the client, unless the client is also a client of a developer, when by Regulation 5 the domestic house developer is the de facto client until the project is completed.

CDM imposed criminal liability onto the client for the first time. In cases where there could be multiple clients, and otherwise, a client can appoint an agent or another client to carry out the client's duties, and then has to make a declaration to the enforcing authority (the Health and Safety Executive) that the transfer of duties has been made.

Under CDM, the client must:

- Appoint a Planning Supervisor and a Principal Contractor in respect of each project, being satisfied that these duty holders are competent and have the resources to perform their duties adequately

- Not permit the construction work to start unless a health and safety plan which complies with Regulation 15(4) is in place for that project
- Provide the Planning Supervisor with information about the state or condition of the premises where the work is to be carried out. This is information which is relevant, and which the client either has or could get after making reasonable enquiries
- Satisfy himself that any designer or contractor he appoints directly is competent for the task and has allocated sufficient resources to it
- Make the health and safety file available for inspection by anyone who may need information in the file to comply with legal requirements. The client will sell or pass on the file to a future owner or a person acquiring the interest in the property of the structure to which it refers.

These duties are amplified in the Approved Code of Practice, and will come to the attention of the client because designers have a special duty under Regulation 13 to take reasonable steps to make a client aware of his duties under CDM. One of the client's major duties is to appoint a competent Principal Contractor for the work, and where CDM applies there will always be one of these. In doing so, the client can rely upon advice from the Planning Supervisor he has already appointed.

Specialist installers and maintainers may well take on other duty holder roles in additional to that of contractor — in some cases acting as Designer and/or Planning Supervisor and Principal Contractor as well. The potential for doing so should be explained to the client at an early stage. The sole limitation on the number and type of roles that an individual contractor can take on is that competence is a basic requirement for each role.

The contractor and CDM

Under CDM, a contractor is defined as any person who carries on a trade or business or other undertaking, whether for profit or not, in connection with which he undertakes to or does manage construction work, or arranges for any person at work under his control (including any employee, where he is an employer) to carry out or manage construction work. This definition includes the self-employed within its scope. A full definition is provided within Regulation 2. Specialist trade work is likely to be classed as 'construction work' within the meaning of the Regulations, so specialists working on 'construction work'

as defined will have duties under CDM if the work is within the scope of the Regulations.

A summary of the CDM Regulations can be found in Part 3 of this book. 'Construction work' means carrying out any building, civil engineering or engineering construction work associated with a structure. 'Structure' is given a very wide interpretation which extends to fixed plant being commissioned, installed or dismantled where anyone could fall 2 m or more. All the CDM Regulations will apply where the work falls within the definition given in the Regulations — non-domestic, notifiable and where five or more persons are to be at work at any one time. Where a project is not notifiable to the enforcing authority and fewer than five persons are involved at one time, only the designer carries CDM responsibilities (Regulation 13). The CDM (Amendment) Regulations 2000 produced a slightly revised definition of 'designer'.

Generally, work will be **notifiable** to the Health and Safety Executive where the work extends for a significant length of time, or involves a significant quantity of work and is done or managed by a contractor. Notifiable projects last longer than 30 days, or involve more than 500 person-days of construction work. The only common exception to this is for the demolition or dismantling of a structure, to which the Regulations **always** apply. The contractor carrying out work has the responsibility of notifying the Health and Safety Executive Area Office before work begins, using Form F10(REV). Copies of the form can be obtained from the Area Office concerned.

A **working day** is one on which construction work is carried out, regardless of the duration of the work and whether a weekend or public holiday is involved. A **person-day** is one individual, including supervisors and specialists, carrying out construction work for one normal working day or shift. Where there is any doubt over qualification for notification, projects should be notified.

Many of the Regulations do not apply when construction work is being done for a **domestic client**. The domestic client of a contractor does not attract any of the duties otherwise given by the CDM Regulations to commercial clients. The following CDM Regulations do apply to work done for domestic clients:

- The contractor must notify the project to the HSE, if it is notifiable under the definition given above
- The duties of the designer apply in full (Regulation 13)

The regulations do not apply to 'minor work', or where it is done in premises normally inspected by the local authority (see below). 'Minor work' is that done by people who normally work on the particular premises. It is either not notifiable, or entirely internal, or carried out in an area which is not physically segregated, normal work is carrying on and the contractor has no authority to exclude people while it is being done.

Premises normally inspected by the local authority include those where goods are stored for wholesale or retail distribution, or sold. Such premises include car windscreen, tyres and exhaust system replacement operations, exhibition displays, offices, catering services, caravan and temporary residential accommodation, animal care except vets, farms and stables, consumer services, launderettes, sports premises and museums.

The Principal Contractor and CDM

The **Principal Contractor** is any person appointed for the time being by the client under Regulation 6(1)(b). There is no restriction on the appointment, except that the duty holder must be competent, be able to allocate adequate resources for health and safety in order to carry out the duties, and the person must be a contractor. The 'person' may be an individual, a company, part of a company — and may already hold other responsibilities and duties under CDM where competent to do so.

The appointment of the Principal Contractor is made by the client as soon as practicable following possession of the necessary information about the candidate's competence and resources. The Principal Contractor's duties are contained in Regulations 16—18. He must:

- Take reasonable steps to ensure that contractors co-operate
- Ensure that everyone complies with any rules in the safety plan
- Take reasonable steps to ensure that only authorised persons enter the construction area
- Display particulars about the site and details about himself and the Planning Supervisor where they can be read by workers
- Take reasonable steps to ensure that the safety plan contains the correct arrangements and details until the end of the construction phase of the work
- Keep the Planning Supervisor up to date with items for the safety file

- Give reasonable directions to contractors, and include rules in the safety plan where necessary
- Ensure so far as is reasonably practicable that every contractor is given information about relevant risks, and ensure that all workers are given training and information by their employers
- Make arrangements to allow discussion of health and safety issues by both employees and the self-employed, and arrangements so that they can offer him advice, and make appropriate arrangements for the co-ordination of the views of the employees or their representatives

These duties are amplified in the Approved Code of Practice. Prior to the start of work on site, the development of the safety plan is the main CDM responsibility of the Principal Contractor. Essentially, the developed plan will include risk assessments and method statements of contractors, the common arrangements in place for welfare, details of how the Principal Contractor will fulfil his CDM duties, site rules and monitoring arrangements. The safety plan is discussed further in Chapter 8.

The Planning Supervisor and CDM

The **Planning Supervisor** is any person appointed to the role by the client, with the aim of helping the client to comply with Regulations 8, 9 and 11. The Planning Supervisor is required, principally by Regulations 14 and 15 to:

1. Ensure as far as is reasonably practicable that the design of any structure in the project includes adequate regard to the 'three needs':
   (a) The need to avoid foreseeable risks to the health and safety of constructors, cleaners and maintainers, or anyone who might be affected by their work
   (b) The need to combat risks to the same categories of people at source
   (c) The need to give priority to measures affecting everyone in those categories rather than measures protecting each individual
2. Ensure so far as is reasonably practicable that the design of any structure in the project includes adequate information about any aspect of the project or structure, or materials, which might affect the health or safety of people in the above categories
3. Take reasonable steps to ensure co-operation between designers to allow each to comply with the three needs. This requires a co-ordinating function, including facilitating the free flow of information

4. Be in a position to give adequate advice to the client and any contractor to enable them to appoint competent designers with adequate resources to allow them to meet the three needs, and be in a position to give similar advice to the client to appoint a competent contractor

5. Ensure that a Health and Safety File is prepared in respect of each structure in a project, which contains adequate information about any aspect of the structure or materials which might affect the health or safety of people in the above categories

6. Review, amend or add to the health and safety file as necessary during the work, and on completion of the work on each structure to deliver the file to the client

7. Ensure that a Health and Safety Plan covering the proposed construction work has been prepared so as to allow a copy to be provided to each contractor before arrangements are made for that contractor to start or manage construction work

8. (Potentially) advise the client that the Health and Safety Plan complies with Regulation 15(4) before the client authorises the construction phase of the work to start.

The Planning Supervisor's duties do not extend to the making of requirements as to the construction methods used, or to carry out any health and safety management function in respect of contractors or the work. These duties may, of course, be carried out by the person acting both as Planning Supervisor and in another capacity on the project.

Competence, qualification and selection under CDM

It is a central feature of CDM that all duty holders (except the client) must be competent, and the Regulations specify that certain steps must be taken to ensure that only competent duty holders are selected. No contractor can be employed to carry out or manage construction work unless the person employing them is satisfied as to their competence. That duty will only be discharged when reasonable steps have been taken, which include making reasonable enquiries or seeking advice where necessary.

A great deal of time and energy has been devoted to the completion and collection of questionnaires which have been used by several of the parties to CDM to attempt to demonstrate competence. The present view of the HSE is that much of this has been unnecessary. The extent of the enquiries and what is reasonable depends on the sort of

work to be done. What will be looked for is the same in all cases, but the depth of enquiry should depend on the circumstances:

- Knowledge and understanding of the work involved
- Ability to manage and prevent risks
- Working to relevant health and safety standards
- The capacity to apply the knowledge
- Availability of adequate resources (including plant, machinery, technical knowledge, trained personnel, time)

Even the smallest **contractors** working under CDM must therefore expect to be asked about their competence and resources. To satisfy questioners, they will need to have available, as a minimum:

- An up-to-date **safety policy statement**, which specifies the means used to audit and monitor the policy and activities on site, and the means of showing compliance with the Management Regulations 1999. For most contractors, this requires recording of the arrangements made to effectively plan, organise, control, monitor and review the preventive and protective measures
- **Risk assessments** covering their common activities to which significant risks are attached. Although there is no requirement for contractors employing fewer than five people to have either a written safety policy or written risk assessments, not having done so is likely to place them at a disadvantage — references from other contractors would be useful in that case
- **Method statements** covering activities where the risks are properly controlled by this means, identifying standard methods and precautions
- **Biographies** for project managers or those in charge of the work, identifying previous experience and/or qualifications

Competence checks of contractors can be made in two stages — one when a general assessment is made as part of a pre-contract qualification process, and secondly as part of a project-specific check. At either stage, competence checks may include:

- Arrangements in place to manage health and safety actively — risk assessments, safety policy, appointment of competent persons, for example
- Procedures the contractor will follow to develop and implement the Health and Safety Plan

- Arrangements for training of site management and employees
- Specific approaches to deal with high-risk activities identified in the pre-tender Health and Safety Plan (which may well not specify detailed control steps)
- The people who will be managing and carrying out the work — qualifications, competence, training, skills
- The time allocated to do the work with minimum risk to health and safety

Making checks by questionnaire is not mandatory — as mentioned above, the HSE indicates that excessive reliance on documentation may be unwise, and that more personal approaches such as interviews, project visits and taking up references are likely to yield better-quality information.

The client is required to make a number of specific checks about the competency of the **Planning Supervisor** before making the appointment, in respect of his knowledge and ability to carry out the duties. Steps which will be considered reasonable by the enforcing authority include asking questions about:

- Knowledge of construction practice relative to the particular project
- Familiarity with and knowledge of the design function
- Knowledge of health and safety requirements
- Ability to work with and co-ordinate the activities of different designers
- Ability to facilitate dialogue between designers and constructors
- The number, experience and qualifications of the employees who will carry out the duties and functions, both internally and externally resourced
- The management systems to be used to monitor the correct allocation of his people and resources
- The time allocated to carry out the various duties
- Technical resources available to aid staff

Similarly, the client is required to make specific checks on the competency of his **proposed Principal Contractor** before making the appointment, in respect of knowledge and ability to carry out the duties. Checks made on other contractors will also be relevant to this appointment — the Principal Contractor is only a contractor with heightened responsibilities. Therefore, the steps taken to make checks which will be considered reasonable by the enforcing authority will also be valid for other contractors:

- The people who will be managing the work — their skills, knowledge, experience, training, numbers, relevant qualifications
- The time allowed to complete various stages of the work
- Knowledge of health and safety requirements and the way compliance will be ensured
- Ability to work with other contractors and co-ordinate them
- Technical and managerial approach to dealing with the risks identified within the Health and Safety Plan
- The technical resources available to aid staff

The Health and Safety File

The Health and Safety File is a collection of documents showing the methods and means of construction of a structure, for the benefit of anyone who carries out construction work on the structure at any time after completion of the current project. It provides information which otherwise might rest only in the minds of those who built or modified the structure, or in documents which might otherwise have been destroyed at the end of work on it. Regulation 14 of CDM requires the Planning Supervisor to amend and add to the file so that it contains the required information when handed over to the client. The client is then placed under a duty by Regulation 12 to make the file available for inspection and pass it to the next owner of the structure.

The contents of the file are detailed in Regulation 14, and expanded within the Approved Code of Practice which gives practical examples of potential contents. The following is an illustrative example.

A sample Health and Safety File

Drawings	Record drawings and plans used and produced throughout the construction process, along with any design criteria
Method details	Construction techniques, methods, materials used and any special characteristics
Equipment	Details of the structure's equipment and its maintenance facilities
Maintenance	Details of the maintenance requirements and procedures for the structure as a whole
Manuals	Those produced by specialist contractors and suppliers in relation to equipment and plant installed as part of the structure, outlining maintenance and operating procedures, schedules and the like
Utilities	Details of the location and nature of utilities and services installed or supplied to the structure, including emergency and fire fighting systems.

5 | The Safety Policy

Previous chapters have discussed the main principles of safety management — but what do they mean in practice? How do they relate to the employee? Communicating the practices and procedures developed by a business requires a powerful tool, and this is the safety policy — the core document in the management of safety anywhere, not just in the construction industry.

Management commitment must be demonstrated by the most effective means. For most contractors, the law requires that this is done by issuing a safety policy statement, a document signed and dated by the most senior member of the management team. The continuing role of management is then to ensure that the requirements of the policy are actively carried out by managers, supervisors and employees alike.

Lack of firm management direction encourages the belief that 'safety is someone else's business'. Without active support at all levels in an organisation, any attempt at the control of accidents and ill-health will be useless — or even worse than useless, since there may be an illusion that environment, safety and health concerns are identified and under control, resulting in complacency. The management of these concerns requires a sustained, integrated effort from everyone — directors or partners, contract and project managers, planners, buyers, site managers, supervisors, foremen and everyone who works on site, in an office or in any other premises managed by the business.

Policies are under constant scrutiny from a variety of interested parties, because only senior management can provide the authority to ensure that safety management is properly co-ordinated, directed and funded. Its influence will be seen in the way the policy is written, the amount of attention it gets and what happens to anyone who decides not to follow it. Most construction employers now have safety policies in place, but they need to be kept up to date. The objective is to reflect what actually happens, and

to make sure the policy does not contain a hopeful list of wishes that are never likely to be fulfilled.

Legal requirements

The original legal requirement for employers to create and publish safety policies is to be found in section 2(3) of the Health and Safety at Work etc. Act 1974. This requires written safety policy statements to be created by all employers, except for the smallest organisations (those with fewer than five employees). The statement itself is an expression of management intention, and as such does not need to contain detail. What is usually referred to as 'the safety policy' will contain this **statement**, together with the details of the **organisation** (responsibilities at each level within the operation) and the **arrangements** (how health and safety will be managed, in detail). These contents are prescribed by the Act, which gives no further advice on the content or the level of detail required.

As the key document pulling all the threads of the management process together, the safety policy will detail the arrangements made to comply with relevant regulations, including the Management of Health and Safety at Work Regulations 1999 (MHSWR), which cover effective planning, organisation, control, monitoring and review of control measures, and the Provision and Use of Work Equipment Regulations 1998. Many regulations contain words like 'adequate', 'suitable' and 'sufficient', which need to be defined in the organisation's own terms. The safety policy is the place to do that.

The policy in practice

Safety policies, as written statements of the intentions of management, acquire a quasi-legal status. Among other things, they serve as a record of the intended standard of care to be provided by the employer. This offers a useful method of evaluating an organisation in terms of health and safety, especially because its standards, beliefs and commitments are on public view and potentially measurable. Therefore, sight of the document will be useful when evaluating another contractor's competence.

In the construction industry, evidence of competence in health and safety terms can be hard to find, although to do so is a legal requirement under the Construction (Design and Management) Regulations (CDM). Any 'safety' paperwork generated by a contractor will be asked for, and may be evaluated by people with little or no training in safety, using a checklist. Some checklists refer to more than 150 items which the senders expect to be contained in a safety policy. Keeping a policy simple and straightforward makes it understandable to the employees — who are the people who have to be able to read and understand it, and comply with it. On the other hand, keeping it simple may lead to omissions, which may in turn result in failure of the policy to pass 'inspection' by third parties.

The most recent guidance from the HSE indicates that passing out questionnaires may be unproductive, and that there are likely to be more effective ways to gauge competence. These include partnering exercises and setting up interviews with prospective contractors where their responses can be gauged more accurately.

Others may wish to use a safety policy for different purposes — to gauge the record and intentions or ability of a possible business partner, and, importantly, for the purpose of establishing an employer's self-assessed standard of care as a prelude to making a **civil claim** for damages. Claimants may be able to use any extravagant wording or undertakings to back up compensation claims, as any failure to achieve whatever commitments are made in a safety policy may be viewed by courts as amounting to negligence. Similarly, the duties of employees set out in the safety policy may define for a court what constitutes taking 'reasonable care' of themselves and others.

Manuals and QA systems

It is important to distinguish between a safety policy and a safety manual — these are not the same thing, but are often found combined. The safety policy will refer to the manual for details on technical points. The main problem is that the likelihood of a document being read is inversely proportional to its length and complexity! Current opinion is that safety policies should be shorter rather than longer, and accompanied by explanatory manuals.

Some employers have devised their safety policy as part of a documentation system set up for BS EN ISO 9000 and 14000 purposes, and there is much to commend an integrated approach to the management of safety, health and quality. However, it must be remembered that possession of an adequate safety policy is a legal requirement, and the law will not be complied with if the safety policy is hidden or hard to find so that it is not effectively 'brought to the attention' of all employees as required by the Act.

Policy contents

The **length** of the document is not necessarily an indication of worth or usefulness, although a policy which occupies only a page or two in total is unlikely to contain the detail required, regardless of the size of the business. What is important is that the employer's general statement is short but comprehensive, and that the policy document describes the organisation and arrangements that are actually in force.

The **general statement** of policy should contain an expression of the employer's intentions. This should not place primary reliance on the safe behaviour of employees. Demands that employees take care of themselves, and comply with what is all too often a long list of 'don'ts' backed by disciplinary threats, are not likely to be met with approval by anyone looking for evidence of management competence. The following should be covered in the general statement:

- A clear statement of intent to provide safe and healthy working conditions
- A clear declaration of intent that work will not adversely affect others or the environment
- Reference to arrangements made for joint consultation with employees
- Commitments to:
  - provide adequate funds and facilities to ensure the success of the policy
  - comply with all applicable legal requirements
  - seek the co-operation of all employees
  - bring the policy to the attention of all employees

At the foot of the general statement, the policy document should be signed and dated by the top person in the business — usually the Managing Director or Chief Executive Officer. Dating the policy should act as a prompt for its regular review and reissue, probably annually, while signing it provides the commitment from the top that is an essential part of safety management.

The part of the policy dealing with the **organisation** is usually considered to require a clear recording of duties and responsibilities at all levels —not just of management but also of the employees. To avoid mistakes and omissions, everybody needs to know exactly what they have to do, and what their personal responsibilities are.

An essential requirement for management involvement at all levels is to define health and safety responsibility in detail within the written document, and then to check at intervals that the responsibility has been adequately discharged. This is **accountability** — the primary key to management action. It is not the same thing as **responsibility**; accountability is responsibility that is evaluated and measured, possibly during appraisal sessions. Final and formal responsibility should be given to a named director or partner, who will be held formally responsible for health and safety matters in the business.

Responsibilities should be defined specifically for:

- The formal appointment of competent person(s) to assist and advise on compliance
- Making risk assessments for all work activities — including manual handling, issue of personal protective equipment, users of display screen equipment and other work equipment, and for COSHH substances
- Assessing safety training needs within the business and providing the training
- Ensuring the competence of subcontractors
- Compliance with CDM requirements
- Monitoring compliance with the policy, including site visits by senior management
- Attending pre-start meetings with other contractors
- Responding to employee initiatives and requests or suggestions
- Maintaining contacts with external sources of advice

The '**arrangements**' part of the policy should describe how things actually get done. All except the most general of the arrangements should be derived from or checked against the risk assessments required by the MHSWR and others. The assessments need to be mentioned in the policy, because someone will have been given the responsibility to carry them out, and so the method by which the assessments are made — the arrangements — should be detailed. MHSWR requires that the significant findings of risk

assessments are to be recorded where there are five or more employees. The practicality of attaching risk assessments in full to the safety policy document is such that it may be simpler to include only the significant risks identified and their controls. A manual of risk assessments can easily be referred to from within the safety policy document, whereas the inclusion of the assessments within the policy itself will make for unnecessary bulk.

One of the functions of the policy document is to serve as a focus for induction training, and so enough information on each topic should be included so that employees understand what happens. Topics which should be considered for inclusion in the 'arrangements' section of the safety policy will depend on the particular business activities being covered by the policy.

The following list of matters which may be considered for inclusion under 'Arrangements' has been compiled from several sources, notably the many detailed checklists used by local authorities to evaluate the safety policies of contractors. It is likely to be impossible to satisfy all those with an interest in the subject, and these suggested main headings are offered as a guide only. It is also possible to argue a case that several of the topics will have been adequately covered elsewhere by risk assessments.

- Role and functions of health and safety professional staff
- Allocation of finance for health and safety
- Systems used to monitor safety performance (not just injury recording)
- Identification of main hazards likely to be encountered by the workforce
- Generic risk assessments — significant findings
- Any circumstances when specific risk assessments will be required
- Arrangements (or cross-references) for dealing with these
- The use of method statements within the organisation
- Safety training policy, and summary details of arrangements
- Design safety policy
- Fire and emergency arrangements for sites and offices
- Responsibilities under CDM — design, safety plan, safety file
- Maintenance of required documentation
- Safe systems of work appropriate to the business, including permit to work systems
- Arrangements for inspecting, testing and maintaining

mechanical and electrical work equipment and systems

- Certification of those working with identified equipment such as dumpers, lifting equipment and transport
- Arrangements for working in occupied premises
- Occupational health facilities, including first-aid
- Environmental impact statement
- Environmental monitoring policy and other arrangements
- Purchasing policy (e.g. on safety of plant, noise, chemicals)
- Methods of reporting and investigating accidents and incidents
- Arrangements for the qualification of subcontractors
- Control of visitors to sites
- Personal protective equipment policy, requirements, availability
- Worker consultation arrangements (e.g. the role and functions of safety committees and employee representatives)
- Revising the policy itself — arrangements, frequency, document distribution

Other safety policy considerations

The inclusion of an **organisation chart** can be helpful, especially in larger organisations, to clarify the reporting lines between different functional areas of the business.

Revision of safety policies should be done at regular intervals, ideally annually. The purpose of this is to ensure that the stated organisation and arrangements are still applicable to the organisation's needs. After changes in structure, senior personnel, work arrangements, processes or premises the hazards and risks may change. After incidents and accidents, one of the objectives of the investigation will be to check that the arrangements in force had anticipated the circumstances and foreseen the causes of the accident. If they had not, then a change to the policy will be required. The revision mechanism should be written into the policy arrangements section.

Circulation/distribution of the policy is important — it must be 'brought to the attention of all the employees'. This means that it must be read and understood by all those affected by it. How this is best achieved will be a matter for discussion.

In some organisations, a complete copy is given to each employee. In others, a shortened version is given out, with a full copy available for inspection at each workplace. Members of the management team should be familiar with the complete document. If it is likely to be revised frequently, a loose-leaf format will be advisable. This is especially true if the names and contact phone numbers of staff are printed in it; these are likely to change frequently.

A growing problem is the increasing use of workers whose native language is not English. It is hard to see how a document can be brought to someone's attention if they are unable to read it, and this implies translation of the policy into relevant languages.

'Off the shelf' safety policies

Experience suggests that policies written by someone else may fail to achieve the desired practical, or maybe even legal, effects. Several commercially-available 'boiler plate' model policies can be bought, including one by the HSE, but they all require adjustment to the particular needs of the organisation. Because of the many uses to which the safety policy is put, it is well to be aware that legal compliance is only the first of the requirements for a successful policy to meet. Successful policies are those which satisfy all the needs of the organisation.

As with risk assessments, the administrative burden of producing a safety policy falls more heavily on the smaller organisation. Nevertheless, there is value for them in carrying out the exercise — small construction businesses tend to have proportionally more accidents to their workers, can engage in high-risk activities and will benefit from a review of the hazards their employees face.

6 Assessing the Risks

There are at least two senses in which risk assessment has been carried out subconsciously over a long period. Firstly, we all make assessments many times each day of the relative likelihood of undesirable consequences arising from our actions in particular circumstances. Whether to cross a road by the lights or take a chance in the traffic is one such. In making a judgement we evaluate the chance of injury and also its likely severity.

The second sense of risk assessment is based on the employer's requirement under many sections of the Health and Safety at Work etc. Act 1974 to take reasonably practicable precautions in various areas to safeguard employees and other parties. This requires the making of a balanced judgement about the extent of a particular risk and its consequences against the time, trouble and cost of the steps needed to remove or reduce it. If the cost is 'grossly disproportionate', as a judge once put it, we are able to say that the steps are not reasonably practicable. Thus, in a very real sense, risk assessments have been carried out at least since 1974 in the UK.

The difference between these assessments and those required by the MHSWR is that the significant results must be recorded by most employers, and information based on them is to be given to employees in a much more specific way than before. When the formal requirement first came into force at the beginning of 1993, those carrying out the statutory assessments found that some activities were being viewed afresh for the first time. The exercise was useful as it forced challenges to long-held assumptions about the safety of traditional work practices — which often did not stand up to the scrutiny.

Benefits

Risk assessment is done to enable control measures to be devised. We need to have an idea of the relative importance of risks and to know as much about them as we can in order to take decisions on controls which are both appropriate and cost-effective.

Types of risk assessment

There are two major types of risk assessment, which are not mutually exclusive. The first produces an objective probability estimate based upon known risk information applied to the circumstances being considered — this is a **quantitative** risk assessment. The second type is subjective, based upon personal judgement backed by generalised data on risk — the **qualitative** assessment. Except in cases of especially high risk, public concern, or where the severity consequences of an accident are felt to be large and widespread, qualitative risk assessments are adequate for their purpose — and much simpler to make. The legal requirements refer to this type of assessment unless the circumstances require more rigorous methods to be used.

Because of the dynamic nature of construction work — with fast-moving, changing hazards, for example — the risk assessment process can be more difficult than in an industry where the work process is essentially static and repetitive. For this reason, many construction employers use **generic** or **model** assessments covering the generalities of particular tasks or activities, which can then be made site-specific relatively quickly. These are most appropriate where there is a similarity to activities, and the hazards and risks associated with them, although they will be carried out in different physical areas or workplaces. In principle, the use of generic risk assessments has much to commend it. It saves reinventing solutions for commonplace tasks where the risks are likely to be much the same each time the task is done. But, where they are used, it is very important to make sure that they have been tailored to the particular situation so as to evaluate the actual risks and not just 'average' or 'hypothetical' ones.

In the construction industry there are likely to be

situations in specific areas or on specific occasions where a generic assessment will not be sufficiently detailed, and those circumstances should be indicated in the safety policy to alert to the need to take further action. There may also be work situations where hazards associated with particular situations will be unique, so that a special assessment must be made every time that the work is done.

To describe the safe system of work which such a risk assessment shows to be necessary, a **method statement** may be required — for example: for demolition work, the erection of steel structures and asbestos removal. Method statements will be discussed later in this chapter, but it is worth observing at this point that risk assessments are used to generate method statements which are themselves a stated sequence of events designed to minimise risks by giving advance warning, knowledge and information to the recipients.

Contents of risk assessments

Significant findings are to be recorded where five or more people are employed by an employer. According to the MHSWR Approved Code of Practice, the record should contain a statement of the significant hazards identified, the control measures in place, the extent to which they control the risk(s) and the population exposed to the risk(s). Cross-reference can be made to manuals and other documents. Any risks which remain uncontrolled should be highlighted; some companies collate them in 'residual risk registers' for easy reference.

As the regulations require review of assessments in specified circumstances, and as it will be necessary to review for changed circumstances over time, it would be prudent to include in the documentation a note of the date the assessment was made and the date for the next regular review. Similarly, it would be wise to include a note to employees reminding them of their duties under MHSWR to inform the employer of any circumstances which might indicate a shortcoming in an assessment.

In 1994 the HSE published guidance entitled 'Five Steps to Risk Assessment', a leaflet setting out the most basic steps in the process and providing a specimen format for recording the findings of assessments. The leaflet has since been updated and reissued. The simplistic approach presented has much to commend it. Comments contained in the text give useful pointers to the approach to risk

assessment which will satisfy minimum requirements in the view of the enforcing authority. For example:

> 'An assessment of risk is nothing more than a careful examination of what, in your work, could cause harm to people so that you can weigh up whether you have taken enough precautions or should do more . . .'

A more sophisticated approach will often be required where hazards and risks are changing frequently, as in the construction industry, and also where clients, work partners and other external organisations seek evidence of more detailed analysis than 'Five Steps to Risk Assessment' is likely to provide. There is no shortage of more complicated commercial systems, and these are often computer-based so as to enable quick updating in changing conditions and rapid review.

The above comments have been directed at general activity risk assessments. It is easy to forget that more specific assessments could be required in order to comply with (for example) the Manual Handling Operations Regulations, the Display Screen Equipment Regulations (which apply to display screens in use in site offices) and the assessment requirements within the Personal Protective Equipment Regulations. These are addressed in more detail later in this chapter.

Hazard evaluation

The hazards to be identified are those associated with machinery, equipment, tools, procedures, tasks, processes and the physical aspects of the plant and the site or premises where the work will be done — everything. Evaluation of the hazards is achieved by assembling information from those familiar with the hazards, such as insurance companies, professional societies, government departments and agencies, manufacturers, consultants and trade unions. Learning from records, including old inspection reports (both internally and externally produced), accident reports and standards is also important.

Even so, some hazards may not be readily identifiable, and there are techniques which can be applied to assist in this respect. These include **inductive analysis**, which predicts failures — failure modes and effects analysis (FMEA) is one of these; job safety analysis (JSA) is another. Inductive analysis assumes failure has occurred and then examines ways in which this could have happened by using logic diagrams. This is time-consuming, and therefore expensive, but it is extremely thorough. MORT (management oversight and risk tree) analysis is an example which is not difficult to use.

Job safety analysis

Once a task has been identified as hazardous, controls need to be applied. For complex tasks, analysis may be required to sort out the component parts of what is actually going to be done. This can result in task improvements and risk reduction. A formal technique for doing this is known as **job safety analysis**, which has developed from work study practices. Essentially, the task is observed, and broken down into steps or stages which are then examined for the level of risk involved (Table 6.1). Control measures are then worked out, and the paperwork is then used to generate a written safe system of work. Naturally, the tasks selected need to be reviewed at intervals to make sure that the analysis and solutions remain valid over time.

Table 6.1: Analysis of potential hazards in complex tasks by job safety analysis (JSA)
Select the job/task to be analysed
Break the job into logical steps
Identify the hazards in each step
Devise risk elimination or reductions for the hazards identified
Record the JSA
Review and update the JSA as necessary

In practice, the most difficult part of job safety analysis is to split tasks up into a sequence of steps. A computer program is a series of steps, but this would be far too detailed for these purposes. One obvious step could be 'Get into the driving position', and some hazard might be then identified such as the potential to slip on bodywork while doing so, leading to the job safety instruction 'Always use the step and handhold provided'. A computer program would be operating at the much more detailed level of 'Lift right foot, extend right leg, place foot on vehicle', and so on. The need is to identify the critical steps which add something to the task. For this reason it is always necessary to begin the analysis by observing the task, possibly making a video as a reference.

It is very important to involve those actually doing the job in the analysis process. They are likely to have most knowledge of how the job is actually done, and also frequently volunteer practical solutions which may not be evident to those without personal experience of the work.

No matter what system is used in job safety analysis, the process is time-consuming and there must be a selection process to avoid unnecessary effort. Factors influencing selection will include known levels of risk (what could go wrong and the extent of loss or damage if it did), any legal requirements, the number of people carrying out the task and whether they are familiar with it. The resulting written safe system of work will be very useful for training purposes, as it will detail all the problems and the selected solutions. It will also be available to managers and supervisors.

Ranking hazards by risk

There may be occasions when a priority list for action will be called for. Techniques for producing a ranking vary from the very simple to the very complex. What is looked for is a priority list of hazards to be controlled, on a 'worst first' basis.

The following system gives a simple way to determine the relative importance of risks (Table 6.2). It takes account of the consequence (likely severity) and the probability of the event occurring. Estimation of the first is easier than the second, as data may not be available for all hazards. Estimates derived from experience can be used. It is possible to carry out ranking using a simple formula, where risk = severity estimate × probability estimate. These estimates can be given any values, as long as they are consistently used. The simplest set of values offers a 16-point scale by multiplying the estimates (Table 6.2).

The categories are capable of much further refinement. Words like 'time' can be defined, increasing if necessary the number of categories. Many organisations increase the categories to take account of numbers exposed to the hazard as well as the duration of exposure. However, the more precise the definitions, the more it will be necessary to possess accurate predictive data.

It should also be noted that ranking systems of this kind introduce their own problems, which must be addressed, or at the least known to the user. One is that longer-term health hazards may receive inadequate evaluation because

Table 6.2: Assessment of the relative importance of risks	
Severity rating of Hazard	**Value**
Catastrophic — imminent danger exists, hazard capable of causing death and illness on a wide scale.	1
Critical — hazard can result in serious illness, severe injury, property and equipment damage.	2
Marginal — hazard can cause illness, injury or equipment damage, but the results would not be expected to be serious.	3
Negligible — hazard will not result in serious injury or illness, remote possibility of damage beyond minor first aid case.	4
Probability rating of hazard	**Value**
Probable — likely to occur immediately or shortly	1
Reasonably probable — probably will occur in time	2
Remote — may occur in time	3
Extremely remote — unlikely to occur	4

of lack of data. Another is that hazards of low severity but high frequency can produce the same risk score through multiplication as high severity, low frequency ones. Although the scores may be the same, the response to them in terms of priority for correction may be very different. Access to good data, and evaluation and categorisation by experts are possible cures.

Decision-making

This process requires information to be available on alternatives to the hazard, also on other methods of controlling the risk. Factors which will influence decision-making are training, the possibility of replacement of equipment or plant, design modification possibilities and the cost of solutions proposed. **Cost benefit analysis** will be used formally or informally at this point.

Cost benefit analysis requires a value to be placed upon the costs of improvements suggested or decided upon. These will include the cost of reducing the risk, eliminating the hazard, any capital expenditure needed and any ongoing costs applicable. An estimate of the pay-back period will be needed. Decisions on action to be taken are often based on

this — a 3- to 5-year period is often associated with health and safety improvements. Using these techniques can direct an organisation's resources to where they will be most effective.

Controlling the risks

It is necessary to be aware that some corrective measures are better able to produce the desired results than others, and that some are very ineffective indeed as controls. The **safety precedence sequence** shows the order of effectiveness of measures:

- **Hazard elimination** (e.g. use of alternatives, design improvements, deciding not to attempt the work)
- **Substitution** (e.g. replacement of a chemical with one with less risk, using a more appropriate (safer) piece of equipment)
- **Use of barriers:**
  - isolation (removes hazard from the worker, puts hazard in a box, e.g. physical isolation of noisy plant)
  - segregation (removes worker from the hazard, puts worker in a box, e.g. a plant driver's cabin)
- **Use of procedures:**
  - limiting exposure time, dilution of exposure
  - safe systems of work, which depend on human response
- **Use of warning systems** — signs, instructions, labels, which depend on human response
- **Use of personal protective equipment** — depends on human response, to be used as a sole measure only when all other options have been exhausted, personal protective equipment (PPE) is the last resort

Regulations require that PPE should only be used if there is no immediately feasible way to control the risk by more effective means, and as a temporary measure pending installation of more effective solutions. Disadvantages of PPE include: interference with ability to carry out the task, the PPE may fail and expose the wearer/user to the full effect of the hazard and continued use may mask the presence of the hazard and may result in no further preventive action being taken.

Communal protection by incorporating fixed barriers and guardrails into designs reduces the need for use of personal protective equipment by contractors and maintenance staff. PPE always requires at least one positive human act before it can be effective — that of making use of it.

Monitoring

Risk assessments must be checked to ensure their validity, or when reports indicate that they may no longer be valid. It is important to remember that fresh assessments will be required when the risks change as conditions change, and also when new situations and conditions are encountered for the first time. Other information relevant to risk assessments will come from monitoring by way of inspection, air quality monitoring and other measurements, and health surveillance.

Health surveillance

Risk assessments will identify circumstances where **health surveillance** will be appropriate. Legal requirements for such surveillance extend beyond exposure to substances hazardous to health. Generally, there will be a need if there is an identifiable disease or health condition related to the work, there is a valid technique for its identification, there is a likelihood that the disease or condition may occur as a result of the work, and the surveillance will protect further the health of employees. Examples where these conditions may apply are vibration white finger and forms of work-related upper limb disorders (WRULDs).

Information to others

MHSWR requires co-operation and sharing of information between contractors working on the same site, and between contractors and clients or owners of property affected by the work. Risk assessments will form the basis of that information. In formal circumstances of contract, there may be a contractual or legal requirement to exchange risk assessment data. The information given to others about risks must include the health and safety measures in place to address them, and be sufficient to allow the other employers to identify anyone nominated to help with emergency evacuation.

Format of risk assessments

There is no prescribed format for risk assessments. Health and Safety Executive guidance is limited to 'Five Steps to Risk Assessment', which is considered to be insufficiently detailed for construction industry needs, and specimen examples in the guidance material associated with the Manual Handling Operations Regulations and the Health and Safety (Display Screen Equipment) Regulations (see below).

It may be helpful to consider the use of generic assessments that can be made site-specific as required; a commercial system marketed by Hascom Network Ltd is shown here by permission (Fig. 6.1). The classification of risks into simple categories avoids the use of numbers as discussed above and the system allows most of the work to be done in advance. All that is required is for any hazards not identified on the front page to be written up together with their control measures on the back page, the identification of 'serious and imminent danger' as required by MHSWR to be made and the document signed off. Space is allowed to draw the user's attention to any circumstances that will require an extra or more detailed examination to be made of the circumstances. The Quick Reference Guide in Part 2 of this book contains details of the sort of information that should be written into the generic section of the form for a range of construction activities.

Project risk assessments

Before work begins it will be necessary to make a preliminary risk assessment of the work to be carried out, in order to draw up a pre-construction phase safety plan to comply with CDM and best practice. The **project risk assessment** is an evaluation of the principal hazards believed likely to be present during the construction phase of the work, allowing due importance to be placed on them and the allocation of necessary resources and planning to reduce the risk of injury to an appropriate and acceptable level. The project risk assessment should take place at an early stage, and its completion is intended to provide sufficient information for the writing or reviewing of the site's safety plan.

Regardless of who will carry out this task, experience shows that the simplest way of doing this is to form a small team of people likely to be associated with the project. Together with copies of the drawings and other relevant information such as specifications and designers' information, the work can be broken down into steps in the same way as for a job safety analysis. If the project is complex, or many contractors are going to be involved, then organisations will find it useful to draw up a checklist to make sure that all the hazards have been included in the discussion. The checklist could be based upon a list of construction hazards known to produce the highest risks, but there are many ways of arriving at the same result.

The output of the exercise should be a list of those activities where contractors' work needs to be co-ordinated or where special training and skills will be required, where a particular method of work must be specified in advance, and where risk assessments and method statements need to be produced by the contractor before work is authorised to start. The exercise will also identify matters that must be addressed in site-specific fire, emergency and environmental plans.

There is no required method of carrying out a project risk assessment. It is suggested that whatever method is used, the deliberations and results should be recorded in some form so that decisions taken can be justified by reference to a 'paper trail' if necessary. Where decisions about what is reasonably practicable are taken, recording the process will 'show the working'. The important issue is to achieve the objective of risk assessment rather than to use any particular methodology. Although the major risk assessment work will be carried out prior to writing the safety plan, it should be remembered that risks change with circumstances, and the assessment as well as the safety plan must be kept under review to enable both to accommodate changes.

The checklist that follows contains some additional features that require explanation. One method of deciding whether a method statement will be required is to put a numerical value on the size of the perceived risk. In this case the parameters used to estimate the size of the risk are the consequences (severity of the outcome), the numbers exposed to the hazard and the probability of harm occurring with the stated control measures in place. Values for the first and last categories can be supplied in advance of the use of the form if desired. The formula used for deriving the risk rating is not based upon any particular piece of research; it shows only one possible view of the relative importance of the three categories selected.

Once the risk rating is established in relation to a particular task or activity, the risk rating can be compared with the scale to see whether a method statement should be called for. Of course, an organisation may wish to set a higher standard than this by revising the formula, or the scale itself. Several companies have decided that they will always ask for a method statement from every contractor; this system is designed simply to offer a decision method which may prove of interest to the reader. Intermediate values can be used — the scoring is not restricted to the numbers shown.

The reader is invited to develop the principle of the checklist to make it relevant for particular needs. It is not implied that the questions used here constitute a complete review of all those activities likely to prove hazardous on a particular site. At the end of the checklist a sign-off sheet

should be added, carrying the names of those who took part in the risk assessment and the date. The work should also be signed off by the person who will be in charge of the work where they were not involved in the assessment process.

Scoring guide

(a) **Consequences** (consider health, property damage, environment, programme loss as well as injury)

Score: 3 — First aid treatment
6 — Lost-time injury
9 — Major injury/permanent disablement
12 — Fatality
15 — Multiple fatality

(b) **Number of workers exposed to hazard**

Score: 2 — 1
4 — 2–5
6 — 6–20
8 — 21–100
10 — 101+

(c) **Probability of harm with control measures in place**

Score: 4 — Happens once or twice annually throughout the country or region
8 — Happens regularly throughout region
12 — Happens once or twice locally on sites
16 — Happens regularly locally on job-sites
20 — Certain or very likely to happen on this site

Upon the **risk assessment charts** (Fig. 6.2), place a bold dot · in the appropriate places in order to obtain a visual impression of the risk areas.

The **risk rating** can be calculated from the formula:

$$\text{risk rating} = a\,(b + c)$$

Determine the **risk category** from the following scale of risk ratings:

18	60	90	130	450
Low	Moderate	Substantial	High	

Risk Categories

PRACTICAL COMPLIANCE

Although this chapter deals with the subject of risk assessment, for the reader's convenience the wider issue of control measures following some types of assessment is also covered here. This is intended to avoid the need for constant reference to other chapters.

COSHH assessments

The Control of Substances Hazardous to Health Regulations 1999 (COSHH) are very detailed, and contain several provisions which will not concern contractors directly. For practical purposes, they can be complied with by following seven basic steps:

- Carrying out risk assessments of 'COSHH substances' — see Part 3 for definitions of these
- Working out what precautions are needed to deal with the hazards identified and introducing control measures to prevent or control exposure to the hazardous substance
- Ensuring that the control measures are used by employees and that they are effective
- Monitoring the exposure of employees, which may involve health surveillance
- Providing information, training and supervision as necessary
- Regular review of the assessments and their validity in the light of experience

1. Assessing the risks

The first step is to make a list of all hazardous substances which are likely to be present at work. The easiest way to do this is to collect the suppliers' information on all the substances which have been bought for use, and read them. There may also be other potentially harmful substances which can be produced by the work process — welding rods which produce fumes for example — and information on these should be obtained from standard reference sources.

COSHH require that employers do not carry out any work which could expose employees to hazardous substances before carrying out a risk assessment. So, for each substance, it is necessary to think about the risks involved. If it is believed that there is no risk, or only a very small one, then no further action is needed.

Section 6 of the Health and Safety at Work etc. Act 1974

requires suppliers to provide the necessary information. It is a good plan to keep all this information together in a file for easy reference. Employers have the responsibility for making the risk assessments; the only requirement is that the person doing the assessment knows what the regulations require, and has the information and knowledge needed to make decisions about risks and controls. Most employers are able to make adequate assessments for themselves, except for complex substances and for complex risks.

Employees and safety representatives should be involved in the process, as they often have practical information on risks at work which may not be known to management. The law requires that they are told about the results of the assessment.

COSHH requirements extend to the assessment of risks to other people and the provision of control measures. This means that visitors to premises, and those employees who work on the premises of others, must have any risks to them assessed by the employer.

In order to judge the level of risk involved, a number of factors must be considered. These include the quantity of the substance used or produced, how often it is worked with, how hazardous it is and whether exposure is likely to exceed the occupational exposure standard (if one is listed). For example, a degreasing agent, paint or an adhesive containing solvent may be relatively harmless when being used in the open air, but in a confined space with poor or no ventilation, significant exposure to it may be life-threatening.

The manufacturer or supplier should give information about these matters in the documentation accompanying the substance (often called hazard data sheets or material safety data sheets (MSDS)). There should be advice about storage, spillage procedures and disposal instructions. Details should also be provided on the 'entry route', which is the way that the substance may get into the body: by absorption through skin contact, by inhaling (breathing it in) or ingestion (eating it knowingly or unknowingly).

Information provided in hazard data sheets is often very detailed, and may not cover the actual conditions of use. **This is why the data sheet by itself cannot be considered as a COSHH assessment.** Many employers provide hazard data sheets to employees, thinking that they have provided an assessment. To be legal and effective, the assessment should be easy to understand, state the hazards and the risk levels, cover the precautions and control measures required and the way in which the substance is to be used, stored and disposed of.

Written assessments must therefore be made as appropriate, setting out the problems and the precautions, and be shown to the employees who are exposed. They need to know what to do, and this forms part of the information employers are required to provide. The crucial point about COSHH assessments is that they must be job-specific, and not merely a general statement about the subject.

Assessments also need to be kept up to date. Experience may lead to revisions in the control measures. Assessments should be dated, so that it is clear when they were made, and they should also contain an automatic review date.

2. *Working out what precautions are needed*

Where a significant risk is identified, COSHH requires employers to take reasonably practicable steps to prevent exposure, either by changing the process or work to eliminate the requirement for the substance to be used or produced, or to replace the substance with another substance with a lower level of risk, or to use the substance in a safer form — by dilution, by using in a solid form rather than liquid, etc. If it is not reasonably practicable to do any of this, then COSHH requires exposure to be adequately controlled by:

- Total enclosure of the process, or if not, then
- Partial enclosure and use of extraction equipment, or if not, then
- Use of general ventilation
- Introduction of systems of work and procedures to minimise the chance of leaks, spills and other escapes of the hazardous substance
- Reduction of the numbers of people exposed, or the duration of their exposure (only where the earlier measures cannot be put into operation)

Use of personal protective clothing or equipment (PPE) is only allowable if none of the above steps can be taken and the desired result is achieved. There is no reason why PPE cannot be provided as an extra safeguard to any of the more effective measures, but its use as the sole defence must be seen as a last resort. The reason for this is simply that none of it works 100% of the time for 100% of the people using it. What the law considers as 'adequate' controls will always depend upon the individual circumstances. A rule of thumb is that it will usually be considered adequate if most people would not suffer any adverse health effects if exposed to the substance at that level day after day.

3. *Ensuring control measures work*

Employees are required by COSHH to make proper use of any control measures the employer provides, and to tell their employer about any defects in them. They need information, training and supervision to ensure that they are aware of what is required of them. It is not sufficient merely to rely upon reports from employees — checks must be made by the employer at appropriate intervals to ensure that the controls are in place and that they work.

There are requirements within the COSHH Regulations on the examination and testing of engineering controls such as dust and fume extraction systems, and records of the tests must be kept for at least 5 years. Usually, this will only apply to construction companies with fixed workshops such as joineries.

4. *Monitoring exposure*

In some circumstances, employers are required to measure the concentration of hazardous substances in the air. These include where the employer cannot be certain that set exposure limits are not being exceeded or that specific control measures are not working. In general, measuring concentrations is a job for specialists, although there are some simple techniques which can be used in-company.

Health surveillance is a more specific examination of employee health, made for a purpose. It has been mentioned elsewhere in this chapter that COSHH requires this to be done where:

- There is exposure of employees and others to a substance known to produce a specific adverse health effect or disease, **and**
- It is reasonable to expect it to occur under the particular working conditions, **and**
- It is actually possible to detect the effect or disease in that way

A relevant example could be dermatitis, a common skin condition, which can be produced by contact with a number of substances commonly used in construction such as cement dust. Doctors or nurses usually carry out health surveillance, but there is no reason why trained supervisors cannot make basic checks and ask appropriate questions. Where health surveillance is carried out, the employer is required to keep a simple record of the results, and to keep that record for 40 years.

5. Providing information, training and supervision

The COSHH Regulations require the employer to give necessary information, training and supervision to employees about the substances they work with and the risks created by exposure to them, and about the precautions to take. They will need to know the control measures decided on and how to use them, about any PPE provided, emergency procedures if appropriate, and also the general results of any health surveillance and monitoring. It is important to remember that it is employees who are at risk so there is no point in doing the assessments and working out the controls without then ensuring that the employees know what they are and how to use them.

6. Review of assessments

Risk assessments and controls should be reviewed at regular intervals. For this reason, it is sound practice to include the date of assessment, and for the review, on the COSHH assessment form. Annual review is usual, although any incident involving the substance should trigger a review as well. Changes notified by suppliers in the contents of data sheets, or the specification of the substance, will also prompt a review of the assessment.

Manual handling assessments

The responsibility for carrying out these assessments rests with the employer in each case, and with the self-employed. This means that contractors and subcontractors must produce their own risk assessments where appropriate.

Under previous legislation, the employer's liability for injuries to employees was limited to cases where the employer had failed to ensure that employees were not required to lift a load 'so heavy as to be likely to cause injury to him (her)'. In addition, the general duty of care required the employer to take reasonable steps to provide a safe system of work, place of work and safe work equipment. The Manual Handling Operations Regulations 1992 are much less specific, and do not contain a Regulation barring their use in civil claims. More information on the

Regulations and technical aspects of manual handling is contained elsewhere in this book.

Any breach of the Regulations may provide grounds for a claim following injury, including failure to make an assessment where one was required; that is, for all manual handling operations, where there is a possible risk of injury and it is not reasonably practicable to avoid moving the load or to automate or mechanise the operation.

It is important to appreciate that the major thrust of the Regulations is to avoid manual handling operations where reasonable to do so, and to limit those which do take place as far as is reasonable. Therefore, for every task, a reduction in the amount and extent of manual handling should be sought, regardless of the loads involved. There is no safe limit; the majority of injuries happen as a result of progressive degradation of an affected area rather than following one particular overload.

The Regulations establish a clear hierarchy of measures to obviate the risk of injury when performing manual handling tasks. To summarise, manual handling operations which present a risk must be avoided so far as is reasonably practicable; if these tasks cannot be avoided, then each task must be assessed and as a result of that assessment, the risk of injury must be reduced for each particular task identified so far as is reasonably practicable.

1. Assessing the risks

What is required initially is not a full assessment of each of the tasks, but an 'appraisal' of those manual handling operations which involve a risk that cannot be dismissed as trivial, to determine whether they can be avoided. If, as a result of the 'initial appraisal', manual handling operations which present a risk cannot be avoided, then a more detailed assessment of them will be a legal requirement.

For guidance purposes only, Figure 1 of the Appendix to the Guidance to the Regulations indicates that a possible risk will arise when lifting any load of more than about 10 kg from the floor or above shoulder height, and a maximum load for the optimum condition (lifting from a bench at waist level using both hands in a stable body position) is put at 25 kg. These values are for males, for females they should be cut by about a third.

These basic guideline figures for lifting and lowering should be reduced if **twisting** of the trunk is required; the values should be reduced by 20% for a 90° twist and 10% for a 45° twist. **Repetition** reduces the values by half where the action is repeated five to eight times a minute, and by 80% for more frequent repetition. **Carrying** loads more than about 10 m will prompt a reduction in the guideline values. **Pushing and pulling** actions carry a guideline figure for further assessment of about 25 kg, where the force is applied through the hands held below shoulder height. Lack of opportunities for **rest and recovery** are significant limiters, as is handling more than about 5 kg while seated.

In any particular case, an assessment should be made if there is even a marginal doubt as to the continued acceptability of an operation. Partly, this is because liability will be hard to disprove in the case of any injury to an employee where an assessment has not been made.

These values are only given as a general indication of risk of injury, and are not limits. It may be acceptable to exceed them if an assessment indicates it is appropriate to do so, given the controls in place. Equally, it may be that the values need to be reduced significantly if there is any twisting of the trunk, repetitive movement, or movement more frequent than about once every 2 minutes. The guideline figures are based on a number of assumptions which are not likely to hold true in the construction environment. Therefore, the guidelines require careful consideration when applying the figures provided. The assumptions include that:

- The handler is standing or crouching in a stable body position with the back substantially upright
- The trunk is not twisted during the operation
- Both hands are used to grasp the load
- The hands are not more than shoulder width apart
- The load is positioned centrally in front of the body and is itself reasonably symmetrical
- The load is stable and readily grasped
- The work area does not restrict the handler's posture, and
- The working environment and any PPE used does not interfere with the performance of the task

Consideration of a series of questions will be useful in completing this stage of the exercise.

1. Is there a risk of injury? A grasp of the basic principles of safety in manual handling will be required here (e.g. an understanding of the types and causes of injury). This understanding will be supplemented with the past experi-

ences of the employer, including accident/ill-health information relating to manual handling and the general numerical guidelines contained in official guidance (see above).

2. Is it reasonably practicable to avoid moving the load? Some work is dependent on the manual handling of loads and cannot be avoided. But, if it is reasonably practicable to avoid moving the load then the initial exercise is complete and further review will only be required if conditions change.

3. Is it reasonably practicable to automate or mechanise the operation? If the manual handling of loads cannot be avoided completely, then automation or mechanisation must be considered. However, introduction of these measures can create different risks (introduction of forklifts creates a series of new risks, for example) which require consideration in this 'initial appraisal'.

The views of staff and employees can be of particular use in identifying manual handling problems. Involvement in the assessment process should be encouraged, particularly in reporting problems associated with particular tasks. Records of accidents and ill-health are also valuable indicators of risk as are absentee records, poor productivity and morale, and excessive product damage.

The aim of the assessment is to evaluate the risk associated with a particular task and identify control measures which can be implemented to remove or reduce it (which may include the use of handling devices and/or training). The assessments made under this requirement will have to be written down for all but the simplest of tasks or where the task is straightforward, of low risk, lasting only a short time and the time needed to record it would be disproportionate. Relevant assessments will have to be kept readily available and where a safety policy has been prepared (to comply with Section 2(3) of the Health and Safety at Work etc. Act 1974) employers should outline their general policy and arrangements for manual handling in it.

For varied work (for example construction and refurbishment) it will not be possible to assess every single instance of manual handling. In these circumstances, each type or category of manual handling operation can be identified and the associated risk assessed. Assessment should also extend to cover those employees who carry out manual handling operations away from the employer's premises (such as site workers and delivery drivers).

Schedule 1 to the Regulations specifies four inter-related factors of which the assessment should take account (the task, the load, the working environment and the individual's capabilities) and further questions relating to each are also identified in the Schedule. Consideration of these factors forms the basis of an appropriate assessment, some of which can be done in advance in a general way by means of generic assessments.

(a) The task

Is the load held at a distance from the trunk? Failure to keep the load close to the body will increase the stress on the lower back and make it less easy to counterbalance it with the weight of the body. Also, the benefits of using the body to support or steady the load will be lost.

Is poor posture experienced? Poor posture introduces additional risks of the loss of control of the load and sudden, unpredictable increases in physical stress. Risk of injury also increases if the feet and hands are not well placed to transmit forces efficiently between the floor and the load.

Does the task involve twisting? Stress on the lower back is increased if twisted trunk postures are adopted or if twisting occurs while supporting a load.

Does the task involve stooping? Stooping reduces the safe capacity that can be lifted significantly whether the handler stoops by bending the back or by leaning forward with the back straight.

Combining risk factors? Safe capacity is reduced if twisting is combined with stooping or stretching.

Does the task involve excessive lifting or lowering distances? Stresses are increased when loads are handled at the extremes of vertical movement. Also, if the load has to be handled through large distances more physical demand is placed on the individual.

Does the task involve excessive carrying distances? In general, if a load can safely be lifted and lowered it can also be carried without endangering the back. However, if the load is carried for an excessive distance (as a rough guide greater than 10 m) physical stresses are prolonged, leading to fatigue and increased risk of injury.

Does the task involve excessive pushing or pulling of the load? Pushing and pulling can present a risk of injury to the handler, with the risk of injury increased if pushing or pulling is carried out with the hands much below waist height or above shoulder height. The risk of slipping should also be considered as this too could result in injury.

Does the task involve a risk of sudden movement of the load? If the load becomes free and the handler is unprepared, sudden unpredictable stresses can be imposed upon the back. The risk can be increased if the handler's posture is unstable.

Does the task involve frequent or prolonged physical effort? The frequency of handling a load can increase the risk of injury. A quite modest load, handled frequently, can create a significant risk of injury. Fatigue can result from prolonged physical stresses, which can be made worse by a relatively fixed posture.

Does the task involve insufficient rest or recovery periods? Insufficient rest during physically demanding work increases ill-health and reduces output.

Handling while seated Handling loads whilst seated imposes physical constraints, such as preventing the use of the legs, with most of the work having to be done by the weaker muscles of the upper limbs. Lifting from below the level of the work surface can result in twisting and stooping, which will increase the risk of injury.

Team handling Handling by two or more people can solve some problems, but can also introduce additional risk factors which will need consideration during the assessment. A formalised system of work will be required for such tasks to be efficiently carried out.

(b) The load

Is the load heavy? Contrary to popular belief, the weight of the load is only one of the factors which can affect the risk of injury. Other factors which need consideration include its resistance to movement and its size, shape or rigidity.

Is the load bulky or unwieldy? The shape of the load will affect the way in which it can be held and may also affect vision. The risk of injury may also be increased if the load is unwieldy or difficult to control. In these circumstances, well-balanced lifting may be difficult to achieve, the load may hit obstructions, or it may be affected by gusts of wind. If the centre of gravity of the load is not centrally positioned, then inappropriate handling may result, increasing the risk of injury.

Is the load unstable, or are its contents likely to shift? Instability of the load can result from a lack of rigidity or the shifting of the contents resulting in unexpected stresses for which the handler is unprepared.

Is the load sharp, hot or otherwise potentially damaging? Sharp edges, rough surfaces and hot or cold surfaces can result in injury as well as impairing grip and discouraging good posture. Protective clothing may help here in reducing such risks.

(c) The working environment

Do space constraints prevent good posture? The working environment can hinder good posture, increasing the risk of injury (restricted headroom will enforce a stooping posture, for example).

Are there uneven, slippery or unstable floors? These will increase the likelihood of slips, trips and falls and can create additional unpredictability.

Are there variations in level of floors or work surfaces? Steps, steep slopes, etc., can increase the risk of injury adding to the complexity of movement, as will excessive variation between the heights of working surfaces.

Are there extremes of temperature, humidity or air movement? Thermal conditions can increase the risk of injury. High temperatures or humidity can cause rapid fatigue and perspiration on the hands, reducing grip. Work at low temperatures may impair dexterity.

Are there poor lighting conditions? Dimness or glare can cause poor posture and contrast between areas of bright light and deep shadow can aggravate tripping hazards and hinder accurate judgement of height and distance. Reliance only on task lighting to illuminate general access routes can result in manual handling problems as well as slips, trips and falls.

(d) Individual capability

Does the task require unusual strength, height, etc.? The ability to carry out manual handling tasks in safety varies between individuals. Generally, the lifting strength of women is less than that of men. However, for both men and women the range of individual strength and ability is large, and there is considerable overlap. Individual capability also varies with age.

Does the task put at risk those who are pregnant or have health problems? Allowance should be made for women who are pregnant as well as any health problem which might have a bearing on the ability to carry out manual handling operations in safety, such as occupational asthma.

Does the task require special knowledge or training for its safe performance? Inadequate knowledge and training can increase the risk of injury when carrying out manual handling tasks. For example, ignorance of safe systems or characteristics of the load can result in injury. If mechanical aids are to be used for such tasks, training may be required in their safe use.

2. *Working out what precautions are needed*

(a) Mechanisation possibilities

Mechanical handling aids which can reduce risk include the use of a lever, which would reduce the force required to move a load. A hoist can support the weight of a load whilst a trolley can reduce the effort needed to move a load horizontally. See page 62 for some examples which may be useful sources of ideas.

(b) Possible housekeeping improvements to minimise risks

- Maintain a tidy site, remove tripping hazards
- Level and maintain ground
- Provide stable walkways on water-logged sites
- Reduce gradients

(c) Possible personal protective equipment for manual handlers

- Overalls
- Safety boots and shoes
- Gloves and gauntlets/arm protectors
- Safety helmet
- Leather apron

(d) Changes in the task

Changes in the layout of the task can reduce the risk of injury by, for example, improving the flow of materials or

---

Examples of mechanisation possibilities

Concrete pump	Lorry offloading — crane
Forklift	HIAB
Hand-powered hydraulic lift	tail lift
Conveyor belt	Tower crane
Pallet truck	Mobile crane
Gin wheel and basket	Hoist
Hand-held power tools — electric	Excavator (lifting certificated)
diesel	Dumper
air	Power wheelbarrow
Mixer	Vibro tamp rails
Hydraulic jack	Scissor lift jack
Wheelbarrow	Hose pipe
Trailer	Trolley for sheet materials
Sack trolley	Ramp
Temporary working platform	Sling
Stretcher	

---

products. Improvements which will permit more efficient use of the body, especially if they permit the load to be held closer to the body, will also reduce the risk of injury. Improving the work routine by reducing the frequency or duration of handling tasks will also have a beneficial effect. Using teams of people and PPE will also contribute to a reduced risk of injury. All equipment supplied for use during handling operations, for example handling aids and PPE, should be properly maintained, and there should be a defect reporting and correction system.

Specific solutions which may offer some ideas are:

- Reduce size and weight of delivery packages
- Palletise material deliveries
- Transfer materials into smaller containers
- Erect rubbish chutes

(e) Reducing the risk of injury from the load

The load may be made lighter by using smaller packages/containers or specifying lower packaging weights. Additionally, the load may be made smaller, easier to manage, easier to grasp (for example by the provision of handles), more stable and less damaging to hold (clean, free from sharp edges, etc.). However, doing this may well increase the frequency of the task!

(f) Improving the working environment

This could be done by removing space constraints, improving the condition and nature of floors, reducing work to a single level, avoiding extremes in temperature and excessive humidity and ensuring that adequate lighting is provided.

(g) Improvements in individual capability

Particular consideration should be given to employees who are, or recently have been, pregnant, or who are known to have a history of back trouble, hernia or other manual handling injury. It is easy to forget that the Regulations force attention to be paid to manual handling in the office as well as in the stores or on site. Typists carrying boxes of paper, or equipment, are equally vulnerable to injury. The health, fitness and strength of an individual can clearly affect their ability to perform manual handling tasks. Selection of suitable handlers is an important tool in selection for manual handling tasks, and is a requirement of the Management Regulations.

3. *Ensuring control measures work*

As with COSHH assessments, for manual handling it is not sufficient merely to rely upon reports from employees — checks must be made by the employer at appropriate intervals to ensure that the controls are in place and that they work. In the construction industry the biggest advances are being made at the design and specification stages, as well as in the increasing availability of handling equipment.

4. Recording the results

The Health and Safety Executive's guidance booklet on the Regulations includes a specimen form which may be used, and also a worked example. There is no fixed requirement for recording assessment results in any particular way, but this format is generally considered to be the most appropriate and its use is commended to the reader.

5. Providing information, training and supervision

Provision of information on the nature and weight of the load, and its centre of gravity, is a requirement of the Regulations. Where this cannot be done easily, training should incorporate giving advice on ways of gauging weights of loads from their shape and size, and their likely density. Manufacturers of products used in the industry are now used to the need to provide weight indications on the packaging and the product itself, where reasonably practicable to do so.

Display screen equipment assessments

Display screen equipment used in the construction industry is normally found in the site or construction office. Details of the requirements of the Regulations can be found elsewhere in this book, where information can be found on the definition of 'user', which will trigger a written assessment of the workstation concerned. The advice which follows is given in an attempt to 'demystify' the process of assessment, focusing on the objectives to be achieved, and is not offered either as 'all you need to do to make an assessment' or as a substitute for training in the requirements and techniques. Assessments are personal to the individual user, so that where equipment is used by more than one person, adjustability and training will be the key to satisfying the legal requirements. The assessment process is an illustration of practical **ergonomics**, the matching of equipment to the people who use it rather than the other way round.

The most useful approach is to encourage the user of the workstation to take an active part in the assessment. The main priorities are to establish that the chair and work surface can, if necessary, be properly adjusted to suit the user. Adjustment should aim to achieve a posture where the forearms can be parallel to the floor and work surface when the hands are positioned at the keyboard. To make this possible, the chair height may need to be altered, and in turn a footrest may be needed.

Sufficient desk space is required, and the keyboard must be positioned in front of the user so that the hands are neither splayed outwards nor inwards when working. The height of the monitor screen should be adjustable so that its top is just below the eyeline of the user.

When adjustments have been made satisfactorily, a check must be made for screen glare. This can come from external light sources, overhead lighting or reflections. Where it is difficult to establish the source of glare, a hand mirror placed on the screen and viewed from the user's position can often identify it. The screen must be adjusted to remove any glare. Positioning it at right angles to major light sources is likely to be a cheap and effective cure, also fitting blinds to external windows. An anti-glare screen may be useful if these simple remedies do not or cannot have any effect.

A full review of aspects of the equipment used, and a more technical discussion of the principles, can be found in the Guidance to the Regulations.

Figure 6.1a: Sample form for generic risk assessment

YOURCO LTD	GENERIC RISK ASSESSMENT		
ACTIVITY COVERED:			

SIGNIFICANT HAZARDS	ASSESSMENT OF RISK		
	LOW	MED	HIGH
1			
2			
3			
4			
5			
6			
7			
8			
9			

ACTIONS ALREADY TAKEN TO REDUCE THE RISKS:

Compliance with:

[Insert relevant Regulations, Codes of Practice, Standards]

Planning:

[Insert pre-planning actions taken, such as design, layout, testing, measurements taken]

Physical:

[Insert general physical controls always required for the task or activity wherever they are done]

Managerial/Supervisory:

[Insert required actions and role of management in relation to the activity]

Training:

[Insert statutory and organisation requirements for training, induction etc related to the activity]

THIS GENERIC ASSESSMENT MUST BE COMPLETED BY THE ADDITION OF SPECIFIC SITE DETAILS ON THE REVERSE	FILE REFERENCE:

Figure 6.1b: Sample form for generic risk assessment (reverse)

SITE-SPECIFIC ASSESSMENT

On each site the generic assessment overleaf must be reviewed to ensure that all significant hazards and their risks are identified and controlled.

SITE/LOCATION:

MAXIMUM NUMBER OF PEOPLE INVOLVED IN ACTIVITY:

FREQUENCY AND DURATION OF ACTIVITY:

ADDITIONAL SPECIFIC HAZARDS IDENTIFIED ON THIS SITE:

ADDITIONAL CONTROL MEASURES REQUIRED ON THIS SITE:

ASSESSMENT OF RESIDUAL (remaining) RISKS: Insignificant / Low / Medium / High

SERIOUS AND IMMINENT DANGERS IDENTIFIED: Yes / No

EMERGENCY ACTION REQUIRED:

NAMES OF COMPETENT PERSON(S) APPOINTED TO TAKE ACTION:

CIRCUMSTANCES WHICH WILL REQUIRE ADDITIONAL ASSESSMENT:

CIRCULATION OF RISK ASSESSMENT

CONTRACTOR		SITE COPY		EMPLOYEES	
SUBCONTRACTOR		OTHER OCCUPIER OF PREMISES		CLIENT	
ON-SITE ASSESSMENT SIGNED:		DATE:		FILE REFERENCE:	

Figure 6.2: Risk assessment charts

YOURCO LTD PROJECT RISK ASSESSMENT

Site or project title / number: _____

Nature of work: _____

No.	Hazard present	YES/NO	Describe the hazards and obvious control or protective measures necessary	Likely consequences of an accident (a)					Number of workers exposed to hazard (b)					Probability of harm (c)					Risk rating and risk category	Extra control measures necessary
				3	6	9	12	15	2	4	6	8	10	4	8	12	16	20		
01	Are there any specific client safety requirements for the work?																			
02	Have site archaeological issues been identified and evaluated?																			
03	Has a geotechnical survey been carried out, and if so do the results indicate hazards which require control measures?																			
04	Is the site adjacent to or over public transport (railways etc.)?																			
05	Is the site adjacent to or over water?																			
06	Is the site adjacent to, over or under any services or drains etc.?																			
07	Is the site adjacent to, over or under any public buildings?																			

No.	Question											
08	Are there any other local hazards such as overhead power cables?											
09	Is the site on contaminated or unstable ground?											
10	Will the ground contours present any construction problems?											
11	Will any demolition works be necessary?											
12	Is there any asbestos removal involved?											
13	Are there any radiation sources involved?											
14	Will any excavation works take place?											
15	Will any works involve 'live' sewers? If YES will sewer entry be required?											
16	Are there any specialist processes required (i.e. exposives, HP fluids, etc.)?											
17	Is there any confined space or tank entry work involved?											
18	Will any piling take place? If YES, what type of piling?											
19	Will this involve any pile entry? If YES who will be required to enter?											

Contd

Figure 6.2: *Contd*

No.	Hazard present	YES/NO	Describe the hazards and obvious control or protective measures necessary	Likely consequences of an accident (a)					Number of workers exposed to hazard (b)					Probability of harm (c)					Risk rating and risk category	Extra control measures necessary
				3	6	9	12	15	2	4	6	8	10	4	8	12	16	20		
20	Will any steel erection works be taking place?																			
21	Will tower cranes be provided or heavy lifting operations taking place?																			
22	Will mobile work platforms, cradles or abseiling be necessary?																			
23	Have site haulage requirements been identified?																			
24	Is the access to the site adequate for vehicles and pedestrians?																			
25	Is there access available to the site for the public?																			
26	Will concrete pumping be carried out?																			
27	Have arrangements been made or co-ordinated for temporary electric supplies?																			
28	Have site lighting needs been identified for all stages of the work?																			

No.	Question									
29	Will any accommodation/office units be located inside a building?									
30	Are floors metal decking?									
31	Are floors pre-cast concrete?									
32	Will there be any RC frame erection works taking place?									
33	What is the type of roof construction? Evaluate fall hazards									
34	Will there be any cladding or curtain walling taking place?									
35	Are pre-fabricated elements to be installed?									
36	Are there any 'Hot Works' to be undertaken?									
37	Are electrical items to be installed?									
38	Will there be any lift installation works?									
39	Will there be any escalators to install?									
40	Have formwork requirements been identified?									
41	Is the project a fire risk?									

Contd

Figure 6.2: Contd

No.	Hazard present	YES/NO	Describe the hazards and obvious control or protective measures necessary	Likely consequences of an accident (a)					Number of workers exposed to hazard (b)					Probability of harm (c)					Risk rating and risk category	Extra control measures necessary
				3	6	9	12	15	2	4	6	8	10	4	8	12	16	20		
42	Have all environmental issues been evaluated and controlled?																			
43	Are there any specific fall protection hazards not already assessed?																			
44	Are there any additional hazards which have been identified as being site specific and which are not covered by the foregoing? If YES, note here:																			

7 Control Strategies for Construction Work

Designing for safety and health

Safety in construction begins with design, where some of the fundamental decisions are taken which can have serious consequences for those who build structures, maintain them and work in them. In some cases, designers may be the only people who can **eliminate** a hazard at source — the best possible control strategy as it eliminates a foreseeable risk. Designers are also well placed to **reduce** unnecessary levels of risk in the industry. All too often, a design decision establishes a hazard, and the contractor, maintenance staff and even premises workers are then left to manage the risk as best they can.

Research carried out by the European Union shows that a majority of construction injuries have their origin at least in part in the preconstruction phase of work. Recent studies have also shown that, while failures can be found in the technical aspects of design (inaccurate or inappropriate calculations, for example), design problems are by their nature related to planning and organisational issues. As a result, concentration upon the purely technical aspects can mislead, and a broader based view is desirable.

A new theory of construction accident causation has been introduced by Suraji and Duff, which includes a much wider range of features than previous models. These features include project conception, design, management and construction. The various deficiencies are classed as proximal (nearby) and distal (further away), which allows the incorporation of the influences of designers and their designs into the construction activity. Fundamental concepts of the theory as it relates to designers are:

- They may introduce factors that lead to accidents directly or indirectly
- Designers work within a variety of imposed constraints — client's decisions, other participants, the designer's own organisation, for example

- The designer's response to those constraints influences construction activity — possibly giving incorrect information leading to an inappropriate process and increased risk
- 'Inappropriate process' includes planning, control, operation and site conditions, plus inappropriate actions by those working on site

Readers interested in the latest views on the role of design in construction safety are referred to the proceedings of a conference recently held on the subject. The reference is given at the end of this chapter.

A review of a design from a safety, health and environmental standpoint is established best practice, especially where errors can be expected to have disastrous consequences for a business after, as well as during, construction — as, for example in some process industries.

Formal analytical techniques such as failure modes and effects analysis, HAZOP and HAZAN are beyond the scope of this book. Contractors should not hesitate to ask their clients for assurance that these techniques have been used where the contractor has not been a party to the design process.

Legal responsibilities of designers

Under the Construction (Design and Management) Regulations 1994 (CDM), the term 'designer' has a specific legal definition — any person who carries on a trade or business in connection with which he prepares a design or arranges for any person under his control to do so. In turn 'design' has a wide definition also; in relation to any structure, the term includes drawings, design details, specifications and bills of quantities.

Designers' duties are mostly covered by Regulation 13 of CDM. The extent of any liability of designers had previously been a 'grey' area within the terms of the Health and

Safety at Work etc. Act 1974, and CDM clarifies the position, at least as far as criminal prosecution is concerned. The issue of civil liability is a matter of debate, but most authorities agree that the setting out of designers' duties within CDM had provided an example of 'reasonable care' which could result in a definitive standard of care in civil law, thus civil liability could result indirectly from failure to carry out the sort of process required by CDM. Designers can still take some comfort from the fact that, in common with the bulk of the CDM Regulations, breach of their duties under Regulation 13 does not confer a **direct** right of action in civil proceedings.

It is appropriate to set out here the CDM designer duties. Designers must:

1. Ensure that any design prepared for construction work pays adequate regard to **three needs**:
   Need 1: to avoid foreseeable risks to health or safety of constructors, cleaners on or in the structure, and third parties affected by their work
   Need 2: to combat at source risks to those classes of people, and
   Need 3: to give priority to measures which protect those classes of people over measures which only protect individuals
2. Ensure the design contains information about any aspect of the project, structure or materials which might affect the health or safety of the above classes of people
3. Co-operate with the Planning Supervisor and any other designer connected with the same project or structure, to enable each of them to comply with their duties, and
4. Ensure the client is aware of the duties of clients under CDM before starting the design

The designer is required to address the first two needs only to the extent that it is reasonable to expect the designer to have done so at the time the design is prepared, and to the extent that it is reasonably practicable to do so. No person can arrange for a designer to prepare a design unless he is reasonably satisfied that the designer has the necessary competence.

Advice on checking the competence of designers is given in the Approved Code of Practice. The Planning Supervisor must be in a position to give advice on competence, as required specifically by Regulation 14(c).

Examples of positive actions which can be taken by designers include:

- Selection of a reversible window type which can be cleaned from the inside
- Detailing a design for items to be lifted to include attachment points to support the load
- Avoiding low level pipe runs in plant rooms
- Specifying early erection of stairways to provide means of escape in case of fire
- Avoiding designs which involve temporary instability during erection
- Incorporating principles of temporary support into the design
- Providing fall protection systems for maintenance which can also be used by constructors
- Designing for a reduced frequency of maintenance

For the contractor, review of the design is closely allied with the selection of appropriate construction methodology. For example, if a particular hazard is identified as present in a large number of similar situations during a project, steps can be taken during the pre-construction phase to design solutions and specify appropriate work systems.

Collaboration between designers, Planning Supervisor and contractor can achieve significant reductions in risks — specification of an erection method for steelwork can lead to more prefabrication, bolting up at ground level and finally from underneath using an elevated working platform — and a new need to check the floor slab loading. Many of these preconstruction activities are closely connected, and it should not be a surprise that as many as 65% of fatalities in the industry can be attributed in whole or part to a failure to recognise and control potential hazards at this stage.

Design choices frequently determine construction methods: failure to recognise this and force the process in the opposite direction leads to muddle and make-do.

Planning the work

Gathering information together must be done before a risk assessment is made for a project, and this in turn is necessary before an adequate safety plan can be written or reviewed. This will contain clear information on all significant hazards likely to be present, and should be sent to subcontractors at an early stage. If tenders and bids are made with knowledge of requirements for safety, health and the environment, contractors are much less likely to complain about lack of advance knowledge, so reducing complaints that the required safety items and precautions have not been priced for.

Relevant information will be available from a variety of sources: the client, design staff, contract documents, equipment and material suppliers, and standards and laws.

Information about the previous uses of the site, and any unusual features, can give guidance on hazards which may be present. Especially important is information about the presence, now or in the past, of:

- Chemical contamination, including the presence of asbestos
- Overhead power lines and underground services
- Unusual ground conditions
- Public rights of way or accustomed access across the site
- Nearby schools, roads, railways and waterways
- Other activities on the site, for example the client's existing business

Common facilities

In recent years there has been a growing awareness of the need to plan for safe **moving traffic** on sites. The basic principle is to keep the movements of vehicles and people physically apart — this may require barriers and designated routes on site. Also, pedestrian crossing points will have to be established, with adequate lighting. So plan for **safe access** to and around the site. How will vehicles be kept clear of pedestrians, especially at site entrances and exits, loading and unloading areas and anywhere where the drivers' vision may be restricted?

Site boundaries should be fenced off and signed suitably. In multiple-occupancy situations, who controls which areas? Site rules may be needed to keep the construction work apart from other activities. What precautions are needed for night-time working? What methods will be used to close the site when no work is being done?

Suitable **welfare facilities** must be available for all workers during all working hours. As a minimum, these will be access to toilet and washing facilities, a supply of clean drinking water, a place to take breaks and meals and store clothing, shelter in bad weather, and first aid facilities. Most, if not all, of these will be covered by local or national regulatory requirements.

Emergency procedures

The most obvious emergency is fire. The safety plan should contain an appropriate emergency plan, written to cover the detailed arrangements on a project. Other potential emergency situations which may require activation of the emergency plan include flooding and multiple injuries from any cause.

A more common form of emergency is the need to evacuate an injured person. It is always necessary to think about how an injured person can be evacuated from the most inaccessible area of the project, and how long that could take. Also consider the full response time — how long it could take between an injury occurring and arrival at a treatment centre. This should always be evaluated and reassessed as construction work proceeds.

Planning for emergencies begins with the goal of minimising their likelihood. The aim of publishing an emergency plan is to ensure that everyone on site can be alerted in an emergency, knows what the emergency signal is and what to do. Emergency routes need to be identified, signed, adequately lit and kept clear.

When planning emergency procedures, routes and exits, the following should be taken into account:

- Site size, and characteristics of the site and the work being done
- How to raise the alarm under those conditions
- Plant and equipment being used (may impede exits, for example)
- How many people are likely to be present (are the exits wide enough?)
- Properties of substances likely to be present
- Location of the nearest emergency services and their capabilities
- Access to the site for emergency services

Fire and fire precautions

There are two methods of dealing with fire in construction work — preventing it happening, and preparing for and controlling the consequences if it should happen. Both require equal attention during the planning process. The three ingredients of fire are fuel, oxygen and a source of ignition. Remove any one of them and there will be no fire.

Much of **fire prevention** takes place at the planning stage, where simple rules apply:

- Use less flammable materials (water-based or low solvent adhesives, for example)
- Keep the quantity of flammables on site to a minimum. Supplies of materials should be enough for half a day or a single shift, and the remainder kept in stores
- Store flammable solids, liquids and gases safely, separated from each other and from oxygen cylinders or oxidising materials
- Arrange for ventilated secure stores or outdoor storage areas for flammables
- Make sure that rubbish is removed regularly
- Ban smoking in appropriate areas
- Require hot work to be done under a permit-to-work system

These can all be dealt with in principle at the planning stage. During construction, ongoing attention is required to prevent fires, and the most important matters to keep under constant review in addition to the above are:

- Close valves on gas cylinders when not in use and check hoses for wear and leaks
- Prevent oil or grease from coming into contact with oxygen cylinder valves
- Do not leave bitumen and other boilers unattended when alight
- Insist that welding or cutting producing sparks is done with a fire extinguisher close at hand
- Check the site at break times and close of work to make sure that plant which might start a fire is turned off
- Hot working should be stopped an hour before close of work

Fire precautions are actions and facilities which allow people to escape safely from a fire if one should break out. The following are the main points to consider:

Means of sounding the alarm A system should be established, which can vary from a fixed alarm unit down to a whistle or word of mouth on a small site. The warning must be distinctive, audible above other noise and known to everyone.

Means of escape Escape routes must be planned. There should be well separated alternative ways of exiting a work area where possible. Escape routes must give access to a safe place where people can assemble and be accounted for. Signs are likely to be needed, also adequate lighting for enclosed escape routes. This may include emergency lighting sufficient to enable safe escape.

Means of fighting fire Fire extinguishers will be required for hot work, but there will also be a general need to have them at identified fire points around the site. Where possible, halon-containing extinguishers should not be used. Fire doors should never be locked, left open or removed once fitted.

Site tidiness Fire precautions include tidiness — rubbish, offcuts and waste materials act as fuels, and they are also tripping hazards and obstructions to emergency routes.

Special fire issues

Storage of 'volatile flammable materials' This term includes LPG and LNG, many solvents, flammable gas and oxygen cylinders. In internal storage areas, good ventilation is necessary to avoid the build-up of dangerous levels of gases. This can be obtained by having high and low level openings in an external wall, but these should not ventilate into the surrounding structure. For flammable liquid storage, the size of the openings should be at least 1% of the total square area of the floor and walls combined. For flammable gases and oxygen cylinders, the opening size should be at least 2.5%.

External stores in the open air must be in well-ventilated areas at least 3 m away from the boundary of the nearest building and any drains or excavations. Leaking gas may collect at lower levels. If stores must be placed alongside buildings, then:

1. There should be a fire-resisting partition between the store and the building (unless the building elevation concerned is fire-resistant 3 m either side of the store and 9 m above it) and
2. Drains and excavations must be sealed unless a spillage retention wall or bund is placed around the store

All stores of volatile flammable materials should have two exits, both unlocked when anyone is inside. The exception is when the store is small, so that nobody has to travel more than 12 m to an exit. Lock all stores except during use. External stores should be protected by a 1.8 m high wire-mesh fence for security.

Small quantities of LPG can be kept safely in a lockable wire cage with a single exit, as long as it is clearly marked and at least 1 m away from any other structure. LPG must not be stored in unventilated huts or metal boxes. Small quantities (up to 50 litres) of flammable materials such as paints and solvents can be kept in lockable steel chests or cabinets.

Oxygen cylinders must always be stored by themselves, and not in company with other gas cylinders such as those containing LPG and acetylene.

Protective coverings Covers are often used to protect final features against damage, but they can add significantly to the potential for fire where ignition sources are common. Their use to protect features in escape stairways should be avoided where possible. Risks can be reduced by using flameproof coverings or coverings coated with flame retardant. At least one escape stairway should have no such coverings within it. Risk can be reduced by installing features needing protection as late as possible, and specifying flame-retardant coverings to be used.

Handling flammable liquids Fires have been caused by failure to limit the likelihood of spills and the release of vapour. Standard precautions are:

- Provide drip trays to contain spillage during decanting and dispensing
- Do not permit the decanting of liquids inside flammable liquid storage areas
- Do not permit contaminated rags to be carried about
- All work operations should be carried out in well ventilated areas
- Flammable liquids should be kept in closed-top containers
- Dispose of contaminated waste safely

Setting up the site

Preparation for construction operations should be based upon an overall project risk assessment, which identifies the major hazards likely to be encountered during the work (see the previous chapter). Physical features of the area will normally dictate the layout of the site, including temporary roads, storage areas and the siting of facilities including offices and welfare facilities.

Preparatory work should include early decisions on removal or diversion of power lines and cables where necessary, land clearance, demolition and access issues. Decisions will be recorded in the pre-tender safety plan, with details to be added by the principal contractor in the finalised construction phase safety plan. One of the aims of site planning should be to minimise the need for manual handling of loads, and the reduction of distances where manual handling is required.

It will also be necessary to consider environmental consequences at this time, and 'good neighbour' policies. Use of checklists enables monitoring to be carried out during internal and external audits. A copy of each completed checklist should be held on file on projects for ease of reference. It is important that all checklists are signed off by project management when completed. Sample checklists are included (Figs 7.1 to 7.4).

Safe place of work

The concept of 'safe place of work' is derived from the law, which has long recognised the duty of the employer to provide a safe place of work and a safe means of access to it and egress from it. But the concept is not just a technicality. To ensure safety, hazards need to be identified and avoided — possibly by changing the design or work method. If this cannot be done, then control is required, and this extends to others who may be affected by the work — other contractors, employees or third parties.

If the 'safe place of work' concept is used as the guiding principle, many of the decisions required on access topics, for example, can be simplified. It can be seen that a safe working platform is preferable to working from a ladder (for an extended period). The person has more space to work, guardrails or barriers to prevent falls, and does not need to carry materials. Third parties can be protected from falling materials. This does not mean that work from a ladder may never be done, but indicates some of the risk factors involved when making a choice about a safe place of work.

Planning and control strategies should be aimed at making decisions about such issues by management, rather than leaving them to site workers who will not normally be aware of all the constraints. For example, a subcontractor may attempt to put up ladders or a temporary scaffold to gain access to the outside of a structure, unaware that the spot chosen is on a major access route that happens to be quiet at the time. Third parties should not have to make decisions about ensuring their own safety — they should be excluded from work areas where they may be at risk.

Another example is the separation of vehicles from people by planning traffic routes to provide a safe means of access and egress. Left to themselves, drivers and pedestrians mostly make reasonable decisions about speed and direction, but they cannot guarantee the control of shifting loads, road conditions and perception which are three major factors in the rising numbers of incidents on site involving transport and people.

Safe systems of work

It has been estimated that at least a quarter of all fatal injuries at work involve failures in systems of work — the way things get done. A safe system of work is a formal procedure that results from a systematic examination of a task in order to identify all the hazards and assess the risks, and which identifies safe methods of work to ensure that the hazards are eliminated or the remaining risks are minimised. Where elements of risk remain, a safe system of work will be required. Some examples where safe systems of work will be part of the controls are:

- Cleaning and maintenance work
- Changes to normal procedures, including materials and methods
- Working alone
- Dealing with plant and equipment breakdowns and emergencies
- Interactions between contractors and clients' workers on the client's premises
- Vehicle loading, unloading and movements

Safe systems of work are often documented in the form of method statements, or formalised into permit-to-work systems. Some safe systems can be verbal only — where instructions are given on the hazards and the means of dealing with them, for short duration tasks. These instructions must be given by supervisors or managers, because leaving workers to devise their own method of work is not a safe system of work. A more formal analysis may be required, and the result can be used to train new workers in the required method. This technique is called job safety analysis, and is discussed elsewhere.

Method statements

The key feature of method statements is that they provide a time-frame and order — a sequence for carrying out an identified task; some work activities must be done in sequence to ensure safety. In such cases, it is necessary not only to know what the control measures are, but also to carry the work out in a particular order. Examples of

activities where the HSE expect method statements to be provided include demolition work, asbestos removal and structural steel erection.

Some principal and management contractors impose their own stricter requirements concerning the production of method statements — the trend is to require them for all activities. They are entitled to do so if they wish, as compliance with the safety management system will be a condition of the contract. Most principal contractors are prepared to assist smaller organisations to prepare their method statements.

Risk assessments can usefully form the basis for method statements, but there will usually be more detail in a method statement than in a risk assessment. Those reading method statements need to know how what is proposed to be done will or could affect them. A method statement that fails to give that information is not adequate.

Method statements normally include the following information:

1. Originator and date
2. A brief description of the work
3. Identification of individual(s) who will be responsible for the whole operation and for compliance with the method statement. Key personnel responsible for particular operations may also be named
4. Training requirements for personnel carrying out tasks which have a competency requirement. (Examples are crane and fork-lift drivers, testing and commissioning staff)
5. Details of access equipment which will be used, safe access routes and maintenance of emergency routes
6. Equipment required to carry out the work, including its size, weight, power rating, necessary certification
7. Locations and means of fixing the stability of any lifting equipment to be used
8. Material storage, transportation, handling and security details
9. Detailed work sequence, with hazard identification and risk control measures, including co-operation between trades which may be required
10. Special considerations, including any limitations following part completion of works, any temporary supports or supplies required or other special circumstances
11. Details of all personal protective equipment and other measures such as barriers, signs, local exhaust venti-

lation, rescue equipment, fire extinguishers, gas detection equipment

12. Any environmental limitations which may be applicable, such as wind speed, rain, temperature
13. Details of measures to protect third parties who may be affected
14. Mention of any site-specific rules that may apply to the work
15. The means by which any variations to the method statement will be authorised
16. Distribution list of those who need the information

Permit-to-work systems

Systems of work often operate under the control of a permit system, known as a permit-to-work. Because of the administration needed to run them, permit systems are relatively uncommon in construction. They are more common in industries where the hazards are associated with high risk levels and their use is well established in many organisations. They require certain precautions to be taken in order to control them. Examples include hot work permits, electrical lock-out systems and vessel or other confined space entry systems.

The main requirements of formal permit-to-work systems are:

- Specification of the control measures
- Planning of job sequences if necessary
- Specified safe means of access and egress to the work area
- Conditions which must be verified before work starts
- Clean-up at the completion of the task
- Adequate communication
- No exceptions permitted
- Training required for management and workers
- Monitoring to ensure the system remains appropriate and is being complied with
- Not a substitute for controls not depending upon people to make them work

These written safe systems of work will be found where the risks are high, the precautions needed are complicated and need written reinforcement, and where the activities of different groups of workers or multiple employers have to be co-ordinated to ensure safety.

Permit systems usually use preprinted forms that list specific checks and/or actions required at specified stages

of the work — the fitting of locking devices to controls, for example. Most permits are designed to cover work lasting up to 24 hours and require an authorisation signature for any time extension. Permit systems that regularly allow work over a longer period than this are likely to be defective.

Key features of permit systems are that experienced, trained and authorised persons will assess the hazards and risks involved in the proposed work, and will then complete and sign a certificate giving authority for the work to go ahead under controlled conditions specified on the permit. Nobody should be in a position to authorise a permit for themselves to do the work.

The permit should specify:

- Details of the work to be done
- Details of all the precautions required
- Emergency procedures
- Any limits on the work, work area or equipment
- Written acceptance by the person who will do the work
- Written signed confirmation that the work has been completed and the area restored to safety
- Any permitted time extension for the work
- How the permit is to be cancelled

It has been established in law (notably *R.* v. *Octel*) that any company which operates a permit-to-work system must accept liability for its adequacy and safe operation. Therefore, whoever issues a permit must accept personal responsibility for ensuring that the required and listed safety precautions have in fact been taken, before issuing the permit against the signatures of those involved. Use of a permit system does not absolve contractors of their responsibilities to provide a safe place and safe systems of work.

Permits-to-work are not alternatives to the provision of other permanent safeguards, as they suffer from the chief disadvantage that they depend upon people to operate them and to work within the conditions they lay down. Purely verbal instructions are never a safe alternative to a permit system. At the heart of a permit system is the basic requirement that a written statement (the permit) confirms that all necessary precautions have been taken to reduce risks and is in the possession of the person directly in charge of the work, before the work begins.

Model permit-to-work system

A model permit system is shown in Table 7.1, containing a suite of permits comprising a general all-purpose format and special applications. The table shows the sequence of actions involved in a permit-to-work system. Where there is expected to be significant use of any or all the permits, it is recommended that a self-carbon pad of each is printed, in order to allow easy sequential numbering and filing of record copies.

General permit — type G (Figure 7.5)

This permit can be used for situations where a more specialised permit is not required. Examples include, but are not limited to:

- Alterations or installation of plant or machinery where mechanical, toxic or electrical hazards may be present
- Work on or near overhead crane installations
- Work on pipelines with hazardous contents
- Work with radiation
- Work on pressurised systems operating at plus or minus 0.5 bar or greater

Table 7.1: Operation of a permit-to-work system	
	TASK and SEQUENCE
	1. Request for work permit*
	2. Assessment of hazard
Permit types	3. Selection of permits
General — **G**	
Confined space — **CS**	
HV electrical — **E**	
Hot work — **HW**	
Excavation — **EX**	
Possible specific precautions	4. Raise permit and state precautions**
Isolation	5. Undertake initial precautions
Protective clothing	6. Verify precautions taken before releasing permit
Special equipment	
Breathing apparatus	
Fire fighting equipment	
Fire watchers	
Removal of waste	
Potential need to test	7. Sampling and testing
Toxic concentrations	
Flammable concentrations	
Lack of oxygen	
	8. Issue permit to competent person in charge of the work
	9. Carry out work in accordance with precautions
Monitoring considerations	10. Monitor conditions
Sampling	
Spot checks	
Shift changes	
	11. Completion of work and hand back
	12. Cancellation of permit
	13. Keep record of permit

* When permits to work are in use on a project or site, their use should be covered as part of the induction training of all workers on site, and during pre-start meetings held with contractors.

** No permit can be issued until the persons issuing and receiving the permit have personally satisfied themselves that the permit is properly filled in, and have visited the scene of the work to inspect the task and verify that the specified precautions are all in place.

Confined space entry permit — type CS (Figure 7.6)

This permit can be used to confirm that necessary precautions have been taken before entry into any confined space, such as a chamber, tank, vat or pipe where there may be insufficient oxygen or toxic fumes from any source (including from the work itself).

Excavation permit — type EX (Figure 7.7)

Where buried cables, sewers or pipes are present, or where work involves breaking into existing sewers or pipes, appropriate measures are required to identify the hazards and specify necessary precautions. These must be identified on the permit.

Hot work permit — type HW (Figure 7.8)

'Hot work' means work involving cutting, welding, lead burning and soldering, or other application of heat using portable equipment. It also includes drilling and grinding where a flammable atmosphere may be present or may be released. Unless the work is to be carried out in a special area prepared for the purpose, the use of a hot work permit system should be considered during the project risk assessment.

High voltage electrical work permit — type E (Figure 7.9)

Equipment over 650 volts is conventionally designated as 'high voltage'. Errors during work on this kind of equipment are likely to result in fatality, and precautions other than the use of competent electricians will usually be required to protect anyone likely to be affected by the work.

Using the model system

The general principles of use are the same for each type of permit. Notes on the use and completion of each type of permit are contained in the following sections.

General permit — type G

The general permit has been designed to ensure adequate isolation of machinery, plant and pipework, and it also covers most general circumstances other than those dealt with by the specialised permits listed above. It may be necessary to raise more than one kind of permit for a particular task, such as a hot work permit, in addition to the general permit.

Accidents can easily occur when plant and machinery are started up during repair and maintenance, and also during installation. Safety devices provided on equipment may often protect both operators and maintenance workers, but on large machines and process plant total isolation may be difficult, construction workers may have to enter the plant to carry out their work, and vision may be restricted. In order to minimise the risk of injury, a written permit-to-work system will be needed.

Other special circumstances where a permit system is appropriate include work on overhead travelling cranes and on chemical plant. Contractors should be aware that, for these cranes, analysis of accident causes shows that an effective safety system must prevent the crane from approaching those working in the area. Complete electrical isolation together with the use of stop blocks on the crane tracks are the precautions required to achieve this.

Completion of the G permit (Figure 7.5)

General

The person responsible for issuing the permits must complete sections A to D inclusive. Additionally, Section E1 must be signed and the time limit completed in E3. The details of the permit must be fully discussed with the person in charge of the work, who will then sign it at section E2, accepting the permit. A copy must then be made and given to the person in charge of the work. This copy must be displayed in the work area, and all details must be fully explained to those doing the work. After completion of the work, the person in charge of the work must sign at section F1 on both copies, following which the issuer signs both copies to cancel the permit. One copy must be kept on file by the issuer, and the person in charge of the work should retain his copy as a record.

Section A

The permit is to be addressed to the person in charge of the work to be done. A clear description is to be given of the area or equipment, and of the work to be done, to ensure accurate identification and no ambiguity.

Section B

Required isolations must be clearly indicated by the person issuing the permit. The person in charge of the work must ensure that each isolation is initialled by the person who has carried it out.

Section C

Adequate information must be given to ensure that the correct type of equipment is used. Details of any atmospheric tests required must be written on the back of the permit, stating the intervals between tests at C5 if required.

Section D

This section is used to identify other permits which may be required to cover different aspects of the same work. If additional signatories are not required, the person issuing the permit must delete D5 and 6. Where they are required, the person issuing the permit must obtain them before issuing the permit.

Section E

E1 must be signed by the person issuing the permit, who must identify the person to whom the permit is issued, and state the time limits in E3. Time extensions are permitted up to 24 hours, at which time a fresh permit must be raised and issued. The person issuing the permit may extend its life for up to 24 hours, but must satisfy himself that it is safe to extend the limit before doing so. Extension details must be written on the back of the permit. E2 must be signed by the person in charge of the work to be done, who must explain the permit to those doing the work and then display the original permit near the place of work. Where there is a change in the person in charge of the work, the person issuing the permit must ensure that the permit is properly transferred and an additional signature obtained at E2.

Section F

F1 must be signed by the person in charge of the work, on completion. F2 must be signed by the person issuing the permit, cancelling it and describing the state of the equipment and/or area concerned.

Confined space entry permit — type CS

Stringent precautions must be taken before anyone is allowed to go into a confined space. Also, rescue must be carried out quickly and efficiently if anything goes wrong. The purpose of this permit is to identify and control the precautions and to confirm the presence of necessary rescue facilities.

A **confined space** is defined as any chamber, room, vat, tower, pipe, duct or similar place, open or closed, into which people may enter and where:

- There is or is liable to be a dangerous liquid, gas, fume, vapour or dust, or
- There is or is liable to be a deficiency of breathable oxygen, or
- There is a fire, explosion or radiation risk, or
- There are mechanical hazards, or
- There are physical hazards such as noise and extremes of temperature.

The following list shows examples of such places.

Examples of confined spaces		
Agitators	Ducts	Settling tanks
Boilers	Flues	Sewers
Bunkers	Gas mains	Steam drums
Culverts	Poorly-ventilated	Stills
Cold stores	tunnels	Tanks
Drains	Towers	Vats

The **hazards** which may be present during entry work include:

- **Gassing by toxic vapours** which may result from material in the confined space, from reaction between the vessel contents and the structure, from gradual releases from sludge or scale, from leakage through interconnected systems because of failure to disconnect or blank off pipelines or ducts, or from the type of work being carried out in the confined space
- **Asphyxiation** — lack of oxygen to breathe, which can be caused by chemicals replacing or displacing oxygen in the space's atmosphere, inadequate ventilation during the work, use of flames, or absorption of oxygen in closed clean vessels by rusting of the wall metal
- **Drowning** — due to ingress of liquids or sludge in sumps, sewers and vessels
- **Fire or explosion** — flammable materials can be produced from the contents of the space or by the work being done, ignited by ignition sources such as static, electrical faults and mechanical sparks
- **Electric shock** — from portable tools and equipment
- **Injury from mechanical equipment** — especially from installed equipment being started inadvertently

- **Corrosive and heat burns** — from opening of steam or other valves, use of welding equipment, contact with introduced or leaked chemicals
- **Physical hazards** — falls, falling objects and collapse of materials or structures

A general checklist of precautions includes:

Isolation and opening — delivery pipes and ventilation ducts into confined spaces should be broken and blanked off or sealed. Manhole covers should always be removed to admit air.

Draining and cleaning — residual products, sludge and other material should be removed from the outside, so that the space is clean, and outlet pipes disconnected.

Ventilation — mechanical means such as portable blowers should always be advocated to remove dangerous fumes and to introduce clean air.

Gas testing — the atmosphere within a confined space must be tested for toxic gases and lack of oxygen, and the test must be repeated each time a change might have occurred in the atmospheric conditions or at such regular intervals as are specified on the permit.

Mechanical hazards — isolators controlling power to any moving parts within the confined space must be locked off (with the keys being held by a responsible person) and warning notices placed on the isolators. Fuses should be withdrawn as an additional precaution.

Provision for rescue — lifting tackle, if necessary on a tripod, should always be rigged for use before entry into the confined space if there is a possibility of this being needed. An evaluation of the means of evacuating a casualty is always required, which must also consider removal to ground level. Rescue teams must be properly equipped with breathing apparatus, and trained to use it and in their duties in the event of an emergency. Rescue watchers must be aware of the action to take in the event of an emergency. There must always be at least one external watcher, and where rescue may be physically difficult or there is more than one lifeline in use, more will be required. At least one watcher must have no other work to do than to give undivided attention to those within the confined space. The extreme end of each lifeline must be tied off and held taut by the watcher. No rescuer is permitted to enter the confined space unless wearing breathing apparatus.

Completion of the CS permit (Figure 7.6)

General

The person responsible for issuing the permits must complete sections A to D inclusive. A clear description of the confined space must be given in A1. Each isolation must be signed off when confirmed by the person carrying out the isolation, in section B. Section E1 must be signed by the issuer, and the time limit completed in E3. The details of the permit must be fully discussed with the person in charge of the work, who will then sign it at section E2, accepting the permit. A copy must then be made and given to the person in charge of the work. This copy of the permit must be displayed in the work area, and fully explained to those doing the work. After completion of the work, the person in charge of the work must sign at section F1 on both copies, following which the issuer signs both copies to cancel the permit. One copy must be kept on file by the issuer, and the person in charge of the work should retain his copy as a record.

Section A

The permit is to be addressed to the person in charge of the work to be done. A clear description is to be given of the confined space, and of the work to be done, to ensure accurate identification and no ambiguity.

Section B

Required isolations must be clearly indicated by the person issuing the permit. The person in charge of the work must ensure that each isolation is initialled by the person who has carried it out.

Section C

Adequate information must be given to ensure that the correct type of equipment is used. Details of any atmospheric tests required must be written on the back of the permit, stating the intervals between tests at C10 if required.

Section D

If additional signatories are not required, the person issuing the permit must delete D5 and 6. Where they are required, the person issuing the permit must obtain them before issuing the permit. Personnel controls will depend upon the conditions within the confined space.

Section E

E1 must be signed by the person issuing the permit, who must identify the person to whom the permit is issued, and state the time limits in E3. Time extensions are permitted up to 24 hours, at which time a fresh permit must be raised and issued. The person issuing the permit may extend its life for up to 24 hours, but must satisfy himself that it is safe to extend the limit before doing so. Extension details must be written on the back of the permit. E2 must be signed by the person in charge of the work to be done, who must explain the permit to those doing the work and then display the permit near the place of work. Where there is a change in the person in charge of the work, the person issuing the permit must ensure that the permit is properly transferred and an additional signature obtained at E2.

Section F

F1 must be signed by the person in charge of the work, on completion. F2 must be signed by the person issuing the permit, cancelling it and describing the state of the equipment and/or area concerned.

Excavation work permit — type EX

The excavation work permit has been designed to ensure that excavations can proceed without putting people or processes at risk as a result of contact with buried services, cables, main drains and other underground obstructions. The permit does not cover the method of excavation and the standard precautions which should be taken.

Issue of an EX permit does not invalidate or alter any other requirements which may be contained in G or CS permits which may apply to a location. Where such permits apply, they must be obtained and issued independently of an EX permit.

Where there may be a hazardous atmosphere or deficiency of oxygen in the area where excavation is to take place, atmospheric tests must be carried out before work starts and if necessary during the course of the work. If the excavation becomes a confined space as defined above, a CS permit will be required and must be issued before entry into the excavation. A hot work permit may also be required in some cases.

During the course of work which is covered by an EX permit, if the programmed work must be changed, or if unantici-

pated cables, pipes or drains are discovered, the existing EX permit must be cancelled and a new one issued to cover the new circumstances.

EX permits cease to be valid after 2 weeks from the date of issue.

Where alterations are carried out or new buildings are erected, a detailed plan showing the location of pipes, cables, hydrants, sewer systems, etc., must be kept in a safe place. The adequacy of a permit system for excavations depends upon the possession of accurate and up-to-date plans.

Completion of the EX permit (Figure 7.7)

General

The person responsible for issuing the permits must complete sections A to D inclusive. Additionally, section E1 must be signed and the time limit completed in E3. The details of the permit must be fully discussed with the person in charge of the work, who will then sign it at section E2, accepting the permit. A copy must then be made and given to the person in charge of the work. This must be displayed in the work area where possible, and the details must be fully explained to those doing the work. After completion of the work, the person in charge of the work must sign at section F1 on both copies, following which the issuer signs both copies at F2 to cancel the permit. One copy must be kept on file by the issuer, and the person in charge of the work should retain his copy as a record.

Section A

The permit is to be addressed to the person in charge of the work to be done. A clear description is to be given of the area, and of the work to be done, to ensure accurate identification. The description of the excavation should include its location, direction, width and depth, and a sketch plan should accompany the permit.

Section B

Known obstructions must be clearly indicated by the person issuing the permit. In each case the section should be initialled by the person providing the information.

Section C

Adequate information must be given to ensure that the correct type of appliances and equipment are used. Details of any atmospheric tests required must be written on the back of the Permit, stating the intervals between tests at C6 if required.

Section D

This section is used to identify other permits which may be required to cover different aspects of the same work. If additional signatories are not required, the person issuing the permit must delete D5 and 6. Where they are required the person issuing the permit must obtain them before issuing the permit.

Section E

E1 must be signed by the person issuing the permit, who must identify the person to whom the permit is issued, and state the time limitation in E3. Validity of permits should not be extended beyond 2 weeks from the date of issue, at which time a fresh permit should be raised and issued. E2 must be signed by the person in charge of the work to be done, who must explain the permit to those doing the work and then (where the nature of the work allows) display the permit near the place of work. Where there is a change in the person in charge of the work, the person issuing the permit must ensure that the permit is properly transferred and an additional signature obtained at E2.

Section F

F1 must be signed by the person in charge of the work. F2 must be signed by the person issuing the permit, cancelling it and describing the state of the area concerned.

Hot work permit — type HW

'Hot work' hazards are associated with many types of construction, maintenance, repair and demolition work. The main sources of danger appear to be obvious, but experience shows that some are not, and those which are recognised are often treated as 'traditional' hazards of the industry.

'Hot work' is defined as work involving cutting, welding, lead burning and soldering, or other application of heat using portable equipment. It also includes drilling and grinding where a flammable atmosphere may be present or may be released.

Arcs and flames from cutting and welding equipment are rarely the direct cause of fire. Other ignition sources are produced in their vicinity as sparks of hot slag which may rain down through openings to start fires in remote locations, and be a cause of eye injuries. Many fires have started inside ducts not completely cleared of combustible residues. Flammable liquid containers, tanks or pipework may explode on contact with a cutting torch, unless effectively cleaned and purged. 'Flammable liquid' means almost any liquid which will burn, because at welding temperatures the liquid will vaporise.

Fires can also be caused by poorly maintained or defective equipment, including faulty valves, poor hoses and badly made connections.

Wherever possible, items requiring hot work should be taken to a safe place in the open, or an exempt area where the permit system does not apply. The need for such an exempt area should have been identified during the project risk assessment. **No work involving vessels, pumps, pipe or plant which contain flammable liquids or residues should be done without a permit.**

The standard precautions to be taken against fire during hot work are listed in section C of the HW permit. All those which are appropriate to the circumstances should be adopted. If hot work is to be done on a metal wall or partition, combustible material on the other side will need protection against conduction and radiation of heat. An additional fire watcher may be required in some circumstances.

Safer alternatives to open hot work include removing the work to a safer place, using cold cutting or electric saws, or using other methods of fixing.

A **fire watcher** can be any competent worker, who has been trained in the use of appropriate fire fighting and protective equipment and is familiar with the means of escape in the area. He should extinguish fires and call for assistance when necessary. He should also be given the authority to stop the hot work if he considers the conditions have become unsafe.

Completion of the HW permit (Figure 7.8)

General

The person responsible for issuing the permits must complete sections A to D inclusive. Additionally, section D1 must be signed and the time limit completed in D3. The details of the permit must be fully discussed with the person in charge of the work, who will then sign it at section D2, accepting the permit. A copy must then be made and given to the person in charge of the work. This copy of the permit must be displayed in the work area, and the details must be fully explained to those doing the work. After completion of the work, the person in charge of the work must sign at section E1 on both copies, following which the issuer signs both copies to cancel the permit. One copy must be kept on file by the issuer, and the person in charge of the work should retain his copy as a record.

Section A

The permit is to be addressed to the person in charge of the work to be done. A clear description is to be given of the area or equipment, and of the work to be done, to ensure accurate identification and no ambiguity.

Section B

The precautions indicated in this section are in addition to any which may be required by other permits in force, or other aspects of the work. Some or all may require approval or authorisation by the client. The person issuing the permits must ensure that each precaution is in place before the permit is issued.

Section C

Fire watchers are extremely important, and have saved many premises from serious damage as a result of fire from smouldering material or structures. In general, at least one of the controls should apply to each permit issued for hot work.

Section D

D1 must be signed by the person issuing the permit, who must identify the person to whom the permit is issued, and state the time limits in D3. Time extensions are permitted up to 24 hours, at which time a fresh permit must be raised and issued. The person issuing the permit may extend the life of a permit up to 24 hours, but must satisfy himself that it is safe to extend the limit before doing so. Extension must be written on the back of the permit. D2 must be signed by the person in charge of the work to be done, who must explain the permit to those doing the work and then display the permit near the place of work. Where there is a change in the person in charge of the work, the person issuing the permit must ensure that the permit is properly transferred and an additional signature obtained at D2.

Section E

E1 must be signed by the person in charge of the work. E2 must be signed by the person issuing the permit, cancelling it and describing the state of the equipment and/or area concerned.

High voltage electrical work permit — type E

Unless all the necessary safety precautions are taken, work on high voltage equipment can result in fatal injuries.

'High voltage' means any voltage in excess of 650 volts. 'High voltage equipment' means any equipment (including cables and switchgear) operated at a voltage exceeding 650 volts.

Those persons who are appointed by contractors to work on high voltage equipment must possess the necessary training, technical knowledge and experience for the work to be done safely, and a high level of personal responsibility. Experience of work on lower voltage systems only is not sufficient.

The type E permit specifies exactly what high voltage equipment it is safe to work on. As with other permits, it is issued to the person in charge of the work. It is essential that all instructions for the safe operation of equipment are unambiguous. The person receiving the instructions should repeat them as a check that the requirements have been understood correctly. Switching must not be carried out on the basis of pre-arranged signals or after agreed time intervals.

Entry to substations, switchrooms and other enclosures containing high voltage equipment must be restricted to workers positively identified as competent, as above, and to persons accompanied at all times by them. All those entering and about to carry out work on any part of high voltage equipment must be fully aware of the nature and extent of the work to be done.

High voltage electrical equipment can only be worked on in safety if **either** the part of the apparatus or equipment to be worked on is permanently and adequately earthed, so that it is electrically dead at all times, **or** the apparatus or equipment is first made safe by earthing, tested for verification, and then released for work by the issue of a E permit. Before any work is carried out on remotely or automatically controlled equipment such as circuit breakers and isolators, the automatic or remote controlled features must first be rendered inoperative.

Special issuing conditions for E permits

The person issuing an E permit is personally responsible for the safe custody of all keys for any locks fitted to prevent danger. The number of E permits issued must be limited, and in particular at any one time there must be only one valid E permit for any specific piece of equipment. When an E permit is to be issued, the person issuing it must accompany the person receiving it to the place of work and ensure that the person receiving the permit has correctly identified the apparatus before any work is started. The person receiving the permit should be asked to read it aloud to the person issuing it. The receiver of the permit must then satisfy the issuer that the apparatus or equipment is electrically dead and safe to work on. The person issuing the E permit must then sign both copies of the permit in the presence of the person receiving it.

E permits are valid only for work under the supervision of the person to whom they are issued, and are not transferable to any other person or work under any circumstances. If the person to whom the permit is issued ceases to have responsibility for the work permitted, the permit must be cancelled immediately and all persons withdrawn from the work until another permit is issued.

High voltage equipment and apparatus can only be made ready for use after the E permit has been cancelled, as this involves removing earthing, locks, danger notices, etc.

The future requirement for work of this type should have been identified in the project risk assessment.

Completion of the E permit (Figure 7.9)

General

The person responsible for issuing the permits must complete sections A, B and C1. The details of the permit must be fully discussed with the person in charge of the work, who will then sign it at section C2, accepting the permit. A copy must then be made and given to the person in charge of the work. This copy of the permit must be displayed in the work area, and the details must be fully explained to those doing the work. After completion of the work, the person in charge of the work must sign at section D1 on both copies, following which the issuer signs both copies to cancel the permit. One copy must be kept on file by the issuer, and the person in charge of the work should retain his copy as a record.

Section A

The permit is to be addressed to the person in charge of the work to be done. A clear description is to be given of the system or equipment to be worked on. Section A is a statement that the system or equipment is safe, and the person issuing the permit must satisfy himself that this is so before issuing it. It also contains details of the means of earthing the system, and where warning notices have been placed.

Section B

Section B is an unambiguous statement of the work to be done, to ensure accurate identification and that those doing the work are aware of the limitations of the safe environment and what precautions have been taken to make it safe.

Section C

C1 must be signed by the person issuing the permit, who must identify the person to whom the permit is issued. E permits should be issued for specific tasks, and will not normally be subject to time limits. The permit will last for the duration of the work or until conditions change. No handovers can be made; a fresh permit must be issued to the new person in charge of the work to be done, who must explain the permit to those doing the work and then display the permit near the place of work. Doing so confirms the supervisor's understanding of the nature of the work, the working conditions and the control measures in place.

Section D

D1 must be signed by the person in charge of the work, only after the workers have been withdrawn. This section indicates that the system or equipment is no longer safe to work on. Cancellation using section E should take place immediately.

Section E

This must be signed off by the permit issuer on both copies, keeping one on file and returning one to the person in charge of the work for record purposes.

References

Gibb, A.G.F. (ed.) (2000) *Proceedings, Designing for Safety and Health Conference*, June 2000. European Construction Institute, c/o Loughborough University, LE11 3TU.

Suraji, A. and Duff, R. (2000) Constraint-response theory of construction accident causation. In: Proceedings, Designing for Safety and Health Conference, June 2000, (ed. A.G.F. Gibb). European Construction Institute, c/o Loughborough University, LE11 3TU.

Figure 7.1: Sample checklist: good neighbour				
Project:				
NO.	**STANDARD**	**YES**	**NO**	**COMMENTS**
1	Has the project Safety Plan taken the following items into account (as they apply to the project)?			
2	Has a 'good neighbour' policy been established with occupiers of premises in the immediate vicinity of the site?			
3	Are perimeter boundary fences to be erected? Are there arrangements to keep them maintained in good repair, well lit as necessary, signed and kept clean?			
5	Is pedestrian footway disruption being considered and kept to a minimum?			
6	Where necessary, are crash barriers, safety nets and pedestrian tunnels provided and maintained?			
7	Has provision been made for public viewing locations?			
8	Is the external level of lighting being maintained as the existing lighting is disturbed (i.e. footpath lighting levels?			
9	Is traffic flow scheduling for material delivery/collection satisfactory?			
10	Is optimum use being made of off-site fabrication to reduce traffic flow to the site?			
11	Are site and delivery vehicles maintained in good condition?			
12	Are security procedures effectively preventing unauthorised access to the site?			
13	Are all adjacent roads/footpaths kept clean?			
14	Is all rubbish being cleared from site regularly?			
15	Are dust control measures effective?			
16	Are the locations of services clearly marked?			
17	Is effluent and waste from the works effectively controlled?			
18	Are passers-by harassed by site work in any way?			
Checklist completed by:		Date:		

Figure 7.2: Sample checklist: noise control

Project:

NO.	NOISE CONTROL	YES	NO	COMMENTS
1	Has a survey of planned site activities been made to establish potential high noise level activities?			
2	Have all contractors been asked to identify and control activities likely to produce high noise levels?			
3	Where such activities have been identified, and it appears likely that acceptable or permissible noise levels will be exceeded, have any of the following actions been instigated by the contractors?			
(a)	Silencing the noise at source?			
(b)	Controlling the spread of noise?			
(c)	Reducing the noise exposure duration?			
(d)	Providing hearing protection and warning signs?			
4	Where feasible has all noisy plant been located away from sensitive areas?			
5	Where this is not possible has the noisy plant been acoustically screened effectively?			
6	Have contractors provided suitable hearing protection for their workers?			
7	Have arrangements been made for regular maintenance of noisy plant?			
8	Has substitution been considered for noisy plant or procedures?			
9	Is the necessary information, instruction and training on noise control techniques given to site workers?			

Checklist completed by: Date:

NO.	POLLUTION CONTROL	YES	NO	COMMENTS
1	**Dust Emissions**			
(a)	Have the operations/processes which create dust emissions been identified?			
(b)	Are they critical activities or can they be deleted?			
(c)	If critical, is there an alternative process?			
2	**Dust Control Measures**			
(a)	Can dust be suppressed at source by either wet processes or forced air extraction?			
(b)	Can the process be contained by localised sheeting or containment within the structure itself?			
(c)	Is ventilation or filtration required?			
(d)	Can the process be relocated to a less sensitive location?			
3	**Site Roads and Exits**			
(a)	Are the site roads a source of dust?			
(b)	Can the surface of the roads be compacted sufficiently to prevent dusting?			
(c)	Is the road surface regularly cleaned (using a dust-free wet method)?			
(d)	Are site roads watered to reduce dust emissions in hot weather?			
(e)	Are wheelwash facilities required at site exit points? If so, are they fully operational and staffed continuously?			
(f)	Do all vehicles leaving the site have to use the wheelwasher?			
(g)	Is the wheelwasher water cleaned appropriately before dispersal?			
4	**Local Roads**			
(a)	Is the status of roads local to the project site monitored for cleanliness and condition? Is this done daily/weekly/less often/adequately?			
(b)	Are facilities available to clear roads of any mud or debris deposited (power sprays, road sweepers etc.)?			
5	**Air Monitoring**			
(a)	Has air monitoring been implemented to establish the amount of dust or other significant contaminants in the air?			
(b)	Have initial readings been taken to establish ambient (background) levels before work commenced?			
6	**Discharges to Watercourses**			
(a)	Are there any surface water courses within, adjacent or downstream of the project/site?			

Figure 7.3: Sample checklist: pollution control

Project:

Contd

Figure 7.3: *Contd*				
Project:				
NO.	**POLLUTION CONTROL**	**YES**	**NO**	**COMMENTS**
(b)	If so, are these vulnerable to pollution by:			
	■ Chemical spillage			
	■ Diesel/oil spillage			
	■ Waste materials			
	■ Soiled water			
(c)	How are these sources of pollution to be avoided?			
(d)	How will the effectiveness of any control measures be monitored?			
7	**Discharges to Drains and Sewers**			
(a)	Are the locations of drains and sewers on or adjacent to the project/site known?			
(b)	Are these vulnerable to pollution by:			
	■ Blockage			
	■ Stored chemical spillage			
	■ Diesel/fuel spillage			
	■ Waste materials			
(c)	How are these sources of pollution to be avoided?			
(d)	How will the effectiveness of any control measures be monitored?			
(e)	Are facilities available on site to clear and clean blocked drains?			
8	**Material Storage**			
(a)	Are storage areas located downstream from watercourses, drains and sewers?			
(b)	Are diesel/fuel oil and chemical storage areas bunded to contain spillages?			
(c)	Are facilities available on site to deal with on site spillage of pollutants?			
9	**Statutory Authorities**			
(a)	Has the relevant local authority been notified of works affecting watercourses, drains and sewers?			
(b)	Has the authority issued any conditions regarding either permitted dust in air levels or watercourse protection?			
(c)	If so, how are these conditions to be complied with and how will they be monitored?			
(d)	Have these conditions been passed to a qualified Safety Manager for comment?			

Checklist completed by: Date:

	Figure 7.4: Sample checklist: waste control			
Project:				

NO.	WASTE CONTROL	YES	NO	COMMENTS
1	Are you aware of local waste disposal requirements which must be complied with by the project?			
2	Have you identified the types of waste that will be disposed? Examples are:			
	■ Wood			
	■ Metals			
	■ Builders' waste			
	■ Chemicals			
	■ Hazardous substances			
	■ Flammable substances (those with flash points exceeding 21°C — check the data sheet)			
	■ Toxic waste			
3	Is it a requirement that waste be sorted by type prior to disposal?			
4	Are contractors' obligations for proper waste disposal clearly identified in the Tender and/or Contract documents?			
5	Are contractors required to:			
(a)	Control and dispose of all their waste?			
(b)	Remove waste to agreed collection points for removal by others?			
(c)	Contribute to a Service Team with necessary labour/funding/equipment?			
6	Where required, are the waste removers and the disposal tips licensed to receive the type(s) of waste being disposed of?			
7	Who controls the issue of transfer or consignment documentation, where this is required?			
8	How long are disposal records required to be maintained? (UK = a minimum of 2 years)			
9	Can waste materials be recycled?			
10	Can the level of waste material be reduced by design/quality control?			
11	Is rubbish burning banned on site?			

Checklist completed by: Date:

Figure 7.5: General permit to work — G

To: . . . .   Company: . . . .

Project: . . . .

EVERY ITEM MUST BE COMPLETED OR DELETED AS APPROPRIATE

A JOB DETAILS

1. Area or equipment to which this permit applies:	2. Work to be done:

B ISOLATIONS (specify where necessary)   **Initials and comments**

1. Circuit breaker locked out/fuses withdrawn/isolator locked off	YES/NO	
2. Circuit tested and confirmed to be dead	YES/NO	
3. Mechanical or physical isolation	YES/NO	
4. Valves closed/locked off/spades inserted	YES/NO	
5. Pipelines drained/purged/disconnected/vented to atmosphere	YES/NO	
6. Documented isolation procedure attached	YES/NO	
7. Other	YES/NO	

C PRECAUTIONS (to be taken as indicated, additional to those specified on other permits)

1. Protective clothing	Yes/No	Type . . . .
2. Respiratory equipment	Yes/No	Type . . . .
3. Protected electrical equipment	Yes/No	Type . . . .

4. State additional precautions required (if none state none)

. . . .

. . . .

. . . .

. . . .

5. Atmosphere tests are not/are required at intervals of . . . . and results must be recorded overleaf.

D ADDITIONAL PERMITS AND SIGNATURES REQUIRED BEFORE WORK STARTS

1. Confined Space Entry — CS	Yes/No	5. In my opinion the engineering precautions are adequate.
2. HV Electrical — E	Yes/No	Signed (Engineer) . . . . Date . . . .
3. Hot Work — HW	Yes/No	6. In my opinion the precautions against special hazards
4. Excavation — EX	Yes/No	within my knowledge are adequate
		Signed (Specialist) . . . . Date . . . .

E ISSUE AND RECEIPT BEFORE WORK STARTS

1. Issue	2. Receipt
I have examined the area specified and permission is given	I have read and understood the conditions of this permit
for the work to start, subject to the conditions hereon,	Signed . . . . Date . . . .
under the control of	3. This permit is valid from . . . . hours to
Signed . . . . Date	. . . . hours (max 24 hours)

F CLEARANCE AND CANCELLATION AFTER WORK

1. Clearance	2. Cancellation	Yes	No	I have notified those affected.
All workers under my control have been withdrawn. The	Work complete			This permit is cancelled.
permitted work is/is not complete.	Isolations removed			
	Area/Equipment			Signed . . . .
Signed	is safe to use			Date . . . .
Time . . . . Date				Time . . . .

Figure 7.6: Confined space permit — CS

To: . . . .   Company: . . . .

Project: . . . .

EVERY ITEM MUST BE COMPLETED OR DELETED AS APPROPRIATE

A JOB DETAILS	
1. Area or equipment to which this permit applies:	2. Work to be done:

B   ISOLATIONS (specify where necessary)   **Initials and comments**

1. Circuit breaker locked out/fuses withdrawn/isolator locked off	YES/NO	
2. Circuit tested and confirmed to be dead	YES/NO	
3. Mechanical or physical isolation	YES/NO	
4. Valves closed/locked off/spades inserted	YES/NO	
5. Pipelines drained/purged/disconnected/vented to atmosphere	YES/NO	
6. Documented isolation procedure attached	YES/NO	
7. Other	YES/NO	

C   PRECAUTIONS (to be taken as indicated, additional to those specified on other permits)

1. Protective clothing Yes/No Type	5. Safety belt/harness Yes/No Type . . . .
2. Respiratory equipment Yes/No Type	6. Lifting tackle Yes/No Type . . . .
3. Protected electrical equipment Yes/No Type	7. Special ventilation requirements Yes/No Specify:
4. Hot work permit Yes/No	. . . .

8. State additional precautions required (if none state none)

. . . .

. . . .

9. Rescue procedures are in place (specify) . . . .

. . . .

10. Atmosphere tests are not/are required at intervals of . . . . and results must be recorded overleaf.

D   ADDITIONAL PERMITS AND SIGNATURES REQUIRED BEFORE WORK STARTS

1. Maximum number of persons allowed in space at any one time	5. In my opinion the engineering precautions are adequate. Signed (Engineer) . . . . Date . . . .
2. Maximum duration each person is allowed for each entry	6. In my opinion the precautions against special hazards within my knowledge are adequate
3. Length of rest pauses to be taken between entries . . .	Signed (Specialist) . . . . Date . . . .
4. Number of external watchers	

E   ISSUE AND RECEIPT BEFORE WORK

1. Issue	**2. Receipt**
I have examined the area specified and permission is given for the work to start, subject to the conditions hereon, under the control of	I have read and understood the conditions of this permit Signed . . . . Date . . . .
	3. This permit is valid from . . . . hours to
Signed . . . . Date	. . . . hours (max 24 hours)

F   CLEARANCE AND CANCELLATION AFTER WORK

1. Clearance	**2. Cancellation**	Yes	No	I have notified those affected.
All workers under my control have been withdrawn. The permitted work is/is not complete.	Work complete			This permit is cancelled.
	Isolations removed			
	Area/Equipment is safe to use			Signed . . . .
Signed				Date . . . .
Time . . . . Date				Time . . . .

Figure 7.7: Excavation — EX	

To: . . . . Company: . . . .

Project: . . . .

EVERY ITEM MUST BE COMPLETED OR DELETED AS APPROPRIATE

A   JOB DETAILS

1. Area to which this permit applies:	2. Work to be done:

B   KNOWN OBSTRUCTIONS — specify and initial

1. Electrical	4. Gas
2. Water services	5. Process pipeline and drains
3. Telephone/cable	6. Other:

C   PRECAUTIONS (to be taken as indicated, additional to those specified on other permits)

1. Personal protective equipment   YES/NO   Type: Helmets/Gloves/Footwear/Clothing . . . .
2. Respiratory equipment   YES/NO   Type . . . .
3. Protected electrical equipment   YES/NO   Type . . . .
4. Hand digging only   YES/NO
5. State additional precautions required (If none state none)

. . . .

. . . .

. . . .

. . . .

6. Atmosphere tests are not/are required at intervals of . . . . and results must be recorded overleaf.

D   ADDITIONAL PERMITS AND SIGNATURES REQUIRED BEFORE WORK STARTS

1. General Permit to Work — G   Yes/No
2. Confined Spaces Entry — CS   Yes/No
3. HV Electrical — E   Yes/No
4. Hot Work — HW   Yes/No
	5. In my opinion the engineering precautions are adequate.
	    Signed (Engineer) . . . . Date . . . .
	6. In my opinion the precautions against special hazards
	    within my knowledge are adequate
	    Signed (Specialist) . . . . Date . . . .

E ISSUE AND RECEIPT BEFORE WORK STARTS

1. Issue I have examined the area specified and permission is given for the work to start, subject to the conditions hereon, under the control of . . . . Signed . . . . Date	2. Receipt I have read and understood the conditions of this permit Signed . . . . Date . . . . 3. This permit is valid until . . . . (date)

F   CLEARANCE AND CANCELLATION AFTER WORK

1. Clearance All workers under my control have been withdrawn. The permitted work is/is not complete. Signed . . . . Time . . . . Date	2. Cancellation	Yes	No	I have notified those affected.
	Work complete			This permit is cancelled.
	Isolations removed			
	Area/Equipment			Signed . . . .
	is safe to use			Date . . . .
				Time . . . .

Figure 7.8: Hot work permit — HW

To: . . . . Company: . . . .

Project: . . . .

EVERY ITEM MUST BE COMPLETED OR DELETED AS APPROPRIATE

A   JOB DETAILS

1. Area or equipment to which this permit applies:	2. Work to be done:

B   PRECAUTIONS (to be taken as indicated) — Client approval/authorisation

1. Isolation of service (simple)	YES/NO	
2. Sprinkler system in service	YES/NO	
3. Extinguishers or hose reels present	YES/NO	
4. Means of sounding fire alarm present	YES/NO	
5. Cutting or welding equipment in good condition	YES/NO	
6. Floors cleared of combustible materials	YES/NO	
7. Combustible floors suitably protected	YES/NO	
8. Flammable liquid containers removed, including empties	YES/NO	
9. If answer to No. 8 is yes, the containers are suitably protected	YES/NO	
10. Wall and floor openings suitably protected	YES/NO	
11. Walls, ceilings, roofs suitably protected	YES/NO	
12. Area on other side free from combustible material	YES/NO	
13. Area to be wetted or fire blankets used	YES/NO	

14. State additional precautions required (if none, state none): . . . .
. . . .
. . . .
. . . .

C   PERSONNEL CONTROLS AND AUTHORISATION

1. Trained fire watcher required throughout duration of work	YES/NO	
2. Trained fire watcher required for 30 minutes following end of work	YES/NO	
3. Fire inspection to be carried out at end of shift following end of work	YES/NO	

D   ISSUE AND RECEIPT BEFORE WORK

1. Issue	**2. Receipt**
I have examined the area specified and permission is given for the work to start, subject to the conditions hereon, under the control of . . . . Signed . . . . Date	I have read and understood the conditions of this permit Signed . . . . Date . . . . 3. This permit is valid from . . . . hours to . . . . hours (max 24 hours)

E   CLEARANCE AND CANCELLATION AFTER WORK

1. Clearance	**2. Cancellation**	Yes	No	I have notified those affected.
All workers under my control have been withdrawn. The permitted work is/is not complete.	Work complete			This permit is cancelled.
	Isolations removed			
	Area/Equipment			Signed . . . .
Signed	is safe to use			Date . . . .
Time . . . . Date				Time . . . .

Figure 7.9: HV electrical permit — E

To: . . . .   Company: . . . .

Project: . . . .

A EQUIPMENT/SYSTEM ON WHICH IT IS SAFE TO WORK	B WORK TO BE CARRIED OUT
It is safe to work on the following HV equipment/system, which is dead, isolated from all live conductors and is connected to earth/ground: LOCATION: EQUIPMENT/SYSTEM:	**Only** the following work is to be done:
EARTHING/GROUNDING DETAILS:	SPECIAL PRECAUTIONS TO BE TAKEN:
POINTS AT WHICH WARNING NOTICES HAVE BEEN PLACED:	

C ISSUE AND RECEIPT BEFORE WORK STARTS	
1. Issue I have examined the area specified and permission is given for the work to start, subject to the conditions hereon, under the control of . . . . Signed . . . . Date	**2. Receipt** I have read and understood the conditions of this permit, and understand that work is restricted to the equipment specified above. Signed . . . . Date . . . .
D CLEARANCE	**E CANCELLATION**
The work for which this permit was issued has been completed/suspended, and all workers under my control have been withdrawn and warned that it is no longer safe to work on the equipment specified in this permit. All tools and temporary earthing/grounding connections have been removed. Signed . . . . Time . . . . Date	This permit to work is cancelled Signed . . . . Date . . . . Time . . . .

8 The Health and Safety Plan

Although a requirement of the Construction (Design and Management) Regulations 1994 (CDM), the development of project- and site-specific Health and Safety Plans is a recognised part of best practice in safety management in the construction industry worldwide. Only a legal requirement in the member states of the European Union and Australia, their use is now encouraged in many other countries and mandated by contractors wishing to achieve maximum levels of control of site work. In this Chapter, 'Safety Plan' is used as an abbreviation for 'Health and Safety Plan'.

The Safety Plan is produced to assist and contribute to the establishment of a safe, healthy and environmentally sound working environment, so as to benefit all workers and management personnel on site, neighbours and third parties, and to minimise the environmental impact of the work. It should include details of the project and its safety, health and environmental requirements, anticipated site hazards and conditions and their means of control. It will also contain rules developed following consideration of all significant hazards and their risks, and other items which are important for the safe and efficient conduct of the work. 'Significant hazards' in this context refers to those hazards which a competent contractor could not be expected to identify in advance.

It must be the responsibility of every contractor to be aware of the Safety Plan and its contents, and to make known to their workers all specific provisions which they are required to comply with. One particular benefit of the Safety Plan is that, when made available in its preliminary form at the bid stage, it allows contractors foresight of conditions which may alter their bids. 'I didn't know you wanted me to allow for doing it that way' has been a frequent complaint of contractors in the past.

The Safety Plan cannot be developed until a project risk assessment has been made, in order to list all the issues and obtain an overview of significant hazards and risks, and if necessary to specify solutions to them. Many contractors have developed systems to do this; early resistance to the concept on the grounds of time lost and difficulty can be removed by the demonstration of its worth. Benefits include the advance identification of problems so that they can be managed by organising rather than by 'fire fighting'.

Pre-tender Health and Safety Plan

A summary of the matters which should be addressed at the pre-tender stage is given in an Appendix to the Approved Code of Practice to the Regulations, which should always be consulted. This shows the topics which should be covered where relevant to the project:

- General description and location of the work, with details of those involved including the client and Planning Supervisor, and any safety requirements of the client (probably in relation to interfacing with the client's own operations)
- Time and programme — including stage start and finish dates
- Existing off-site conditions — adjacent land use, ground instability and contamination, traffic systems and restrictions, potential for trespass or vandalism, restrictions on noise and other nuisances
- Existing on-site conditions — status and location of buried and overhead services, traffic conditions and restrictions, site investigation reports, underground obstructions, ground contamination and ground water conditions
- Hazards from existing structures which might arise from demolition and refurbishment — presence of asbestos, fragile materials, fire damage, post-tensioned concrete structures and structural instability
- Existing records — available drawings of existing structures, the health and safety file if available,

previous site investigation reports, historical maps if relevant

- The design — principles and assumptions of design for structures, including suggestions for methods or sequence of erection/assembly, specific inherent risks where the contractor will be required to state how he will avoid or control them, significant hazards not eliminated by the design
- Site layout and management — access egress, storage and unloading arrangements, site offices and welfare facilities
- Site rules and procedures — security arrangements, permit-to-work procedures, site rules from statutory bodies and emergency procedures
- Procedures for review of the Safety Plan itself — procedures for managing design changes
- Information required for the health and safety file
- Arrangements and format for the Safety Plan
- Arrangements for communication and liaison between all parties
- Issues which will arise when the project is located in occupied premises

Developing the construction phase Health and Safety Plan

The principal contractor's role is to take the pre-construction plan and breathe life into it. In so doing, the principles of prevention must be followed, as required by Schedule 1 of the Management of Health and Safety at Work Regulations 1999:

- Where possible, avoid risks altogether — by doing the work a different way without introducing new hazards
- Evaluate risks that cannot be avoided by risk assessment
- Combat risks at source, for example by removing a problem rather than posting a warning sign
- Adapt work to the individual, and give them control over speed of work where possible, rather than making the individual adjust to the work
- Make use of technological progress — mechanical handling, for example, replacing manual handling
- Have a policy of active risk prevention based on continuous improvement
- Give priority to solutions protecting the whole workplace rather than individuals — guardrails rather than safety harnesses
- Make sure everyone knows what they need to do to remain safe and healthy

- Achieve a positive organisational safety culture, where risk control is accepted as a normal part of life rather than an add-on.

The level of detail required in the Safety Plan should be in proportion to the risks involved. As with other control documents, the intention is for it to describe the practical arrangements made to control those risks.

The contents of the plan are not specified within the CDM Regulations, but guidance is given on them in the Approved Code of Practice. The list of topics given in Table 8.1 illustrates the comprehensive nature of the plan, but those that are not relevant to a particular project can be left out. Many of the topics are discussed elsewhere in this book.

Project safety management commitment statement

It has been found useful to require the most senior member of the project management team to sign the Safety Plan, thus demonstrating commitment to it. In some cases, principal contractors require the production of a project-specific statement of commitment to be placed in the front of the plan. For example see page 99.

Roles and responsibilities of project staff

The numbers and levels of management on a project will be proportional to its size and complexity. The following suggested duties and responsibilities of senior management may be found useful. Roles can be combined where appropriate on smaller projects.

Project Director / Senior Project Manager

1. Understand and be responsible for the implementation of the company safety, health and environmental policy, and policies, practices and procedures developed in compliance with it to meet the needs of the project
2. Co-operate in the co-ordination of safety, health and environmental activities between all contractors and clients, and any other contractors who may be working on the project
3. Appoint one or more members of staff to monitor the standards of safety, health and environment awareness and action on the project
4. Ensure that all relevant legislation is observed on the project

Table 8.1: Safety Plan contents	
Topic heading	**Examples of detail that may be appropriate**
Project introduction	Description of the work, project safety policy or statement of intent, restrictions on work such as traffic flows and noise limits, work activities of the client, contractors' layout for the site including storage areas
Project management	Management structure and responsibilities of the principal contractor's project team, arrangements for co-ordination meetings
Safety, health and environmental standards set for the project	Any special codes and conditions to be complied with, possibly set by the client
Information for contractors	Information that must be passed to on to lower tier contractors
Principal contractor's selection procedures for contractors, designers and suppliers	This may be relevant on larger projects where substantial procurement continues throughout the project's life
Communications	Examples are: how information is to be communicated to project teams and workforce, how co-operation is to be achieved between contractors, how design changes will be carried out and authorised
Hazardous activities	Arrangements including method statement requirements for dealing with the effective management of hazardous activities, of which examples are: ■ Storage and distribution of materials ■ Waste control and disposal ■ Means of access and places of work ■ Mechanical plant ■ Temporary services ■ Temporary support structures ■ Commissioning ■ Permit-to-work arrangements ■ Fall prevention and protection ■ Protection against falling materials ■ Exclusion of unauthorised people from work areas ■ Protection of the public
Emergency Plan or procedures	Dealing with injuries, fire and other dangerous occurrences, including remediation — see below
Reporting of incidents and other information	Arrangements for reporting incidents under RIDDOR, and statistical information
Welfare	Responsibility for provision and maintenance of welfare facilities, statement of what is available
Information and training for workers	Induction training and site orientation requirements, toolbox talks or task instruction
Consultation arrangements	Communication with workers and/or their representatives
Site rules	Any site rules established by the principal contractor
Health and Safety File	Requirements for the collection and handling of information to enable the file to be prepared
Monitoring arrangements	Compliance with the law, the site rules, procedures and standards
Project review	System for applying lessons learned, including continuing evaluations of contractor competence

Example of a project-specific statement of commitment

[YOURCO] is committed to the provision of a safe and healthy working environment for all employees and contractors, as well as other people affected by our work. These may include employees of the client or owner, our own visitors and those of contractors, and third parties such as the general public.

We recognise that the project work will result in the creation of a potentially dangerous work environment, and our aim is to achieve our objectives by eliminating all potential life-threatening and disabling situations, and to minimise the occurrence of all incidents which could lead to personal injury, damage to or loss of property and plant, and damage to the environment.

To achieve these objectives, we are committed to the following:

■ Identification in advance of potential hazards, and organising the work activities so as to minimise the risks arising from them
■ Setting and achieving essential basic standards for safety, health and the environment, and aiming to achieve best construction practice through a process of education and development

■ The presence of a nominated person on site whose role will be to manage safety on site, including the identification and rectification of any potentially dangerous situations
■ Establishing the site so that all necessary safe working procedures and equipment are implemented and used
■ Ensuring that all workers are given induction training prior to access to the site, and making appropriate arrangements to brief visitors
■ Establishing a site safety committee with regular meetings and follow-up actions, to which each contractor is to nominate a representative
■ Keeping our client advised in advance of any works that will affect their operations

As the senior Project Manager for this project, I have reviewed the contents of this Health and Safety Plan, and will ensure the necessary support and resources to achieve the above objectives and the provision of a safe and healthy workplace.

. . . .

Project Manager

---

5. Ensure the prompt reporting of injuries suffered on the project, and the investigation of pollution incidents, significant incidents and other matters which may be required by company policy to be investigated or recorded
6. Reprimand any member of staff failing to discharge satisfactorily the responsibilities allocated to him/her
7. Ensure that adequate information is received regarding matters which might affect safety, health and the environment in order to determine at the planning stage:
   (a) The most appropriate order and method of working
   (b) Allocation of responsibilities with contractors and others
   (c) Facilities for welfare and sanitation
   (d) Necessary fire precautions
   (e) Hazards which may arise from overhead or underground services and other situations which may lead to unnecessary improvisations on site
   (f) Provision of adequate power supplies and lighting

8. Ensure that pre-start kick-off meetings are held with contractors and that adequate safety and health plans or method statements have been prepared by them which establish work methods and sequences of operations, together with the precautions to be adopted
9. Act as Chairman of the project Safety and Health Committee and ensure that safety and health are discussed and recorded as part of the agenda of all project and progress meetings held with contractors
10. Be responsible for the planning, implementation, publication, operation and review of the written Project Major Emergency Plan, and ensure that the proposed Emergency Plan meets the objectives of company policies on safety, health and the environment
11. Appoint a senior manager to act as Fire and Emergency Procedures Co-ordinator
12. In the event of an emergency, act on advice from the Fire and Emergency Procedures Co-ordinator and

activate the Project Major Emergency Plan where appropriate

13. In relation to company and statutory environmental requirements:

   (a) Ensure the removal of waste in accordance with statutory requirements, or best practice

   (b) Implement measures necessary to control ground, river and coastal water pollution

   (c) Implement measures necessary to control noise pollution

   (d) Implement a 'good neighbour' policy

   (e) Prevent the atmospheric discharge wherever possible of ozone-depleting gases such as CFCs and halons

   (f) Protect wherever possible and as appropriate wild life, flora including trees, archaeological and heritage remains

   (g) Promote sustainable development, minimise materials wastage, promote recycling options and conserve water, paper and energy

   (h) Investigate and report on environmental incidents, taking preventative action against repetition

   (i) Operate vehicles where possible on unleaded fuels to minimise exhaust pollutants

14. Set a personal example

Assistant Project Manager(s)

1. Organise the project and supervise to ensure that work is carried out to agreed standards of safety and quality, with minimum impact on the environment and risk to workers, equipment and materials

2. Co-ordinate the activities of contractors, and monitor that legislation, site rules and other instructions are being complied with

3. Ensure that first-aid and welfare facilities are available as planned, and that their location is known to all workers

4. Be familiar with the details of the Project Major Emergency Plan, and management's role in it

5. Set a personal example

Appointed site safety person

1. Supervise to ensure that work is carried out to agreed standards of safety and quality, with minimum impact on the environment and risk to workers, equipment and materials

2. Monitor that legislation, site rules and other instructions are being complied with

3. Carry out weekly formal inspections of the project and record the results, giving copies to the contractors concerned and keeping a file record

4. Investigate all injuries reported, to identify the cause and determine responsibility

5. Accompany visiting company and contractors' safety staff and local enforcement officials during their visits to the project

6. Set a personal example

All employees

1. Develop a personal concern for their own safety and health whilst at work, and for that of others, especially new employees

2. Report defects in equipment or plant, and any obvious health or environmental risks, to their superior

3. Be aware of the principles of the company policy on safety, health and the environment, and promote these

4. Comply with any company and project rules on safety and health as they affect their work, particularly in regard to the wearing or use of personal protective clothing and equipment

5. Set a personal example

Project Major Emergency Plan

The objective of a Project Major Emergency Plan (PMEP) is to ensure that all members of the project's management are able to respond to a major emergency quickly and systematically, by following a sequential plan of action. There is no statutory requirement to include a PMEP within the Health and Safety Plan, but their value has been proved beyond doubt. The remainder of this chapter offers suggestions on the format and development of a PMEP.

A **major emergency** can be defined as a situation where, in the opinion of the senior member of the project management, significant loss or damage has been caused to the project, persons working on the project in any capacity, or the surrounding area, and where external rescue, emergency or regulatory agencies and the media are likely to become involved.

Examples of major emergencies include (but are not restricted to) multiple injury incidents, fires which cannot be controlled by site facilities, natural disasters such as floods, earthquakes and exceptional wind speeds, and significant structural failures.

The PMEP will contain details specific to the project, and information under the following suggested headings:

Emergency planning   Before physical work begins on the project, and at intervals throughout its duration, the following should be reviewed: exposure to risks arising from fire, structural collapse, terrorist attack, risks to the general public outside the site perimeter and the likelihood of trespass or other unauthorised access to the premises. At the same time, the major work tasks involved should be reviewed to identify those with high risks which, in combination with the nature, size and location of the project structures, will make it necessary for special emergency arrangements to be put in place. The plan should begin with a statement of the project's exposure to these and other risks, and the significant work tasks.

Details of notifications made and the dates of notifications   The local police and fire authorities should be advised in writing of project start and estimated end dates, and key personnel contact telephone numbers during and outside normal working hours

Notification priority list   Who needs to know and in what order?

Posting of emergency telephone numbers   Outside normal working hours contact telephone number for notification of an emergency/incident, and details of project management outside hours numbers should be supplied to contractors' representatives and posted on noticeboards as well as included in the PMEP.

Evacuation plan, including emergency escape procedures and routes. It is important that everyone working on a project knows what actions they are expected to take in emergency situations that may require evacuation of the premises. The main points of the PMEP will therefore need to be covered during induction training.

The plan should include:

- Floor plans or maps clearly showing the emergency escape routes and assembly areas — these may need to be updated frequently
- Location of safe or refuge areas in high-rise buildings and other premises where speedy evacuation may not be possible
- Procedures to be followed by any workers allowed to

remain behind to look after essential operations until their evacuation becomes absolutely necessary
- Details of who is responsible for making emergency notifications for situations that may occur outside normal working hours
- Safe or refuge areas offering safety to occupants pending evacuation. Such an area is created by using barriers that either fold across or shut, to provide a space protected from fire and smoke, such as in enclosed stairways

Arrangements for personnel head count after emergency evacuation has been completed. Generally, evacuated staff should be directed to a known assembly point a safe distance and location away from the evacuated premises. Supervisors should then make sure that all their staff are accounted for, and confirm this to the Emergency Procedures Co-ordinator or person deputising.

Visitors to the project should be accompanied at all times by a responsible person in the organisation being visited, who must ensure their safety at all times and account for them at the head count. This requirement should be covered during the kick-off briefing for contractors.

Emergency evacuation drills   The need to practise emergency evacuation cannot be over-emphasised. Without such drills, it is unlikely that an adequate response in a real emergency will be achieved. The PMEP should contain appropriate arrangements for them, but their frequency will depend on individual needs, and these are likely to alter at different stages of a project. Accordingly, their scope and frequency must be kept under constant review and revisions given appropriate publicity.

Rescue and medical duties for those/any employees who are to carry them out. It is important to make sure that only those who have a detailed knowledge of the project layout and any processes involved are allowed to undertake rescue operations. These people should be in possession of current first-aid qualifications where possible, and be properly equipped with personal protective equipment so as to minimise the risks to themselves.

Security   Effective security arrangements are important in emergency planning. The security unit should be located away from potential disaster areas, in a commanding position. It should be fire-proof and contain a copy of the emergency contact numbers, the PMEP, and plans showing fire points, emergency exit routes and assembly areas, dry/

wet fire main inlet points and access route for emergency services.

It will normally be necessary to detail the arrangements made for project security within the PMEP. Security staff have a valuable role to play, especially outside normal working hours, and they should be involved in any rehearsal of arrangements. It is also possible that they may themselves be involved in an incident, and so the personnel head count must also include security staff.

Site fire precautions   Summary details of the project fire precautions should be included in the PMEP, and will be useful as a basis for decisions on what situations may require activation of the plan.

Out of hours incidents   The PMEP should provide for emergency notification for situations that may occur outside regular working hours, including nights, weekends and holidays. Security staff should be given specific instructions to call the senior person in charge of the project. In the event of a fire or serious injuries, the appropriate emergency services should be alerted before calling the person in charge of the project.

Personnel

Project major emergency plans should be co-ordinated by someone other than the senior person in charge of a project, designated the **Emergency Procedures Co-ordinator**. That person should, in conjunction with colleagues, write the PMEP and keep it up to date. The PMEP may nominate people to take various roles and act as the Emergency Management Team, or the group can be nominated at the scene.

The composition and numbers of the Emergency Management Team will depend on the nature, size and complexity of the project. They should all be familiar with the layout and nature of the project and any processes involved, and should be given the opportunity to practise their actions in association with fire drills. Members should be sufficiently senior in position to allow them to give instructions where necessary, and have experience which allows them to make appropriate assessments of the situation and judgements as to what actions will be most appropriate under the changing circumstances.

In the event of a major emergency, the Emergency Procedures Co-ordinator's job is to:

1. Assess the situation, determine whether the PMEP should be activated and advise the senior member of the project management accordingly
2. Take charge of the work of the Emergency Management Team
3. Direct all efforts in the area, including taking charge of the evacuation process and minimising property damage and loss
4. Ensure that outside emergency services are called in where necessary, and liaise with the senior person controlling those services, especially concerning the number and locations of any persons not accounted for, and for handover of control where appropriate
5. Direct the shutdown of project operations as necessary

Immediate action following activation of the PMEP

The decision to activate the PMEP will be taken by the senior member of management. The decision will be taken on the basis of advice given by the Emergency Procedures Co-ordinator or person deputising. The guiding principle should be that the safety of site personnel and any person likely to be affected by the emergency always takes priority. When the plan is activated, the following steps should be taken without delay:

1. Evacuate the premises according to the detailed PMEP, under the control of the Emergency Procedures Co-ordinator or person deputising
2. Arrange for a site personnel head count to be taken, and advise the senior external emergency services controller if anyone is unaccounted for
3. Establish an Emergency Management Team for the immediate control of the situation, led by the Emergency Procedures Co-ordinator or person deputising. Ensure that members of the Emergency Management Team take prompt steps to advise their own next of kin of their personal safety
4. Initiate discussion with external emergency services to establish who will be in overall control of the situation. Control should normally be relinquished to external emergency services, except where they decline to accept this
5. Notify company management as soon as possible by telephone
6. Pass information to the client, insurers and contractors affected by the emergency.

The following tasks should be carried out by the Emergency Management Team:

1. Confirm that full evacuation has taken place
2. Confirm that the PMEP has been properly followed
3. Establish a preliminary view of the nature and scale of the emergency and report this to the senior member of management
4. Decide what needs to be done to stabilise the situation. For example, is there a need for demolition to remove unstable materials; is there a safe method for re-entry into the premises; is there an emergency evacuation signal and system in place before re-entry is attempted; is there a safe temporary exit route?
5. Send non-essential personnel home, having decided upon the need for and size of a multi-service crew and clean-up gang
6. Alert specialist service contractors of any immediate needs, such as temporary equipment, generators, pumps, lighting and lifting equipment
7. Brief the senior member of management on what needs to be done, so that liaison can be maintained with the client, insurers and contractors, and the necessary actions agreed
8. Co-ordinate investigations to determine the cause of the emergency
9. Obtain clearance to re-enter the premises from rescue service and enforcement agencies, where necessary (for example, where there is any question of structural weakness having resulted)

Management information requirements

Information required by senior management at the beginning of an emergency is best addressed by the use of a checklist as a scratchpad memory aid (Figure 8.1).

Figure 8.1: Emergency information checklist	
What happened?	
What was the probable cause?	
When did the incident happen?	
How many casualties are there?	
Who is the employer of any casualties?	
What is the total number of people normally working on the project?	
How many were present at the time of this incident?	
Have all those who were present at the time of the incident been accounted for?	
Has the project site been evacuated?	
Is anybody trapped?	
Which emergency services have been involved?	
How long did it take for them to arrive?	
What are the details of any injuries?	
Which hospital (if any) received casualties [address required]?	
What is the current status of the project in terms of damage and delay to work?	
Have there been other previous similar incidents on the project?	

9 Training

Training for health and safety is not an end in itself, it is a means to an end. Talking in general terms to employees about the need to be safe is not training; workers and management alike need to be told what to do for their own health and safety and that of others, as well as what is required by statute. A knowledge of what constitutes safe behaviour in a variety of different occupational situations is not inherited but must be acquired, either by trial and error or from a reputable source of expertise. Trial and error methods extract too high a price in construction work, where the consequences of forced and unforced errors may be very serious, even catastrophic.

Experience and research also shows that knowledge of safe behaviour patterns, gained by instruction, films, videos, posters and booklets, does not guarantee that safe behaviour will be obtained from individuals. Training is therefore never a substitute for safe and healthy working conditions, and good design and planning. Because humans are fallible, the need is to lessen the opportunities for mistakes and unsafe behaviour to occur, and to minimise the consequences when it does.

Three conditions need to be present for any safety training to be successful: the active commitment, support and **interest of management**, necessary **finance** and organisation to provide the opportunity for learning to take place, and the availability of suitable **expertise** in the subject. The support of management demonstrates the presence of an environment into which the trained person can return and exercise new skills and knowledge. The management team also demonstrates support by setting good examples; **it is pointless to train workers to obey safety rules if supervisors are known to ignore them.**

Trainers must not only be knowledgeable in their subject, they must also be qualified to answer questions on the practical application of the knowledge in the working environment, which will include a familiarity with work practices, procedures and rules.

Reinforcement training will be required at appropriate intervals, which will depend on observation of the workforce (training needs assessment), on the complexity of the information needed to be held by the worker, the amount of practice required and the opportunity for practice in the normal working environment. Assessment will also be needed of the likely severity of the consequences of behaviour which does not correspond with training objectives when required to do so. If it is absolutely vital that only certain actions be taken in response to emergencies, then more frequent refresher training will be needed to ensure that routines are always familiar to those required to operate them. A construction industry example of this could be the need to practise the client's imposed evacuation routine at regular intervals, where premises are jointly occupied with the client during construction work.

Training needs

Three levels of training can be identified within the construction industry. **Craft and skills** training will involve the acquisition of skills needed to do the particular work, and training in environment, health and safety must be part of the package. This is recognised in the existence of the certification schemes discussed below. The second level is that provided by the **employer** to new employees upon joining, when the employer complies with legal obligations to provide formal health and safety training as required by the Management of Health and Safety at Work Regulations 1999. This will take place as part of the employer's induction programme. Training must also take place when work conditions change and result in exposure to new or increased risks. It must be repeated periodically where appropriate, and be adapted to any new circumstances. By law, no health and safety training can take place outside working hours. The third level of training is that given in relation to the particular site or project the person is working on, usually by means of an **on-site** induction process.

Craft and skills training

Certification by an appropriate body should provide confirmation that the holder is fully trained, competent and authorised to do particular types of work. In the UK there are many competence-based registration schemes that can be relied on to provide recognised evidence of competence, as is required by the Construction (Design and Management) Regulations 1994. The most general of these is the Construction Skills Certification Scheme (CSCS), which at present covers the following occupations:

Bench joiners	Painters and decorators:
Bricklayers	■ decorative
Built-up roofers	■ industrial
Carpenters and joiners	Partition fixers
Ceiling fixers	Piling operatives
Construction operatives:	Plant mechanics
■ general	Plasterers
■ concreting	Roof slaters and tilers
■ drainage	Sheeters and cladders
■ paving	Shopfitters:
■ street working	■ site fixers
Cornish masons	■ bench joiners
Dry liners	Single-ply roofers
Façade cleaners	Stonemasons
Formworkers	Wood machinists
Mastic asphalters	

Unskilled labourers are not covered by the scheme. The CSCS is administered by the Construction Industry Training Board and controlled by a Management Board of employer organisations and trade unions, with observer members from other interested parties including the HSE. Membership of the scheme is confirmed by the issue of a record card. Full details of CSCS can be obtained from PO Box 114, Bircham, Kings Lynn, Norfolk PE31 6XD.

Other recognised competence-based registration schemes and their organisations include:

- Certificate of Training Achievement for construction plant operators (CTA)
- BICS Operators Proficiency Certification Scheme (cleaning)
- Registration and Certification Scheme for Window and Curtain Wall Installation (administered by Bath University)
- UK Register of Electricians

- British Locksmiths Association
- Leadworkers Certificate
- Engineering Construction Skills Database
- EITB/EMTA (Engineering and Marine Training Authority) NVQ3 in lift maintenance/installation engineering (or apprenticeship)
- UK Register of HVCA Operatives
- UK Accreditation Service for Gas Installers (BCOP) — replacement for CORGI
- Building Engineering Services Scheme (CITB)
- Gas Distribution Record Scheme
- JIB Plumbing Registration Scheme
- JIB Gas Grading Scheme
- JIB Electricians Scheme
- Scaffolding Registration Scheme
- Steeplejack and Lightning Conductor Fitters Record Scheme
- Tunnel Miners Record Scheme
- Scheme for the Certification of Competence of Demolition Operatives (Topman and Mattockman)
- Concrete Repair Operatives Record Scheme
- Spray Concrete Registration Scheme
- Drilling and Sawing Operatives Registration Scheme
- Street Works Excavation and Reinstatement
- Construction Skills Register (Northern Ireland)
- Scottish Construction Operatives Registration Executive (SCORE)
- International Powered Access Federation MEWP Operators 'Type Specific' Training Programme
- Fencing Industry Skills Scheme

The foregoing is not intended to represent a complete list of schemes; mergers and additions are likely to alter the list over time.

Apart from specific legal requirements which may exist, the need for **first-aid training** in the workplace depends upon a number of factors. These include the nature of the work and the hazards, what medical services are available in the workplace, the number of employees and the location of the workplace relative to external medical assistance. Shift working may also be taken into account, also the ratio of trained persons present to the total number of workers.

Driver training and certification may be a requirement for particular classes of vehicle, according to national or local regulations in force, and in these cases the detail of the training programme may be defined in law.

A common cause of death both at work and away from work

is the road traffic accident. As the loss of a key worker can have a severe impact upon the viability of an organisation, training for all workers who drive should be considered. Defensive driver training has been found effective and cost-beneficial in reducing numbers of traffic accidents, and has been extended to members of workers' families, particularly those entitled to drive the employer's vehicles.

Fire and emergency training should be given to all employees, and included in induction and on-site induction training. Everyone needs to know the action to take when fire alarms sound. Knowledge of particular emergency plans and how to tackle fires with equipment available may be given in specific training at the workplace. At whatever point the training is given, the following **key points** should be covered:

- Evacuation plan for the project or premises in case of fire, including assembly point(s)
- How to use fire-fighting appliances provided
- How to use other protective equipment, including sprinkler and other protection systems, and the need for fire doors to be unobstructed in occupied premises
- How to raise the alarm on projects and in premises from call points
- Smoking rules
- Awareness of housekeeping practices which could permit fires to start and spread if not carried out, such as waste disposal, flammable liquid handling rules
- Any special fire hazards peculiar to the project or premises

Fire training should be accompanied by practices, including regular fire drills and evacuation procedures. No exceptions for senior management should be permitted during these. Evacuation drills on large projects have been found to improve the response time dramatically in subsequent real emergencies.

The employer's induction training

The safety policy of each employer should define the standard of induction training that will be given to employees as they join the organisation. Ten **key points** which should be covered in the employer's induction training are:

- Review and discussion of the employer's safety policy
- The employer's safety philosophy; safety is as important as production or any other organisational value

and activity; accidents have causes and can be prevented; prevention is the primary responsibility of management; each employee has a personal responsibility for his or her own safety and that of others

- That site, national and organisational health and safety rules or regulations will be enforced, and that those violating them may be subject to some form of discipline
- That the health and safety role of supervisors and other members of the management team includes taking action on and giving advice about potential problems, and that they are to be consulted if there are any questions about the health and safety aspects of work
- That where required, the wearing or use of personal protective equipment is not a matter for individual choice or decision — its use is a condition of employment
- That in the event of any injury, no matter how trivial it may appear, employees must seek first aid or medical treatment and notify their supervisor immediately
- That for any work involving repetitive, awkward, heavy physical or timed movements, workers are specifically instructed to report any adverse physical symptoms immediately. (These will need to be recorded, and investigated without delay)
- Fire and emergency procedure(s)
- Welfare and amenity provision
- Arrangements for joint consultation with other employees and their representatives should be made known to all newcomers

On-site induction training

Project-specific training is often arranged by the principal contractor, who will provide a training room facility on larger projects, and may also provide the trainer. Some clients with existing businesses may require that the induction process includes knowledge about the processes carried on adjacent to the project, and also about evacuation and other rules that must be observed by construction workers. This training may be carried out by a supervisor, but it should be properly planned and organised by the use of checklists.

Typically, on-site induction takes between 30 minutes and 2 hours, depending on the involvement of the client and other employers in joint-occupied premises. Induction training should be recorded in a log or register, and may be connected to the issue of identity badges. Newcomers to projects should receive induction before they start work, as it has been found that new arrivals are statistically the most likely to be injured, and soon after starting work.

Key points for site induction programmes should include:

- The general requirements of the principal contractor and the client
- Explanations of applicable safety regulations and organisation rules and procedures
- Review of any necessary skills applicable to the work, and training in them, such as a demonstration of any personal protective equipment which may be required and provided for the work (including demonstration of correct fit, method and circumstances of use and cleaning procedures)
- Hand-over of any documentation required, such as permit-to-work documents, safety booklets and chemical information sheets
- Review of applicable aspects of emergency and evacuation procedures
- Review of risk assessment findings. (This is proving popular and worthwhile as a training aid, and by this means employers can also fulfil the requirement to bring risk assessment findings to the notice of those affected by them)

Training for management

Supervisory and general management training at all levels is necessary to ensure that responsibilities are known and the organisation's policy is carried out. The legal requirement for suitable and sufficient training is not confined to site workers. Management failures which have come to light following investigations into disasters, plant accidents and other health and safety incidents have been concentrated in the following areas:

- Lack of awareness of the safety management systems in use, including their own job requirements for health and safety
- Failure to enforce health and safety rules adequately or at all
- Failure to inspect and correct unhealthy or unsafe conditions
- Ignorance of environmental requirements or opportunities for improvement of conditions
- Failure to inform or train workers adequately
- Failure to promote health and safety awareness by participation in discussions, motivating workers and setting an example

It is not sufficient simply to tell site managers and foremen that they are responsible and accountable for health and safety; they must be told the extent of the responsibilities and how they can discharge them.

Ten **key points** to cover in the training of managers are:

- The employing organisation's safety policy, organisation and arrangements
- The legal framework and the legal duties of the organisation, its management and the workforce
- Specific laws and rules applicable to the organisation's work, and its targets and commitments
- The causation and consequences of accidents
- Basic accident prevention techniques
- The control of hazards likely to be present in the work, (for example) roof work, confined spaces, scaffolding requirements, fire, materials handling, hazards of special equipment related to the industry and the use of personal protective equipment
- The main environmental concepts relating to the organisation's work, targets and solutions
- Safety inspection techniques and requirements
- Techniques for motivating employees to recognise and respond to organisational goals in health and safety
- Disciplinary procedures and their application

Senior managers should be given essentially the same information, as this gives them a full appreciation of the tasks of subordinates, makes them more aware of standards of success and failure, equips them to make cost-beneficial decisions on health and safety budgeting, and allows them to play an important part in setting organisational measures and targets.

External assessment of the training given to management at all levels is desirable. This can be done by training to the appropriate syllabi of national or international professional organisations, and encouraging those trained to take the relevant examination. In the UK, these are the 'Managing Safely' and 'Working Safely' courses moderated by the Institution of Occupational Safety and Health (IOSH), and for professional training courses leading to the two-part Diploma which is administered by the National Examinations Board in Occupational Safety and Health (NEBOSH). The IOSH course 'Safety for Senior Executives' is particularly significant as it aims to secure the commitment of the most senior management to the organisation's safety programme.

Environmental, health and safety specialists

All employers need access to competent advice on the environment, health and safety, in order to comply with their duties under the Management of Health and Safety at Work Regulations 1999. Whether this is achieved by training an existing member of staff, employing a full time or part-time person or the use of consultants will depend upon an assessment of the needs of the organisation in terms of its size and the nature of the hazards likely to be encountered in the work. Evidence of professional competence should be obtained, and advice sought from one or more of the professional organisations involved, which include the Institution of Occupational Safety and Health (IOSH) which has a Special Interest Group for its construction industry members, the Royal Society for the Prevention of Accidents (RoSPA) and the British Safety Council (BSC). The most demanding of current membership requirements are those of IOSH. Full membership is accorded to those who can satisfy membership requirements based upon experience and academic achievement via examination at Diploma level or via the NVQ system. Professionals who join the Register of Safety Practitioners (RSP) are required to demonstrate their commitment to continuing professional development through a peer review system.

Legal requirements

Many specific pieces of health and safety legislation contain requirements to provide training for employees engaged in certain tasks, and the most commonly applicable of these have been selected for discussion elsewhere in this book. As noted above, there is a general duty placed on employers to ensure that all employees are trained (and provided with information, instruction and supervision in addition) as necessary to ensure their health and safety so far as is reasonably practicable. This duty is to be found in section 2(2)(c) of the Health and Safety at Work etc. Act 1974. More specific requirements are contained in the Management of Health and Safety at Work Regulations 1999, and in specific sets of regulations covering tasks and conditions, such as confined space entry.

10 Meetings

Almost everyone has been present at meetings where nothing seemed to be accomplished, and they were generally agreed to have been a waste of time. They can easily turn into rambling unfocused discussions, arguments, or monologues from one or more participants, often the person leading the meeting. To achieve their objective, meetings should be structured communication sessions, with an agreed goal and outcomes. People have to be led in order to function effectively and reach decisions appropriate to the organisation. There should be an agenda, to make sure nothing is forgotten and make best use of the time of those attending. There should be a record made of what was decided and the actions to be taken.

In addition, meetings must be seen as relevant and important by those who attend them. If the focus wavers, or the task to be achieved is perceived as unimportant, key people will not make time to attend. Six important rules for successful meetings are:

- Keep them short and to the point
- Make sure everyone knows what will be discussed in advance — if necessary, circulate a written summary of important information for people to read before the meeting. This minimises unnecessary questions and comments
- Keep a record of what was discussed and what will happen as a result
- Set a timetable for the discussions and follow it
- Invite only those people whose presence is absolutely necessary
- Provide a method of following-up actions between meetings

Meetings with contractors

Clients and principal contractors have legal duties under the Construction (Design and Management) Regulations 1994 (CDM) to ensure that only competent contractors are engaged, and that an adequate Safety Plan is in place before the start of work is authorised. One way of making sure that these duties are complied with is to hold meetings between the selected contractors at which safety, health and environmental aspects of the work are reviewed. Each contractor should provide a detailed Safety Plan, including where appropriate a health, safety and environmental risk assessment.

A **pre-start meeting** should be held by the principal contractor with each contractor to review their plan, and the overall site Safety Plan, in advance of commencement of work on site. The meeting can also be used to explain the principal contractor's role in safety, health and environmental management, together with any specific requirements related to the particular project. A follow-up meeting may be required.

The objective of the pre-start safety kick-off meeting is thus to ensure that the client's and principal contractor's policies are known to each contractor, and to allow contractors adequate time to revise their own plans and procedures where necessary to secure conformance.

The meeting should be held as far as possible in advance of the start of work on site to allow revision of the plan(s) where necessary and to enable problems to be identified and resolved in good time. When completed and accepted, each contractor's plan will be included within the overall safety plan for the site or project.

The pre-start meeting should be chaired by the organiser, who may be the client, the principal contractor's Package Manager or Senior Project Manager, with a qualified safety professional in attendance where this is practicable. The agenda should be advertised in advance of the meeting, as the intention is to make sure all necessary details are reviewed; the purpose of the meeting is not to catch anybody out. Minutes of the meeting should include the response of the contractor together with any problems identified. If a further meeting is required to resolve any outstanding issues, the future meeting date should be agreed and minuted. Any variations to the client's or principal contractor's standard safety procedures should be noted.

A copy of the completed agenda should be signed off and retained on file for record and audit purposes. It would also be advisable to record the holding of the meeting in the weekly or monthly progress log. Also, where circumstances require the commencement of work by the contractor without a pre-start meeting having been held, those circumstances should be detailed in the progress log.

The following suggested agenda is necessarily lengthy, but its purpose is to cover the major issues which are likely to be found on a major project, and which must be acknowledged by the client, the principal contractor and individual package contractors involved. For smaller work, or less complex projects, the agenda can be modified to meet the needs of those involved. The recommended topic headings are followed by a detailed checklist giving more information on the questions that may need to be asked and answered (Figure 10.1).

Weekly safety meetings

Weekly progress meetings are often the main opportunity for liaison between contractors and the co-ordinating principal contractor. On larger or complex projects a separate meeting may be held weekly to discuss safety, health and environmental liaison. In either case the following agenda can be used as a checklist to make sure that nothing relevant is overlooked.

Purpose

Discussion between the principal contractor's site management staff and contractor management of standing items and new matters (new issues and future planning items) concerning safety, health and environmental practice on site.

Agenda for contractor pre-start meeting

1. General introduction for contractor
2. Information to be provided by contractor: organisation chart, names of staff, safety plan, risk assessment(s), method statement(s)
3. Work method review and agreement
4. Details of required attendance at review and progress meetings
5. Project safety meeting details and attendance requirements
6. Contractor's own safety meetings
7. Contractor's monthly safety report: contents requirements
8. Safety training (management training, induction, skills training, toolbox talks)
9. Contract subletting (arrangements for the control of subcontractors)
10. Injury and data reporting requirements: reporting of injuries and dangerous occurrences, recording of hours worked, major incident procedure
11. Welfare and first aid facilities
12. Inspection requirements
13. Access arrangements
14. Fall prevention and protection
15. Lighting arrangements
16. Personal protective equipment
17. Crane and hoisting facilities
18. Site transport and mobile plant: establish training and certification requirements
19. Noise control
20. Materials storage
21. Waste disposal: project policy on waste clearance
22. Housekeeping
23. Permits to work
24. Fire precautions: Emergency Plan review
25. Site access and security: limitations on visitor entry and movement
26. Public protection

Specimen agenda for weekly safety meeting

1. Recorded attendance and absence
2. Review of minutes of previous meeting
3. Standing items:
   (a) Access routes on site
   (b) Public safety
   (c) Environment
   (d) Confined space entry
   (e) Control of substances hazardous to health
   (f) Crane safety and slinging
   (g) Electrical safety
   (h) Excavations/trenching
   (i) Fall protection
   (j) Falling objects
   (k) Fire prevention and protection
   (l) Gas cylinders
   (m) Tools, plant and equipment
   (n) Hot work, welding and burning
   (o) Access, ladders and stepladders
   (p) Laser equipment
   (q) Manual handling
   (r) Mechanical handling and lifting devices
   (s) Personal protective equipment
   (t) Scaffolding
   (u) Traffic routes
   (v) Waste disposal
   (w) Welfare facilities (site accommodation, first-aid)
4. Presentation and review of:
   (a) Accident/injury statistics
   (b) Injury investigation reports
   (c) Corrective Action Reports issued to contractors
5. Tabled information — includes contractors' safety reports, audit and inspection reports, method statements provided since last meeting
6. New issues arising not previously raised
7. Future planning of work — long range, middle range and short range (to be commenced before the next meeting)
8. General remarks agreed to be minuted and dates for future meetings

Monthly safety meetings

A monthly safety meeting is normally held at a higher level on larger projects, in order to decide strategic issues rather than the detail of day to day work.

Purpose

Discussion between the most senior management of the participators in the work, to review reports and recommendations and to review minutes of the weekly meetings held on site to identify trends and points where executive action is required to implement necessary changes.

Specimen agenda for monthly safety meeting

1. Recorded attendance and absence
2. Review of previous minutes and follow-up reports on outstanding items
3. Auditors' reports since the previous meeting
4. Review of accident/injury experience since the previous meeting and overall
5. Review of any investigation reports received since the previous meeting
6. Review of response to Corrective Action Reports issued
7. Review of environmental concerns
8. New issues arising not previously raised
9. Future planning of work — long range, middle range and short range (to be commenced before the next meeting)
10. General remarks agreed to be minuted
11. Dates for future meetings

ITEM NO	ITEM	YES	NO	COMMENTS
1	**Introduction for contractor**			
1.1	Explain the main principles of the client's or principal contractor's safety policy			
1.2	Overview of the client's or principal contractor's general requirements for contractors			
1.3	Review of contractor's safety responsibilities			
1.4	Provide details of client's or principal contractor's safety staff and inspection routines, supply copy of specimen safety report and agree methods of follow-up action			
1.5	Explain the need generally to co-operate with the client or principal contractor and advise of any penalties which may apply in the event of failure to do so.			
1.6	Supply list of external contacts, including emergency services			
1.7	Supply list of client's or principal contractor's emergency contacts			
1.8	Review of any rules or procedures required by the client, including security			
2	**Information to be provided by the contractor**			
2.1	Names of contractor's Safety Manager, site safety adviser if appointed, and the site safety supervisor			
2.2	Receive copy of contractor's organisation chart			
2.3	Receive copy of contractor's safety plan, including appropriate method statements			
2.4	Receive copies of contractor's risk assessments (where appropriate)			
3	**Work method to be reviewed in detail and agreed**			
4	**Review and progress meetings**			
4.1	Advise that attendance is required and state the frequency of meetings			

Figure 10.1: Sample checklist: agenda for a contractor pre-start meeting

Contd

Figure 10.1: *Contd*				
ITEM NO	**ITEM**	**YES**	**NO**	**COMMENTS**
4.2	Advise that safety, health and environment must feature as the first item on the agenda at every future meeting			
5	**Project safety meeting**			
5.1	Advise that attendance is required at these meetings and state their frequency			
6	**Contractor's safety meetings**			
6.1	Advise that the contractor must hold any such meetings as required by the safety plan. Also, sufficient prior notice must be given to allow the client's or principal contractor's safety staff to attend			
7	**Contractor's monthly safety report**			
7.1	Advise that the contractor is required to provide a monthly written safety report and statistics, as may be detailed in the contract and safety plan			
8	**Safety training**			
8.1	Explain the client's or principal contractor's policy on training generally			
8.2	Obtain details of management and supervisory training provided to site staff			
8.3	Obtain details of induction training to be provided. (All workers should go through induction training before they start work on site)			
8.4	Skills training — advise requirements, especially for certification where required			
8.5	Advise of any specialist training requirements of the client			
8.6	Toolbox talks — state requirements for contractors			
9	**Contract subletting**			
9.1	Disclosure of subcontracts			
9.2	Obtain details of the arrangements made to control safe working conditions for subcontractors			

Contd

Figure 10.1: *Contd*				
ITEM NO	**ITEM**	**YES**	**NO**	**COMMENTS**
9.3	Advise that all subcontractor(s) must comply fully with the contract EH & S requirements			
9.4	Obtain contractor's confirmation that the competence of subcontractors has been checked as regards safety, health and environmental aspects of the work			
10	**Injury and data reporting requirements**			
10.1	Explain the client's and/or principal contractor's requirements			
10.2	Explain that all hours worked by the contractor's and any subcontractors' workers are to be recorded and notified to the client's or principal contractor's project staff			
10.3	Advise that all injuries requiring first-aid are to be reported to the project staff and details recorded			
10.4	Advise that the contractor must notify the client or the principal contractor immediately of any injury where the injured person is absent for more than one shift			
10.5	Advise that fatalities and major injuries are to be reported immediately with full details to project staff			
10.6	Explain that the contractor is to take all necessary steps to notify the circumstances to the various authorities as required by RIDDOR			
10.7	Advise that the contractor must not disturb the scene of any major injury or fatality until authorised to do so by the client's or principal contractor's project staff			
10.8	Advise that the contractor must provide facilities and information to enable the proper and full investigation of any injuries			

Contd

Figure 10.1: *Contd*				
ITEM NO	**ITEM**	**YES**	**NO**	**COMMENTS**
10.9	Advise the procedure in the event of a fatality or major incident/injury occurring to anyone engaged by the contractor or any of his subcontractor(s)			
11	**Welfare and first-aid facilities**			
11.1	Explain the facilities provided by the client or principal contractor for general use. The contractor is normally required to provide his own first-aid facility			
12	**Inspections by client's or principal contractor's staff**			
12.1	Explain the steps taken by the client's or principal contractor's staff to monitor safe working conditions on the project, and the action which is to be taken on receipt of a copy of the principal contractor's weekly and monthly inspection reports, and other Corrective Action Reports			
13	**Access arrangements**			
13.1	Explain that the contractor must provide safe and clear access throughout the project at all times for workers and rescue services			
13.2	Explain the scaffolding and other forms of access to be provided on the project, and the contractor's duty to ensure these are used and used properly			
13.3	Review the arrangements made by the contractor to inspect scaffolding and other access equipment regularly			
14	**Fall protection**			
14.1	Emphasise that falls are the main cause of fatalities on site, and explain specific requirements for fall prevention			
14.2	Explain the project's requirements regarding fall protection generally, including edge and hole protection, the erection of barriers, fans and nets, the availability and wearing of safety belts and harnesses, and the penalties for not wearing the equipment			

Contd

ITEM NO	ITEM	YES	NO	COMMENTS
Figure 10.1: *Contd*				
15	**Lighting arrangements**			
15.1	Review the standard and nature of the lighting required for access areas and for the contractor's work area(s), and how it will be provided			
16	**Personal protective equipment**			
16.1	Explain the law regarding the wearing of safety helmets at all times indicated, and the penalties for non-compliance			
16.2	Explain the policy with regard to the provision and wearing of safety footwear			
16.3	Explain the policy with regard to provision of other protective equipment for the tasks to be carried out, with special emphasis on eye protection			
16.4	Explain arrangements made for the storage of non-work clothing worn to the project, and for the issue of protective equipment as required			
17	**Crane and hoisting facilities**			
17.1	Explain operational requirements and arrangements made for craneage and/or hoisting facilities on the project			
17.2	Obtain details of any such facilities to be used or brought onto the project by the contractor or his subcontractor(s), obtaining details of the plant item(s)			
17.3	Explain the general requirement for third-party certification of all lifting equipment brought to the project before its first use			
17.4	Summarise project requirements for operators of lifting and hoisting equipment to hold training and competence certificates			
17.5	Explain requirements for regular checks and maintenance that must be recorded by the contractor			
18	**Site transport and mobile plant**			
18.1	Obtain details of all such plant to be operated by the contractor			

Contd

Figure 10.1: *Contd*				
ITEM NO	**ITEM**	**YES**	**NO**	**COMMENTS**
18.2	Explain any local or client training requirements for operators to hold certificates of training and competence			
19	**Noise control**			
19.1	Explain any local Codes or requirements which may limit the project noise output			
19.2	If relevant, obtain noise level output estimates and establish that the contractor is aware that measurement may be required			
20	**Materials storage**			
20.1	Explain project requirements for materials storage and establish any special needs of the contractor in this regard			
20.2	Explain project requirements for materials delivery, including any hours or other vehicle movement restrictions			
20.3	Explain storage and use requirements for LPG/LNG, oxygen and acetylene, including minimising hose lengths and use of flashback arrestors			
21	**Waste disposal**			
21.1	Explain the client's and principal contractor's environmental policy to restrict the potential for environmental damage caused by unplanned waste disposal			
21.2	Explain the project policy on waste clearance as it relates to the contractor			
22	**Housekeeping**			
22.1	Explain and agree arrangements for storing materials and clearing away rubbish. Explain that if rubbish is not regularly removed, the client or the principal contractor may at its option arrange for its removal and charge the contractor for the costs incurred			

Contd

Figure 10.1: *Contd*				
ITEM NO	ITEM	YES	NO	COMMENTS
23	**Permits to work**			
23.1	Will the contractor be issued with permits by other contractors or the client? if YES, obtain the procedures in writing and issue to the contractor			
23.2	Does the contractor need to establish his own permit-to-work system for himself and/or other contractors? If YES, set out the required procedures and confirm them in writing			
23.3	Establish who is to accept and sign off permits			
24	**Fire precautions**			
24.1	Obtain details of, or instruct on the siting of, the contractor's proposed site offices, huts and storage areas, together with details of the proposed fire-fighting equipment for these			
24.2	Review the emergency plan with the contractor and ensure that it is fully understood with respect to raising the alarm, rescue, and evacuation procedures in the event of an emergency			
24.3	Review project requirements for the display of emergency information			
24.4	**Assembly points**			
24.5	Establish and review the contractor's method for accounting for workers and visitors in the event of an emergency			
24.6	Explain the responsibility for maintaining escape routes clear and free from obstruction at all times, and the need to make regular inspections to verify this			
24.7	Consider the contractor's specific arrangements for storage of explosive and/or flammable materials			
25	**Site access and security**			
25.1	Identify safe access to the project for vehicles and pedestrians			

Contd

Figure 10.1: *Contd*				
ITEM NO	**ITEM**	**YES**	**NO**	**COMMENTS**
25.2	Describe the requirements for hoarding or otherwise barring access to work areas to those not entitled or required to be there			
25.3	Explain any limitations to the entry of visitors and their movement on site			
26	**Public protection**			
26.1	Establish whether the contractor or his subcontractor(s) need to provide fans, netting, full edge protection, covered walkways or protective barriers below work areas to safeguard the public — or other workers			
26.2	Review the means of preventing public access to work areas			
26.3	Explain the requirements and responsibilities for the protection of the public in respect to holes, voids and edges			

Project . . . .

Client . . . .

Principal Contractor . . . .

Signed . . . .   Signed . . . .   Date . . . .
For Client/Principal contractor   For contractor

11 Understanding People

Accidents in construction (and elsewhere) are 'people problems' at least as much as any other kind of problem. At some point, immediate or distant, people and decisions were involved. 'To err is human'; we all make mistakes. It is the task of modern safety management systems to recognise that fact, and therefore to minimise the opportunities for mistakes, and to minimise the harm that can arise when they are made. Awareness of our limitations is needed before we can set up systems successfully which take those limitations into account and maximise safety efforts on site.

Behavioural factors, sometimes called 'human factors', which affect human performance include:

- Perceptual, mental and physical capabilities of people
- Interaction of people with their organisation, jobs and working environment
- Influence of equipment and systems design on human performance
- Organisational characteristics which influence safety-related behaviour
- Social and inherited characteristics of people

Some of these issues have been discussed previously: modern techniques of managing safety and health incorporate best practice within them so as to produce and influence a positive **safety culture**. That means fostering positive attitudes to safety and health, and this can be done by:

- Effective communication — passing information to and from, and regular consultation with, the employees
- Achieving a positive commitment to safety by senior management that is both real and visible
- Maintaining good training standards
- Maintaining good working environments with a high potential for safe working conditions to be achieved

All these can be found within organisations committed to a high standard of excellence in safety and health at work.

A full discussion of the interactions between people and their organisation, jobs and working environment, and the influence of equipment and systems design on human performance, is beyond the scope of this book. The aim of this chapter is to provide an insight into some of the basic concepts involved, and to discuss some of the solutions currently used to tackle them. A concise and readable summary of human factors in safety can be found in the excellent HSE publication 'Reducing error and influencing behaviour'.

Why people fail

'Human error' as a simple catch-all explanation for accidents is now discredited. The term, if it means anything at all, does not provide an adequate description of the many ways in which the failure of people at all levels in organisations can contribute to the complex phenomenon we call an accident. It is more useful to think about '**human failure**', which involves both errors and violations, and also to distinguish between active failures and latent failures.

Active failures have the potential to trigger an incident very quickly, when, for example, a machine operator misjudges an action or someone takes a 'calculated risk' by ignoring rules or training. **Latent** failures are those involving people not immediately involved in the activity, such as managers and designers. Latent failures set up the conditions for things to go wrong.

Errors are actions or decisions which were not intended and which involved a deviation from an accepted standard, rule or procedure. They can be divided into slips, lapses and mistakes.

Slips are simple 'failures to do it right'. From misplacing the decimal point in a calculation to moving a lever the wrong way or too far, they can happen when attention is diverted or when a procedure is too complicated.

Lapses involve forgetting to do the right thing, usually because of a distraction, and they can occur in tasks that take a long time to complete. The individual can remain unaware of both slips and lapses, and so may fail to report them for this reason alone during investigations. The aim of safety management techniques is to minimise the opportunities for them to occur, by improving the design of equipment and procedures.

Mistakes are more complex — when making a mistake we do something wrong but believe it to be right. This can happen because the reasoning behind the action was wrong, or because the wrong procedure was followed. Slips, lapses and mistakes can happen to experienced and trained people, and poor communication is often a factor.

Violations happen when people deliberately do the wrong thing. Removing machine guards, driving too fast and not wearing protective equipment are obvious examples, but the reasons behind them are not simple.

Routine violations happen when it becomes standard practice in a group or even an entire organisation to break the rule or procedure. Lack of knowledge and failure(s) to enforce the rule make the situation worse, and this is one reason why rules which are actually unnecessary and are not enforced should be eliminated.

Situational violations happen when the demands of the particular circumstances are seen as more important than compliance with the rule — pressure of work, not enough time or equipment to do the job. Better planning and supervision can assist.

Exceptional violations happen when people believe that something has gone wrong and a chance has to be taken, even though it means breaking the rule. Improvisation and the belief that the benefits outweigh the risks are key factors. Preventive action against exceptional violations includes the provision of more training for abnormal and emergency situations.

We know that accidents are caused by a combination of factors rather than simply by carelessness and/or 'unsafe things'. Multiple causation is an accepted principle, which must be responded to in kind. There is rarely a 'simple fix'

available, although some of the solutions may be easier to implement than others. Solutions which avoid considering and acting on the human factors and the organisation are unlikely to be effective in the longer term.

Ergonomics

Ergonomics is the applied study of the interaction between people and the objects and environment around them. In the context of work, the 'objects' include tools and equipment, chairs, tables and steps. Ergonomics is not limited to the design of chairs — it promotes well-being at work by addressing all aspects of the work environment. The subject may be hidden within Standards and Codes, but the main emphasis is on designing and specifying tools, equipment and work places so that the job fits the person rather than the other way around. Ergonomics is concerned with the application of scientific data on human capabilities and performance to the design of workplaces, tools, equipment and systems.

Other aspects of the wide scope of ergonomics include organisational arrangements, which aim to limit the potentially harmful effects of physically demanding jobs on individuals. They include selection and training, matching personal skills to job demands, job rotation and the setting of appropriate work breaks.

In the construction industry, examples of the positive use of ergonomics include the design of tools and the limiting of weights in sacks and bags. Studies carried out by the former Swedish construction and insurance organisation Bygghalsan, together with manufacturers and employers have shown that considerable productivity gains can be achieved by optimising working conditions using ergonomic solutions. One such study investigated musculoskeletal disorders in the necks and shoulders of trade workers, including painters and plasterers.

The survey showed that almost half of all construction workers spend more than 10 hours a week with their arms above shoulder level, leading to increased risk of problems in neck and shoulders. Sickness absence due to these problems increases for individuals over 30 in all trade categories in the industry, indicating that work-related problems in the neck and shoulders manifest themselves after 10 to 15 years of exposure.

Recognising that it is impossible to eliminate totally the need to work with the hands above shoulder height, ergonomists at Bygghalsan looked at alterations in the way the work was organised, work methods, improved equipment, and techniques to make the work easier. One of the discoveries made was that the use of micro pauses — very short breaks — reduced muscular load and resulted in faster work. Screws were tightened into a beam at eye level, and those who took a 10-second pause after every other tightening did the work 12% faster than those who did not. Interestingly, the workers themselves thought it took more time to carry out the work with micro pauses than without.

Designing work and work equipment to suit the worker can reduce errors and ill-health — and accidents. Examples of construction problems which can benefit from ergonomic solutions include:

- Hand tools which impose strain on users
- Poor layout of controls on plant and equipment, making them hard to use or steer
- Control switches and gauges which are hard to reach or read
- Jobs reported to be found excessively tiring

An ergonomic approach is now being used in the Manual Handling Operations Regulations 1992. The Regulations begin by forcing a review of the needs to handle loads manually at all, and permit handling only following specific assessment of the risks associated with the task. The contrast with the former approach could not be more marked: we no longer train people to lift heavy weights as the sole solution to the problem.

Stress

One definition of stress is the reaction people have to excessive pressures or other types of demand placed on them. While there may be some disagreement over an exact definition, most people believe that it is a serious problem in their organisations. Surveys carried out by the TUC and employers' organisations show that stress is a significant issue at work. Nearly a fifth of managers surveyed in 1997 said they had taken time off work in the previous year because of stress. Generally, stress is thought to pose a comparatively high risk and yet to be the least well controlled of all risks at work.

Pressures exist in all areas of people's lives, and there is some evidence that we need pressure to be able to function at maximum effectiveness. But the responses to pressure, physical and mental, can be damaging if required to con-

tinue beyond the short term. Psychological and physical illness can result — anxiety, depression and heart disease may be work-related, although they can develop from other causes as well.

At the time of writing, discussion has been initiated by the Health and Safety Commission about the extent to which stress should be regulated. Its Discussion Document contains a very useful review of knowledge and opinion on the subject, and the reader interested in pursuing the subject will find this a useful starting point. It is available on the HSE website at http://www.open.gov.uk/hse/condocs/.

Factors which contribute to stress in the construction industry include:

- The physical work environment — noise, cold, heat, etc.
- Working conditions, including pay, long hours, travel needs
- Change in practices and techniques
- Volume of work — overload or underload
- Work design and pace, and the amount of control over the work that people can have
- Roles within the organisation, including tension between demands of safety and production
- Relationships with other people at work
- Organisational style, and 'office politics'
- Lack of job security
- Poor communication and involvement in decision-making

Isolating one or more of these factors as a cause of a particular stress problem is difficult, especially because most are inter-related. Also, the reactions of individuals to stress vary widely. Because the causes of stress are likely to include non-work components, it is thought that the subject is best considered as a single issue rather than trying to separate work-related stressors from home or life stress. As a result, introducing or revising legislation which can only be directed at the work environment is likely to meet with only partial success. This is why partnership efforts are being encouraged, to include other government departments, voluntary and professional groups, aiming at promoting a positive health culture at work.

Nevertheless, an attempt is required at identifying and managing stress by the employer. Stress at work is believed to reduce the individual's effectiveness, increase absenteeism and labour turnover, and it may increase the likelihood of injury. In 1999, a former Birmingham City Council housing officer forced to retire on the grounds of ill-health was awarded compensation of £67000 after the council admitted liability for the stress she was caused. This was the first time that an employer admitted liability for causing stress. Major factors were said to be lack of experience and qualifications for the work, with training promised though not given.

There are many aspects of current criminal law on safety and health at work which have a bearing on stress levels, and which lay duties of various kinds on employers. These duties begin within section 2 of the Health and Safety at Work Act 1974, with its reference to the health and welfare of employees, and continue through the Management Regulations with their requirements for defined management arrangements and risk assessments. Proving that criminal breaches have occurred in relation to stress, and that reasonably practicable (or other) steps were not taken, may be another matter.

Communication

Verbal and non-verbal communication govern our personal relationship with the outside world — with everybody else. Until a reliable form of telepathy is developed, whatever message we want to send or receive to others has to be by word of mouth directly, written, or by gesture and appearance — involving the five senses. Sometimes there can be conflicts between the forms of the message. A verbal instruction from a foreman to wear safety shoes, when the foreman is not doing so himself, causes confusion.

Fortunately, communication skills can be defined and learned, so that the right messages are received by the right people with no distortion in their meaning. After the accident, 'I thought you knew what I meant' is an admission of a failure to communicate. The fact is that no matter how bright and knowledgeable a person is, if necessary knowledge and advice cannot be passed to others in an acceptable form then the person's abilities are of no value at all.

Managing safety and health requires a good ability to communicate, and to be personally aware of the potential causes of failure to do so. Verbal skills can be improved by attending short courses on presentation techniques which involve role playing and practice. Often these courses include advice on non-verbal communication — unconscious gestures and habits, wearing appropriate clothes to relate to the target audience, and ways to secure acceptance of the message being given. Trainers need training themselves at intervals; knowledge of techniques needs to be acquired and used.

Reports

Report and procedure writing skills focus on the clear expression of problems and solutions, and are founded on well tried principles.

For example, **reports** should have a clear structure to aid comprehension. There should be an introduction, containing the reason for writing the report and the scope, followed by a summary. If there is no summary, it will be necessary to write a conclusion section. Recommendations, if called for, are placed as an Appendix to the report, as are lists of photographs, plans and drawings, references and material used or consulted in the preparation of the report. At some point, probably the introduction, the status and qualifications of the writer to hold the opinions expressed should be given. The body of the report will contain all the details, divided into sections where this will aid comprehension. A logical sequence should be followed, so that the reader is guided and carried along by the narrative and not left to skip about the report looking for clues.

Accident investigation reports are likely to be disclosed to a wide audience, and for this reason should be purely factual in content. Speculation as to causation and outcome should be avoided, as should the allocation of blame where possible. Whether the accident report is made on a standard form, or specially written, it should contain the following:

- A short summary of what happened
- An introductory summary of events prior to the accident
- Information gained during investigation
- Details of witnesses
- Information about the injury and any other loss sustained
- Conclusions
- The date, and the signature of the person(s) involved in the investigation
- Appendices — supporting material (photographs, diagrams, plans to clarify)
- Recommendations

Significant questions that should receive an answer in any report include:

- What was the immediate cause of the accident, injury or loss?
- What were the contributory (secondary) causes?
- What is the necessary corrective action?
- What system changes are necessary or desirable to prevent a recurrence?
- What reviews are needed of policies, procedures and risk assessments?

The use of propaganda about safety and health issues is a special type of communication.

Getting the message across

Various forms of propaganda selling the 'health and safety message' have been used for many years - posters, flags, stickers, beer mats and so on. They are now widely felt to be of little measurable value in changing behaviour and influencing attitudes to health and safety issues. Because of the long tradition of using safety propaganda as part of safety campaigns, however, there is a reluctance to abandon them. Possibly, this is because they are seen as constituting visible management concern whilst being both cheap and causing minimal disturbance to production. In contrast, much money is spent on advertising campaigns and measuring their effectiveness in selling products.

How can safety messages be effective? The advice which follows is based upon a limited range of studies — despite the huge sums spent on advertising products of all kinds, remarkably little work has been done on the effectiveness of traditional safety campaigns.

Avoid negativity — studies show that successful safety propaganda contains positive messages, not warnings of the unpleasant consequences of actions. Warnings may be ineffective because they fail to address the ways in which people make choices about their actions; these choices are often made subconsciously and are not necessarily 'rational' or logical. Studies also show that people tend to make poor judgements about risks involved in activities, and are unwilling to accept a perceived loss of comfort or money as a trade against protection from a large but unquantified loss which may or may not happen at some point in the future.

'It won't happen here' — the non-relevance of the warning or negative propaganda needs to be combated. The short-term loss associated with some (but not all) safety precautions needs to be balanced by a positive short-term gain such as peace of mind, respect and peer admiration.

Safety propaganda can be seen as management's attempt to pass off the responsibility for safety to employees. Posters, banners and other visual aids used in isolation without the agreement and sanction provided by worker participation in safety campaigns can easily pass the wrong message. The hidden message can be perceived as 'The management has done all it can or is willing to do. You know what the danger is, so it's up to you to be safe and don't blame us if you get hurt — we told you work is dangerous.'

Management can have high expectations of the ability of safety posters to communicate the safety message. This will only be justified if they are used as part of a designed strategy for communicating positive messages.

Expose correctly — the safety message must be perceived by the target audience. In practice, this means the message must be addressed to the right people, be placed at or near to the point of danger and have a captive audience. Site accommodation is therefore a good place to give the safety message.

Use 'attention-getter' techniques carefully — messages must seize the attention of the audience and pass their contents quickly. Propaganda exploiting this principle too readily can fail to give the message intended — sexual innuendo and horror are effective attention-getters, but as they may be more potent images they may only be remembered for their potency. Other members of the audience may reject the message precisely because of the use of what is perceived as a stereotype, for example 'flattering' sexual imagery can easily be rejected by parts of the audience as sexist and exploitative.

Strangely, the attention-grabbing image may be too powerful to be effective. An example to consider is the 'model girl' calendar. If the pictures are not regarded as sexist and rejected, they may be remembered. But who remembers the name of the sender? This is, after all, the point of the advertising.

Comprehension must be maximised — for the most effect, safety messages have to explain problems either pictorially or verbally in captions or slogans. To be readily understood, they must be simple and specific, as well as positive. Use of too many words or more than one message inhibits communication. Use of humour can be ineffective; the audience can reject the message given, because 'only stupid people would act as shown' and this would have no relevance to themselves or their work conditions.

Messages must be believable — the audience's ability to believe in the message itself and its relevance to them is important. Endorsement or approval of the message by peers or those admired, such as the famous, enhances acceptability. The 'belief factor' also depends upon the perceived credibility of those presenting the message. If the general perception of management's attitude is that health and safety has a low priority, then safety messages are more likely to be dismissed because the management motivation behind them is questioned.

Action when motivated must be achievable — safety propaganda has been shown to be most effective when it calls for a positive action, which can be achieved without perceived cost to the audience and which offers a tangible and realistic gain. Not all of this may be possible for any particular piece of propaganda, and the major factor to consider is the positive action. Exhortation simply to 'be safe' is not a motivator.

Safety propaganda

There is little evidence for the effectiveness of health and safety propaganda. This is mainly because of the difficulty of measuring changes in attitudes and behaviour which can actually be traced to the use of propaganda. For poster campaigns, experience suggests that any change in behaviour patterns will be temporary, followed by a gradual reversion to previous patterns, unless other actions such as changes in work patterns and environment are made in conjunction with the propaganda.

Limited experimental observations by the author show that the effectiveness of safety posters, judged by the ability of an audience to recall a positive safety message, is good in the short term only. One week after exposure to a poster, 90% of a sample audience could recall the poster's general details. 45% could recall the actual message on it. Two weeks after exposure, none of the audience remembered the message, and only 20% could recall the poster design. In both cases the attention of the audience was not drawn specifically to the poster when originally exposed to it.

Safety propaganda can be useful in accident prevention, provided its use is carefully planned in relation to the audience, the message is positive and believable, and it is used in combination with other parts of a planned safety campaign. Safety posters which are not changed regularly become part of the scenery, and may even be counter-productive by giving a perceived bad image of management attitude ('All they do is stick up a few old posters').

References

HSG48 Reducing error and influencing behaviour
HSC   Managing stress at work (discussion document)

12 Joint Consultation

The desirability of a co-operative approach to health and safety has been recognised for many years. The construction industry is unusual because the co-operative approach required for success extends not only to employees of a single employer, but also to all the people working on a site or project, regardless of their employment status. Consultation in this wider sense is anticipated within the Construction (Design and Management) Regulations 1994 (CDM), but the major statutory requirements to consult have an older history.

A main recommendation of the Robens Committee in 1972 was that an internal policing system should be developed, whereby workforce representatives would play an active part in drawing hazards to the attention of workers and management, and play a positive role in explaining health and safety requirements to employees. Provision for these regulations was made in the Health and Safety at Work etc. Act 1974, which originally contained two subsections dealing with the subject of employee rights to consultation by the appointment of statutory representatives with whom the employer would be required to enter into dialogue.

One of the subsections allowed for the election of such representatives from among the workforce; the other gave the right to nominate safety representatives to trade unions. The provision for the general election of such representatives was repealed by the Labour Government's Employment Protection Act 1975, and the right to appoint safety representatives is now restricted to recognised independent trade unions. The Safety Representatives and Safety Committees Regulations 1977 (SRSCR) provided the detail of the general entitlement contained in section 2 of the Act, and came into effect on 1 October 1978. They were further amended by the Management of Health and Safety at Work Regulations 1992 and 1999.

In 1996 the Health and Safety (Consultation with Employees) Regulations (HSCER) came into force. The European Court of Justice had ruled that consultation with the employer on health and safety issues cannot be limited to consultation with trade unions and their appointed safety representatives. Compliance with the Framework Directive requires more general consultation, and so these Regulations require employers to consult, where there are employees who are not represented by safety representatives appointed by recognised trade unions under the SRSCR provisions, either employees directly or representatives elected by them.

In general, the SRSCR entitlements and functions given to safety representatives are more extensive than those given under the HSCER. The SRSCR and their accompanying Approved Code of Practice provide a set of entitlements to consultation to nominees of recognised independent trade unions. They are given the right to make a number of kinds of inspection, to consult with the employer, and to receive information on health and safety matters. The Regulations also provide for training and time off with pay to carry out the functions of safety representation.

The Safety Representatives and Safety Committees Regulations 1977

The right to appoint safety representatives is restricted to trade unions recognised by the employer for collective bargaining or by the Arbitration and Conciliation Advisory Service (ACAS). The presence of only one employee belonging to such a union is sufficient to require the employer to recognise that person (upon application by his/her union) as a safety representative. The Regulations place no limit on the number of such representatives in any workplace, or any construction site, although the associated Approved Code of Practice and Guidance Notes observe that the criteria to consider in making this decision include total number of employees, variety of occupations, type of work activity and the degree and character of the inherent dangers. These matters should be negotiated with the appropriate unions and the arrangements made for representation should be recorded.

Unions wishing to make appointments of safety representatives must make written notification to the employer of the names of those appointed, who must be employees of the employer except in extremely limited circumstances. Upon appointment in this way, safety representatives acquire statutory functions and rights, which are set out in Regulation 4. The employer cannot terminate an appointment; the union concerned must notify the employer that an appointment has been terminated, or the safety representative may resign, or employment may cease at a workplace whose employees he or she represents (unless still employed at one of a number of workplaces where appointed to represent employees).

Safety representatives are not required to have qualifications, except that whoever is appointed should have been employed for the preceding 2 years by the employer, or have 2 years' experience in 'similar employment'. The right to time off with pay during working hours for safety representatives in order to carry out their functions and to undergo 'reasonable' training is given in Regulation 4(2). What is 'reasonable' in the last resort can be decided by an employment tribunal, to which the representative may make complaint (see below).

The Management of Health and Safety at Work Regulations 1992 (MHSWR) inserted a further regulation, requiring employers to consult safety representatives in good time, in respect of those employees they represent, concerning:

- Introduction of any measure at the workplace which may substantially affect health and safety
- The employer's arrangements for appointing or nominating 'competent persons' as required by MHSWR
- Any health and safety information the employer is required to provide to employees
- Planning and organisation of any health and safety training the employer is required to provide
- The potential health and safety consequences of the introduction of new technologies into the workplace

Employers must provide such facilities and assistance as safety representatives may reasonably require to carry out their functions of representation and consultation with the employer as provided by section 2(4) of the Health and Safety at Work etc. Act 1974. The facilities required include

independent investigation and private discussion with employees. The following functions are specifically mentioned in the Regulations:

- Investigation of potential hazards, dangerous occurrences and causes of accidents at the workplace
- Investigation of complaints by employees represented on health, safety or welfare matters
- Making representations to the employer on matters arising from the above
- Making representations to the employer on general matters of health, safety or welfare
- Carrying out inspections of the workplace regularly, following notifiable accidents, dangerous occurrences or diseases, and documents
- Representing employees in workplace consultations with inspectors of the appropriate enforcing authority
- Receiving information from those inspectors in accordance with section 28(8) of the main Act
- Attending safety committee meetings in the capacity of safety representative in connection with any function above

Regular routine inspections of the workplace or site can be carried out by entitlement every 3 months, having given reasonable previous notice in writing to the employer, or more frequently with the agreement of the employer. Where there has been a 'substantial change' in the conditions of work, or new information has been published by the Health and Safety Executive (HSE) relative to hazards of the workplace, further inspection may take place regardless of the time interval since the previous one. Defects noted are to be notified in writing to the employer, and there is a suggested form for the purpose. Employers may be present during inspections.

Safety representatives have a conditional right to inspect and copy certain documents by Regulation 7, having given the employer reasonable notice. There are restrictions on the kinds of documents which can be seen, which include commercial confidentiality, information relating to an individual (unless this is consented to), information for use in legal proceedings, that which the employer cannot disclose without breaking a law, and anything where disclosure would be against national security interests.

The employer has to make additional information available so that statutory functions can be performed, and this is limited to information within the employer's knowledge. The Approved Code of Practice contains many examples of the kind of information which should be provided, and the information which need not be disclosed.

Although the Regulations give wide powers to the safety representative, they specifically impose no additional duty. Representatives are given immunity from prosecution for anything done in breach of safety law while acting as a safety representative. It has been suggested that circumstances where this immunity might apply would include agreement during consultation on, for example, a system of work proposed by the employer which later turned out to be inadequate and became the subject of prosecutions of individuals involved in the decision to use the system.

Safety representatives may present claims to an employment tribunal if it is believed that the employer has failed to allow performance of the functions laid down by the Regulations, or to allow time off work with pay to which there was an entitlement (Regulation 11). If the tribunal agrees, it must make a declaration, and can award compensation to the employee payable by the employer. There is no right of access to the tribunal for the employer who feels aggrieved or who wishes to test in advance the arrangements he proposes to make.

Despite the title of the Regulations, their only reference to safety committees is in Regulation 9, which requires that a safety committee must be established by the employer if at least two safety representatives request this in writing. The employer must post a notice giving the composition of the committee and the areas to be covered by it in a place where it can be read easily by employees. The safety committee must be established within 3 months of the request for it. Apart from these, the Regulations contain no stipulations about size, composition and other practical details, which are a matter for the employer.

The Approved Code of Practice and the Guidance Note contain information and advice concerning the structure, role and functions of safety committees, which should be taken into account by non-statutory safety committees as well.

The Health and Safety (Consultation with Employees) Regulations 1996

Regulation 3 requires that, where there is no representation by safety representatives under SRSCR, the employer must consult employees in good time on matters relating to

their health and safety at work. In particular, employers are directed to do so about:

- Introduction of any measure at the workplace which may substantially affect the health and safety of those employees
- His/her arrangements for nominating 'competent persons' in accordance with MHSWR to assist the employer on health and safety matters, and to take charge of measures to combat identified serious and imminent danger at the workplace
- Any statutory health and safety information which he has to provide
- Planning and organisation of any health and safety training he has to provide
- Health and safety consequences for those employees of the introduction of new technologies into the workplace

Regulation 4 requires the consultation to be either with the employees directly, or with representatives elected by any group of employees. Those elected are referred to as 'representatives of employee safety' (ROES). It is important to appreciate that the choice of which form of consultation to adopt is left to the employer. If the employer decides to consult ROES, he must inform the constituents of their names and the group of employees represented. The employer must not consult an individual as a ROES in four circumstances, which are where:

- The person has notified the employer that the person does not intend to represent the group
- The person is no longer employed in the group the person represents
- The period for which the person was elected has expired without the person being re-elected
- The person has become incapacitated from carrying the functions under the Regulations

If the consultation is discontinued for any of these reasons, the employer must inform the employees in the group concerned that it has been. Paragraph 4 of Regulation 4 requires the employer to inform employees and ROES if he decides to change the consultation method to one of direct consultation.

Regulation 5 obliges the employer to make necessary information available to employees consulted directly to enable them to participate effectively, and where ROES are consulted as well as this the employer is required to make available records he must keep in compliance with RIDDOR.

As with the SRSCR, the employer does not have to disclose information which would be against national security, which could contravene any (statutory) prohibition on the employer, which relates to any individual without their consent, which would cause substantial injury to an undertaking, or where the information was obtained by the employer in connection with legal proceedings against him. Inspection rights are not available over documents not related to health or safety.

Consultation at site and project level

Exchanging information between employers and employees about work practices and likely future hazards is a requirement of MHSWR. Guidance from the HSE indicates that site safety committees are an effective way of doing this. Where there are more than about 250 people working on a project, such a committee should be considered by the management team as an aid to joint consultation. In some cases it may be necessary to co-ordinate a committee with existing arrangements made by a client, so that those working on site can be updated on the client's own working methods and safety controls.

The major functions of a site safety committee are to allow a full exchange of information, discussion of safe working practices, review information on the accident record and receive reports from safety staff and others. Written arrangements for safety committee meetings and constitution should form part of the site's Safety Plan. The principal contractor may require every contractor to be represented at meetings, which can be conducted as a part of the regular progress meeting.

Variants on the traditional safety committee can be found on some larger projects, especially where the numbers of contractors working would make the size of a committee unwieldy. Safety and quality circles have been found useful in these cases.

13 Access to Information

There is a great deal of information available on construction safety topics. Unfortunately the information sources are mostly unco-ordinated, in many places, and often written by specialists so that it cannot easily be understood by those people who have to work with it. Alternatively, the information may be presented in such a simple way that it cannot easily be applied to particular instances. Information technology is moving towards the production of solutions, for example by facilitating interaction between authors and enquirers. Discussion forums on websites (see below) are allowing one-to-one exchange of specific information. A familiar problem still persists; information quickly goes out of date. Because of these factors, the search for knowledge cannot be a 'once only' effort. Getting up to date imposes another equally important requirement: keeping up to date.

Systems for information provision designed since about 1980 recognise the need for 'one-stop' information shopping for answers to problems, and avoid the temptation to cross-reference to a potentially large number of other sources.

Most technical information on safety is still provided on paper, and this is not likely to change despite the introduction of new technologies. Some of these have new problems associated with them — photocopies and facsimile transmissions on waxed paper rolls are both affected by ultraviolet light, so the image gradually disappears over time. Other photographic record systems for holding information such as microfiche are more permanent, but can still be damaged. They are inconvenient to read and copy, although they carry a large quantity of complete material in a small space.

Books, leaflets and other paper-based material are likely to remain the most common information format for the foreseeable future, despite attempts to create the 'paperless' office. HSE Books is the UK's largest supplier of safety technical publications, either direct or through a growing national network of major booksellers. Other paper information providers, such as Croner, offer an updating library resource with a large number of specialist titles in their range. Fortunately, the UK's best construction reference, the two-volume manual *Construction Safety*, is also updated annually to subscribers.

Computer files held at the workplace are widely used as information storage, although there is already evidence that new generations of computers are unable to decipher material stored on the older systems. Computer networks, set up between offices using telephone or fibre-optic lines, are able to share much information, which is often stored centrally on a mainframe or server. Larger companies have developed 'intranets', which can be accessed via password from remote terminals and provide a world-wide core of information. Access to such a system enables a large amount of information to be accessed at short notice at a local site. Disadvantages are the cost, and the need to monitor and service the system regularly.

The expense of using computer data resources can be minimised by gaining access to someone else's information by using an electronic database through a modem connection, and there are now a number of computer resources which accept worldwide connections through the Internet. The ability to carry full text and graphics is spreading. NIOSHTIC and HSELINE are the best known English-language databases.

It is also possible to access versions of these and other databases at the workplace by installing a CD-ROM reader. Compact disc technology allows at least 280 000 pages of information to be stored on a single compact disc, which can be read by a personal computer. Recording onto CD-ROM by the non-specialist can now be done at reasonable cost. The advantages of doing so are cheapness, and the volume of material which can be stored and quickly accessed. CD-ROM constructed databases are available for purchase and rental through providers such as Silver Platter (OSH-UK), which also update and extend their products regularly. International data resources mostly store information on US requirements. A major supplier of chemical data sheets and other health and safety information on CD-ROM is the Canadian Centre for Occupational Health and Safety (CCINFO).

Information can also be stored in learning programmes and combined with video into an interactive system, available on CD-ROM and played through specially-adapted visual display screens. The use of multimedia, which combines still and moving images with sound and text information, is increasing rapidly as it can be produced easily and accessed through many relatively simple computer configurations.

Searching the Internet for health and safety information can be extremely rewarding. There are now several thousand 'home pages' on the World Wide Web giving access to safety, health and environmental information, but it is important to remember that not all of it can be relied on. Just as trade associations can be expected to publish information and views favourable to the members of the association, it should not be forgotten that many interest groups have websites and are anxious to promote a particular view in the guise of unbiased information. A list of websites which have been reviewed for their technical content and usefulness can be found through the Canadian Centre (http://www.ccohs.ca). Other uses of the Internet include the newsgroups, which enable questions to be asked and answered in front of an international audience, and the exchange of software. Though mainly for professionals, the Chat Forum run by the Institution of Occupational Safety and Health is open to anyone with a question, through www.iosh.co.uk and following the links.

Unless a unique home page has been given as a reference, the searcher will have to use one of an estimated 1550 available 'search engines' to track down a particular subject. Examples of search engines which are suitable for this purpose include AltaVista, Lycos, HotBot and Excite. When the word being searched for is entered, the search engine will return a list of 'hits' — web pages which contain the word.

Some refinement of the search is usually required to avoid being swamped by data. The search engines are quite likely to offer a selection of several hundred thousand web pages to the searcher, because too vague a search word (key-word) has been used. The search engines attempt to place the most useful hits at the front, but it can still take significant time to find relevant material without a very precise set of significant keywords to search with.

For example, a search for all references to 'construction safety' will generate a much bigger selection of pages to be checked out than a search for pages carrying the phrase 'UK construction plant safety'. In most search engines, the key to doing this is to use an 'and' function between the keywords, to make sure that the search does not provide all the pages with the word 'UK' plus all the pages with 'construction' and so on. Much time can also be saved by not viewing any web page hit dated more than about 3 months back from the current date. This is because search engines do not distinguish between live and derelict pages.

It is also necessary to remember that the majority of web pages are US-based, and that there are language and terminology differences round the world. The Australian word for 'banksman' is 'dogman', for example.

Computers are also convenient to use to store specific in-company information such as injury records, and to extract information from data using relational databases. They can also take nearly all the drudgery out of preparing injury returns, comparative tables and the like.

Safety professionals

The role of the occupational health and safety practitioner (OHASP) — the term preferred nowadays to 'safety officer', at least by those doing the job — has enlarged considerably since the early 1960s when regulations first required larger employers to appoint a 'site safety supervisor'. The change from the prescriptive set of construction safety rules to a regime where the employer has to assess all hazards and their risks and then take appropriate steps to control them has resulted in a significant increase in demand as well as a wider role.

Today's senior construction safety professional is likely to be professionally qualified by examination or by acquiring vocational qualifications, and a corporate member of the professional body (IOSH). Following established syllabi for self-development equips the professional to set up control systems for others to follow. Safety professionals whose role is solely one of inspection and monitoring are often not so highly qualified, and typically hold a lower grade of IOSH membership. Nevertheless, the intention has always been to provide a clear progression upwards on the career path, and the principles of self-development and continuing professional development are now well established.

The function of the professional is to advise management on the appropriate handling of hazards and risks, and to establish systems to enable that to be done effectively. Responsibility for giving sound, competent advice rests with the professional; responsibility for taking appropriate action based on that advice remains with management.

The role the professional has will usually be defined by management, and set out within the safety policy. Generally, the professional's advice will be available to all employees, as will information on technical issues. Attendance at safety committees, carrying out inspections and audits and formulating responses to new legislation normally occupy the professional's time. When accidents occur, the professional will be expected to carry out a thorough and unbiased investigation.

Consultants in occupational safety and health have multiplied in recent times, many of them professionals taking early retirement from business and the regulatory bodies. Organisations unable or unwilling to support a full-time position for an OHASP frequently find it convenient and cost-effective to subcontract the role. Too often, they also try to subcontract the responsibility. Consultants who are members of IOSH are bound by the Institution's code of ethics. IOSH maintains a register of safety consultants, which allows a prospective user to select those with potential for interview. Using the *Yellow Pages* may be quicker, but a successful relationship is less likely.

The HSE leaflet 'Choosing and using a safety consultant' offers excellent advice and should be followed. Experience shows that 'choosers' often have little idea of what they need, and so it will be well worth spending some time to define the extent of the particular role beforehand. Insurance of a minimum of £1 million cover against professional errors should be held by anyone acting as a consultant, in everyone's interests.

Other advice

Many trade associations operate safety advisory services for their members, as do chambers of commerce. Unbiased advice can be obtained from the HSE's Infoline, which refers the more difficult or policy questions to the HSE's regional offices. IOSH also operates a free advisory service to members and non-members alike.

Part 2
Environment, Health and Safety Issues

14 Construction and the Environment

The construction industry, as a major part of the economy in the UK and elsewhere, has a significant impact on the environment. An estimated 198 000 companies of all sizes make up the industry, with an annual output of £58 billion. The industry produces 29% of the controlled waste in the UK (70 million tonnes annually), of which 12 to 15 million tonnes are currently recycled. About 17% of waste going to landfill sites is directly related to construction work. Good management of the environment makes economic sense because many of the environmental aspects of construction work carry financial cost. Examples are energy use and waste, where taxation on landfill and a proposed climate change levy will add to costs.

Economic benefits include improved tender opportunities and reducing the cost of wastage (disposal, handling, transport, taxation, for example). Environmental benefits include reduced damage to the environment by control of emissions and adverse impacts on ecosystems, and reduced demands on natural resources.

In this chapter, the role and objectives of the contractor are considered. Contractors as constructors can make only a limited difference to construction's environmental impact, and need to work with others. Clients, for example, as procurers of construction work, can call for 'greener' buildings and thus affect the entire supply chain.

'Sustainable development' calls for a balanced approach, pursuing neither 'perfection' in environmental performance nor abandoning it in the pursuit of profit. It requires the prudent use of natural resources, moving towards the 'triple bottom line' of environmental, social and economic accountability. A strategy document was published in the UK in April 2000 to promote more sustainable construction. A full discussion of the issues involved is beyond the scope of this book, but the references quoted at the end of the chapter should be consulted for a wider scope than is presented here.

The short-term objectives for the reduction of environmental impact of any process are:

- Reduction of the consumption of resources
- Reduction of emissions and other byproducts of the process to air, water and land, and
- Reduction of production and increased recycling of waste

In achieving these goals, compliance with relevant environmental legislation will be required. There is now a complex framework of environmental legislation covering a broad range of issues, such as noise and dust nuisance, strict liability ('the polluter pays' principle), and duty of care in waste management. Further European legislation will cover air quality, climate change and waste recycling.

In the longer term, contractors should examine their activities to seek out opportunities for reducing the consumption of energy sources and water use throughout their business, including in offices and other static premises. This will require an assessment of the use of energy resources and water, which in turn requires measurements to be made. Then, policies and programmes can be established which set practical targets to be reached.

The initial steps towards achieving short-term objectives require the identification of materials used in all waste-generating activities, and encouraging contractors and suppliers to co-operate in the management of waste, beginning with reducing its production.

On each site, and within each Safety Plan, there should be specific reference made to environmental impact and the steps taken to minimise it. Safety Plans should also contain notes on the identified opportunities to work towards environmental objectives. For example, each site and office has many potential opportunities to prevent or reduce the generation of waste materials, which can save costs as well as minimise the amount of waste entering the environment.

The checklists at the end of this chapter (Figures 14.1 and 14.2) have been structured so as to be useful to those preparing Safety Plans, and to site-based staff seeking to evaluate their current situation.

Waste management and pollution control

The most common methods of pollution control are source reduction and recycling. Waste treatment and disposal are not classed as components of a construction pollution prevention programme, although they should be considered for inclusion within the scope of the environmental section of the site's Safety Plan.

The **hierarchy of waste management** is a set of stepped alternatives which provides guidance on best practice in minimising or preventing the generation of waste materials. It also applies to pollution in general:

- Waste should be **prevented** or reduced at the source by designing out whenever feasible
- Waste that cannot be prevented or reduced should be **reused** whenever feasible
- Waste that cannot be prevented, reduced or reused should be **recycled** in an environmentally safe and friendly manner, whenever feasible
- Waste that cannot be prevented, reduced, reused or recycled should be **treated** in an environmentally safe and friendly manner and energy should be derived from it, whenever feasible
- Disposal, landfill or other **release** into the environment should be used only as a last resort, and should be carried out in an environmentally safe and friendly manner

Source reduction is any method or technique used at or before the point of generation that reduces or eliminates the creation or use of hazardous substances or all controlled wastes so as to reduce the risks to health and the environment. Source reduction can be achieved through:

- Improvements in housekeeping
- Better and more frequent preventive maintenance
- Upgrading storage and materials handling techniques
- Discussion with, and eventually approving selected suppliers, to encourage their environmental efforts

- Substitution of chemicals
- Improved worker training

Source reduction emphasises conservation and the more efficient use of hazardous and non-hazardous materials, energy, water and other resources.

Recycling is the process of using or reusing a material or residual component of a material. It aims to minimise waste, and can also be achieved through reclamation, which is the process of treating a material to recover a usable product. Examples of recycling are waste management activities which separate recyclable material for collection, and the use and reuse of suitable demolition products. For both projects and offices, the aim should be to put recyclable materials in designated containers and arrange for recycling collection and processing.

Air pollution control involves limiting the emission of pollutants into the atmosphere. Six major air pollutants generally recognised as significant are carbon monoxide, ozone, sulphur oxides, nitrogen dioxide, particulate matter and lead. There are also air emission standards for friable asbestos during a number of activities including, but not limited to, building renovation and demolition, asbestos removal and encapsulation, and the disposal of asbestos and asbestos-containing materials. Other potentially significant pollution sources include the use and maintenance of motor vehicles and releases from stationary sources such as boilers, furnaces and the burning of waste.

Water pollution control is concerned with the restoration and maintenance of the chemical, physical and biological integrity of water resources. In many jurisdictions this is done by establishing limits on the discharge of pollutants through the implementation of a discharge permit programme. Such a system is likely to limit the permissible concentration of toxic pollutants or conventional pollutants discharged to open waters. There may also be:

- Standards for non-domestic sources which discharge into sewers leading to publicly-owned treatment works; these may incorporate pH, temperature, flammable materials and pretreatment requirements
- A requirement for a spillage control and countermeasure plan to minimise the risk of oil spills if the site stores large quantities of oil, and especially the use of bunded containments around fuel stores
- A requirement for storm water permits for certain types of industrial facilities or processes

For both sites and offices, water pollution goals can be met by working towards the following:

- Identification of all substances used by contractors in their work and the provision by each contractor of appropriate information which includes the means of disposal and containment of spillages where appropriate
- Prompt identification of the nature of any effluent from contractors' work areas, which must neither be harmful nor cause deposits or contamination in drains and sewers

Initial ground contamination

Initial ground contamination and its extent should be established before work begins on site. Where contamination exceeds recognised standards, specific measures will be required in order to treat or remove the contaminated ground, and to train and equip workers likely to be exposed to the contamination. Usually, the latter will involve the use of established techniques for the maintenance of occupational health, including limitation of exposure, provision and use of personal protective equipment, availability of welfare facilities including washing facilities and changing rooms, availability of cleaning facilities and storage for work clothing which may be contaminated. Geotechnical surveys must always be checked at the planning stage, so as to identify any conditions which may cause these measures to be put into effect. Attention must be paid to new regulations, most recently the Pollution Prevention and Control (England and Wales) Regulations 2000, SI 2000 No 1973.

A reference quoted at the end of this chapter gives guidelines for the classification of contaminated soils as well as guidance on risk assessment, a range of precautions and examples of information for employees.

Remediation

Remediation is a term which includes all actions necessary to return a polluted site to a condition suitable for its intended future use. Both current and past owners and operators of contaminated property may be found liable to bear the cost of remediation, in addition to all parties who generated, transported or arranged for the disposal of the materials that contaminated the property. Clearly, the cost implications of remediation can be significant.

Spillage control

The accidental release, spill or leak of oil from a facility or transportation vehicle can pose a significant threat to both public and occupational health as well as to the environment. A variety of sources can contribute to release, including:

- Loading and unloading fuel operations
- Ruptured hydraulic or fuel line or tank on heavy equipment
- Releases from petroleum/gas/oil tanks
- Leaking underground storage tanks
- Manhole contamination

To eliminate or reduce this threat, a response in good time is necessary. In addition to physical containment, there may also be notification requirements imposed by local or national regulatory agencies. The advice of the fuel supplier should be sought immediately by telephone in the event of significant spillage, or where knowledge of the necessary corrective actions is not immediately available. Notification to the emergency services may also be necessary. The potential requirement for whatever action is appropriate should have been identified at the outset, and the procedure included in the Safety Plan.

Pesticide use and control

Insecticides, fungicides and rodenticides are types of pesticide that may be encountered in offices as well as during construction work. In many jurisdictions there are requirements concerning the introduction and uses of these in the marketplace, and on how some products are manufactured, distributed, sold and used. Often, the user aspect is controlled by operator certification. In the UK the Control of Pesticides Regulations apply.

Many pesticides have effects on other life forms than those they are used against, including humans. It may be necessary or advisable to apply these products after normal working hours to avoid exposure of staff to them.

The major control considerations are:
- Correct evaluation of the need to use a pesticide, which includes its potential effect on all non-target organisms, including humans and animals
- Correct selection of an appropriate product suitable for the conditions
- Control of storage and handling of the product,

including recording dates of use, placement sites and quantities used

- Its application by trained operators where this is called for by law or the manufacturer
- Protection of workers and the public against the application method and any overapplication
- Appropriate housekeeping standards to minimise the need for pesticide use
- Removal of debris after pesticide application

Hazardous waste management

UK law imposes a system controlling the management of hazardous wastes, which places requirements on producers and carriers of the waste as well as owners and operators of storage and disposal facilities. There are many possible definitions of 'hazardous waste', but the term usually includes any solid, liquid or contained gas which is either ignitable, corrosive, reactive or toxic (or any combination of these) and which is at the end of its useful life.

The term 'hazardous waste' normally applies to any mixture containing a hazardous waste. Common exclusions from the definition of hazardous waste are:

- Material such as domestic sewage
- Any mixture approved for discharge through a sewage system for treatment in a sewage works
- Some industrial wastewater point source discharges
- Waste controlled under other more specific laws (common examples are lead and asbestos)
- Some (but not all) wastes destined for recycling and
- Some kinds of empty container

Ignitable hazardous wastes are those which have the ability to cause a fire during transport, storage or disposal. Examples include waste oils, used solvents and oxidisers.

Corrosive hazardous wastes are those able to deteriorate standard containers, damage human tissue and/or dissolve toxic components of other wastes. Strong acids and alkalis are examples.

Reactive hazardous wastes have the tendency to become chemically unstable during normal conditions, or to react violently when exposed to air or mixed with water, or to generate toxic gases. Examples include some compounds of phosphor, pure sodium, and sulphuric acid in storage batteries. Although these are unlikely to be produced by the construction process, they can be found during demolition.

Toxic hazardous wastes have the potential to leach from landfills and contaminate soil and groundwater. Some common examples are lead, benzene, carbon tetrachloride, mercury and cadmium.

Control of substances hazardous to health

There are many hazardous substances which are commonly used or produced at work. Examples are solvents in glues and paints, silica dust produced during grinding and chasing concrete on site and fixative agents used in enclosed office areas. These substances should be identified and their risks assessed. Designers should eliminate hazardous materials from their designs, or specify the least hazardous products that perform satisfactorily. The control of substances hazardous to health is governed by the COSHH Regulations 1999, which are summarised elsewhere in this book. Special mention is made of the control of these substances in this chapter because of their potential to harm the environment, and other people such as neighbours and other contractors. The substances can be generated by the work, as well as arrive on site as purchased products.

Assessment of the risk is required by law. This is done by looking first at the way in which workers and others are exposed to the substance in the particular job to be done. In construction work, harm is normally caused by:

- Breathing in fumes, vapours and dusts
- Direct contact with skin and eyes
- Swallowing or eating contaminated material

The results of the assessment must be written down, and anyone exposed to hazardous substances should be shown a copy of the significant findings of the assessment as part of induction training.

Prevention is the best solution when reducing risk — remove the hazard. This can be achieved by doing the job in a different way or by using a less hazardous substance. Controls for hazardous substances include:

- Reducing exposure time
- Provision of good ventilation
- Using as little of the substance as possible
- Changing the method of application — brushing is better than spraying, for example
- Using equipment fitted with exhaust ventilation or water suppression to control dust
- Improving personal hygiene

- Health surveillance for workers (for example, those working with lead and silica)
- Proper use of appropriate personal protective equipment

It is important to be aware that personal protective equipment is **not** the solution of choice. It should not be used unless exposure cannot be adequately controlled by any other, or any other combination, of the above measures.

Measurement and reduction of energy consumption

The prevention of pollution and the conservation of energy are complementary activities. Nearly all forms of energy generation consume raw materials and create wastes. By integrating energy conservation activities into construction operations, the quantity of wastes produced during energy generation can be reduced. Construction companies should be committed to promoting and improving the efficient use of lighting, heating, air conditioning and ventilation during operations, and to improving fuel efficiency and economy in transport fleets.

Some opportunities which may be identified for the reduction of energy consumption in offices and on sites are:

- Locating air intakes and air conditioning units in cool, shaded locations
- Turning off all equipment and lights when not required to be in use
- Using more efficient heating and refrigeration units
- Improving lubrication practices for motor-driven equipment and recording consumption of oils and fuels
- Using energy-efficient lighting
- Installation of timers and thermostats to control heating and cooling better

Environmental objectives and targets

The general objectives which could be set are obviously a matter for individual companies; the following are offered as suggestions:

- Reduce energy consumption in offices and motor fleets
- Reduce resource consumption in offices and sites
- Reduce emissions to air, water and land
- Reduce production, and increase recycling, of waste

- Maintain effective environmental management systems throughout the organisation
- Promote environmental activities with relevant external voluntary bodies
- Report on environmental performance in annual reports

More detailed, specific targets can be set on an annual basis. It is advisable to set targets which are likely to be achievable in the time period. The following are suggested for consideration:

- Maintain current knowledge of the legal requirements on environmental protection in relation to the type and location of construction work normally carried out
- Assess and record energy consumption and water use levels for offices and/or identified sites
- Assess and record consumption of energy sources for motor fleet activities
- Provide written information and training on methods of identifying and minimising emissions, with initial emphasis on site safety plans
- Identify opportunities for and encourage recycling of materials in the construction process and offices
- Include clear commitments to environmental goals in publicity material
- Prepare an environmental report on the achievement of these targets at intervals, and at least annually, to be circulated to the board of directors.

Environmental policies

Increasingly, contractors are being asked to produce an environmental policy for their business. This is not a legal requirement, but clients and others working to the ISO 14000 standard seek to establish the environmental awareness of their suppliers and business partners. It is suggested that the layout should follow the same format as the employer's safety policy, and the following topics should be considered for inclusion in a detailed environmental policy:

- **General statement of policy,** signed and dated by the chief operating officer of the organisation
- **Organisation/responsibilities**
  - ☐ Environmental legislation
  - ☐ Environmental management
  - ☐ General staff responsibilities
  - ☐ Specific responsibility for:
    environmental risk assessment
    control of waste

environmental cost assessment

ensuring staff awareness of new and developing technology

energy policy

environmental emergency plan

incident reporting

environmental aspects of acquisitions and property transactions

environmental audits

■ **Arrangements**

☐ Training and development of all staff

☐ Acquisition and distribution of environmental information

☐ Specific arrangements for assessing and controlling:

   waste materials

   substances and processes likely to have an environmental impact

   noise

   vehicle exhaust emissions

   vehicle cleaning

The size of the business is not a reliable indicator of the amount of detail which may be needed in an adequate environmental policy. The question is rather whether all the significant environmental risks have been identified and adequate control of them demonstrated. For small businesses where these risks are well understood and control is relatively simple, a single paragraph included within the safety policy may be sufficient to meet most needs. In these circumstances the following sample statement may be found useful as an indication:

> We recognise that our business activities may have environmental implications, and we therefore follow a policy designed to minimise environmental impact and damage. Our operations will be managed and organised so as to minimise so far as is reasonably practicable environmental damage caused by noise, dust and damage to groundwater and drainage systems. In pursuance of this policy, environmental considerations will be taken into account in tendering.

References

Control of Pollution Act 1974 (as amended)

Control and Disposal of Waste Regulations 1988

Special Waste Regulations 1996

Builders' Skips (Marking) Regulations 1984

Environmental Protection Act 1990

Environmental Protection (Duty of Care) Regulations 1991

Environmental Protection (Prescribed Process and Substances) Regulations 1991

HS(G)66   Protection of workers and the general public during the development of contaminated land, HSE Books

BR351   Green guide to specification, Building Research Establishment, available from CRC Ltd, 151 Rosebery Avenue, London EC1R 4GB

Construction Best Practice Programme, CBPP, PO Box 147, Bucknalls Lane, Garston, Watford WD2 7RE. Website: www.cbpp.org.uk

CIOB: 228   Environmental Management in Construction (model forms for smaller companies, focused on on-site construction activities)

CIRIA: SP135   Waste management and recycling in construction — boardroom handbook
(Gives guidance on waste minimisation measures for policy-makers and board members of construction organisations, including clients, designers, contractors and suppliers)

CIRIA: SP096   Environmental assessment

CIRIA: SP120   A Client's Guide to Greener Construction

CIRIA: C502   Environmental good practice

CIRIA: C503   Environmental good practice — working on site

CIRIA: C513   The Reclaimed and Recycled Construction Materials Handbook

Figure 14.1: Sample checklist — environmental opportunity audit				
Project:				

NO.	OPPORTUNITY IDENTIFIED IN SAFETY PLAN	ACTION TAKEN	ACTION NOT TAKEN (state if not applicable	COMMENTS
1	Dust control — wetting road surfaces			
2	Dust control — use of wet cutting methods			
3	Reporting vehicle emissions to contractors			
4	Ban on open fires and waste burning			
5	Controls on all asbestos work			
6	Welding, burning and cutting require hot work permits			
7	Ban on CFCs in extinguishers and refrigeration units			
8	Identification of all substances used by contractors and supply by them of information			
9	Spillage control and countermeasures plan in place			
10	Storm water permits required			
11	Identification of any contractors' effluents			
12	Spillage controls: drip pans, tunnels used etc.			
13	Pesticides: training, data sheet production, application under permit after hours			
14	Hazardous wastes: method statements to cover disposal			
15	Study of geotechnical survey			
16	Hazardous substances identified and controlled by contractors			
17	Recycling carried out, using designated containers, collection and processing			
18	Energy consumption measured and reviewed, examples available showing positive steps taken			
19	Water consumption measured and reviewed, examples available showing positive steps taken			
20	Other opportunities identified for positive environmental action not described in original safety plan:			

Checklist completed by:	Date:

Figure 14.2: Sample checklist — environmental audit				
Project:				
Auditor:			Date:	
NO.	**QUESTION**	**YES**	**NO** (state if not applicable	**COMMENTS**
1	Has an environmental risk assessment been carried out?			
2	Has project or site environmental plan being written, based on the risk assessment?			
3	Do contract and subcontract documents refer to or include the environmental plan?			
4	Was an environmental impact survey carried out for the project at the planning stage?			
5	If so, do the subcontract documents refer to it?			
6	Has a 'good neighbour' policy been established with occupiers of premises in the near vicinity?			
7	Are hoardings and signs kept clean?			
8	Is the level of existing lighting being maintained at the site perimeter?			
9	Has pedestrian disruption been minimised?			
10	Are public viewing locations provided?			
11	Is there a written project traffic and logistics plan, referenced in the subcontract documents?			
12	Is best use made of off-site fabrication to reduce traffic flow?			
13	Are delivery vehicles maintained in good condition?			
14	Is unauthorised access effectively prevented?			
15	Are all roads and footpaths external to the site boundaries kept clean?			
16	Are all site roads and footpaths kept clean?			
17	Are all adjacent rivers, streams, lakes, etc., kept clean?			
18	Have offices and welfare facilities been located sympathetically?			
19	Is there a 'no burning of rubbish' policy in force?			
20	Is there a 'no radios' policy in force?			

Contd

Figure 14.2: *Contd*				
Project:				
Auditor:			Date:	
NO.	**QUESTION**	**YES**	**NO** (state if not applicable	**COMMENTS**
21	Are the locations of existing underground services known and clearly signed?			
22	Are dust-creating activities identified and controlled?			
23	Is effluent properly controlled?			
24	Has an initial background noise survey been carried out (where necessary)?			
25	Has the local authority issued consents or restrictions under the Control of Pollution Act?			
26	Have the listed control measures been implemented?			
27	Have noisy plant items been identified and marked with expected noise levels?			
28	Have noise zones been established?			
29	Have all contractors provided COSHH assessments?			
30	Are the employees aware of the listed control measures?			
31	Are substances and materials adequately stored?			
32	Are first aid facilities readily available and adequate?			
33	Are all carriers of waste from site registered?			
34	Is the tip being used licensed?			
35	Have consents been obtained from the Environment Agency for permanent or temporary discharges into water courses?			
36	Are diesel and other fuel and oil stores bunded?			
37	Have any endangered species been identified on site?			
38	What control measures have been put in place?			
39	Have any trees been listed?			
40	Have they been adequately protected?			
41	Is timber used on the project from proven (certificated) sustainable sources?			
Audit completed by:			Date:	

15 Construction Hazards and Solutions

This chapter discusses a wide range of selected topics in more detail than can be given in the Quick Reference Guide which follows. Where appropriate, further reading suggested for each topic is listed at the end of the chapter. Except where stated otherwise, all the documents listed can be obtained from HSE Books. Many reference documents have been withdrawn from HSE listings because the advice given does not relate to current legislation or best practice. It is unfortunate that many have not been replaced. The excellent two-volume updating publication *Construction Safety* is considered by most authorities to be the single best reference work, and is highly recommended. Details can be obtained from the Construction Confederation, London. Many construction subjects are covered by British and ISO Standards, but as the reference numbers change frequently the references have generally not been included in this book, apart from within the Quick Reference Guide.

Access equipment

Many injuries result from failures or falls involving access equipment which has been incorrectly selected, erected, used or maintained. Access equipment is frequently used for short duration and emergency work without full consideration of a safe method of work. Each task should be assessed, and a suitable means of access chosen based upon an evaluation of the work to be done, the duration of the task, the working environment (and its constraints) and the capability of the person or people carrying out the task.

There are many different types of access equipment. This section covers the following:

- Ladders, stepladders and trestles
- General access scaffolds
- Scaffold towers
- Suspended cradles
- Personal suspension equipment (abseiling equipment and boatswain's chairs)
- Mast-elevated work platforms
- Mobile elevating work platforms (MEWPS)

Other, highly specialised equipment is available, and the general principles will apply to their use. Usually, they have been specially designed for particular tasks and manufacturers' information should be used in operator training.

General principles

Accidents using access equipment occur because one (or more) of the following common problems have not been controlled in advance, or was thought to be an acceptable risk under the circumstances:

- Faulty design of the access structure itself
- Inappropriate selection where safer alternatives could have been used
- Subsidence or failure of base support
- Structural failure of suspension system
- Structural failure of components
- Structural failure through overloading
- Structural failure through poor erection/inspection/ maintenance
- Structural failure through overbalancing
- Instability through misuse or misunderstanding
- Overreaching and overbalancing
- Climbing while carrying loads
- Slippery footing — wrong footwear, failure to clean
- Falls from working platforms and in transit
- Unauthorised alterations and use
- Contact with obstructions and structural elements
- Electrical and hydraulic equipment failures
- Trapping by moving parts

Ladders, stepladders and trestles

The key points to be observed when selecting and using this equipment are as follows.

Ladders

1. See whether an alternative means of access is more suitable. Take into account the nature of work and duration, the height to be worked at, what reaching movements may be required, what equipment and materials may be required at height, the angle of placement and the foot room behind rungs, and the construction and type of ladder.
2. Check visually whether the ladder is in good condition and free from slippery substances.
3. Check facilities available for securing against slipping — tied at top, secured at bottom, or footed by a second person if no more than 3 m height access is required.
4. Ensure the rung at the step-off point is level with the working platform or other access point, and that the ladder rises a sufficient height above this point (at least 1.05 m or five rungs is recommended), unless there is a separate handhold.
5. A landing point for rest purposes is required every 9 m.
6. The correct angle of rest is approximately 75° (corresponds to a ratio of one unit horizontally at the foot for every four units vertically).
7. Stiles (upright sections) should be evenly and adequately supported.
8. Ladders should be maintained free of defects and should be inspected regularly.
9. Ladders not capable of repair should be destroyed.
10. Metal ladders (and wooden ladders when wet) are conductors of electricity and should not be placed near or carried beneath low power lines.
11. It is important to ensure that ladders are positioned the correct way up. Timber pole ladders often have stiles thicker at the base than at the top, and should have metal tie rods underneath the rungs. Metal ladders often have rungs with both flat and curved surfaces — the flat surface is the one on which the user's feet should rest.

Stepladders

1. Stepladders are not designed to accept side loading.
2. Chains or ropes to prevent overspreading are required, or other fittings designed to achieve the same result. Parts should be fully extended.
3. Stepladders should be levelled for stability on a firm base.
4. Work should not be carried out from the top step, and preferably not from the top third.
5. Overreaching should be avoided by moving the stepladder — if this is not possible, another method of access should be considered.
6. Equipment should be maintained free from defects. Regular inspection is required.
7. No more than one person should use a stepladder at one time.

Working platforms and trestles

1. Trestles are suitable only as board supports.
2. They should be free from defects and inspected regularly.
3. Trestles should be levelled for stability on a firm base.
4. Platforms based on trestles should be fully boarded, adequately supported and provided with edge protection where appropriate.
5. Safe means of access should be provided to trestle platforms, usually by a stepladder.
6. Working platforms in construction work must by law be no less than 600 mm in width, so many older trestles may no longer be suitable to support such platforms as they will be too narrow.

General access scaffolds

There are three main types of access scaffold commonly constructed from steel tubing or available in commercial patented sections. These are:

1. Independent tied scaffolds, which are temporary structures independent of the structure to which access is required, but tied to it for stability.
2. Putlog scaffolds, which rely upon the building (usually under construction) to provide structural support to the temporary scaffold structure through an arrangement of putlog tubes (with special flattened ends) placed into the wall.
3. Birdcage scaffolds, which are independent structures normally erected for interior work which have a large area and normally only a single working platform.

The key points to be observed when specifying, erecting and using scaffolds are:

1. Select the correct design with adequate load-bearing capacity.

2. Ensure adequate foundations are available for the loads to be imposed.

3. The structural elements of the scaffold should be provided and maintained in good condition.

4. Structures should be erected by competent persons or under the close supervision of a competent person, in accordance with any design provided and with applicable Regulations and Codes of Practice.

5. All working platforms should be fully boarded, with adequate edge protection, including handrails or other means of fall protection, nets, brickguards and/or toeboards to prevent materials or people falling from the platforms.

6. All materials resting on platforms should be safely stacked, with no overloading.

7. Adequate and safe means of access should be provided to working platforms.

8. Unauthorised alterations of the completed structure should be prohibited.

9. Inspections of the structure are required, prior to first use and then at appropriate intervals afterwards, which will include following substantial alteration or repair, after any event likely to have affected stability, and at regular intervals not exceeding seven days. Details of the results should be recorded on an inspection form.

Scaffold towers

Scaffold towers are available commercially in forms comparatively easy to construct. They may also be erected from traditional steel tubing and couplers. In either form, competent and trained personnel are required to ensure that all necessary components are present and in the right place. Many accidents have occurred because of poor erection standards; a further common cause is overturning.

The key points to be observed in the safe use of scaffold towers are:

1. Erection should be in accordance with the manufacturer's or supplier's recommendations.

2. Erection, alteration and dismantling should be carried out by experienced, competent persons.

3. Towers should be stood on a firm level base, with wheel castors locked if present.

4. Scaffold equipment should be in good condition, free from patent defects including bent or twisted sections, and properly maintained.

5. The structure should be braced in all planes, to distribute loads correctly and prevent twisting and collapse.

6. The ratio of the minimum base dimension to the height of the working platform should not exceed 1:3 in external use, and 1:3.5 in internal use, unless the tower is secured to another permanent structure at all times. Base ratios can be increased by the use of outriggers, but these should be fully extended and capable of taking loads imposed at all times.

7. Free-standing towers should not be used above 9.75 m unless tied. The maximum height to the upper working platform when tied should not exceed 12 m.

8. A safe means of access should be provided on the narrowest side of the tower. This can be by vertical ladder attached internally, by internal stairways or by ladder sections designed to form part of the frame members. It is not acceptable to climb frame members not designed for the purpose.

9. Trapdoors should be provided in working platforms where internal access is provided.

10. Platforms should be properly supported and fully boarded.

11. Guardrails, toeboards and other appropriate means should be provided to prevent falls of workers and/or materials.

12. Mobile scaffold towers should never be moved while people are still on the platform. This is a significant cause of accidents.

13. Ladders or stepladders should not be placed on the tower platform to gain extra height for working.

Suspended access (cradles)

A suspended access system includes a working platform or cradle, equipped with the means of raising or lowering when suspended from a roof rig.

The key points to be observed in the safe installation and use of this equipment are:

1. It should be capable of taking the loads likely to be imposed on it.

2. Experienced erectors only should be used for the installation.

3. Supervisors and operators should be trained in the safe use of the equipment, and in emergency procedures.

4. Inspections and maintenance are to be carried out regularly.

5. Suspension arrangements should be installed as designed and calculated.

6. All safety equipment, including brakes and stops, should be operational.
7. The marked safe working load must not be exceeded, and wind effects should also be considered.
8. Platforms should be free from obstruction and fitted with edge protection.
9. The electrical supply is not to be capable of inadvertent isolation, and should be properly maintained.
10. Adverse weather conditions should be defined so that supervisors and operators know what is not considered acceptable.
11. All defects noted are to be reported and rectified before further use of the equipment.
12. Safe access is required for the operators and unauthorised access is to be prevented.
13. Necessary protective measures for those working below and the public should be in place before work begins.

Personal suspension equipment (boatswain's chairs and abseiling equipment)

A boatswain's (or bosun's) chair is a seating arrangement provided with a means of raising or lowering with a suspension system. This should only be used for very short duration work, or in positions where access by other means is impossible.

Abseiling equipment is used by specialists to gain access where the duration of work is likely to be very short indeed and the nature of the work lends itself to this approach.

The key points to be observed in using personal suspension equipment are:

1. The equipment must be suitable and of sufficient strength for the purpose it is to be used for, and the loads which are anticipated. A specific risk assessment should be made in every case.
2. The equipment must be securely attached to plant or a structure strong and stable enough for the circumstances.
3. Suitable and sufficient steps must be taken to prevent falls or slips from the equipment.
4. The equipment must be so installed or attached as to prevent uncontrolled movement.

Mast-elevated work platforms

Generally, this equipment consists of three elements:

- Mast(s) or tower(s) which support(s) a platform or cage
- A platform capable of supporting persons and/or equipment
- A chassis supporting the tower or mast

The key points to be observed in the erection and use of this equipment are:

1. Only trained personnel should erect, operate or dismantle the equipment.
2. The manufacturer's instructions on inspection, maintenance and servicing should be followed.
3. Firm, level surfaces should be provided, and outriggers are to be extended before use or testing, if provided.
4. Repairs and adjustments should only be carried out by qualified people.
5. The safe working load of the equipment should be clearly marked on it, be readily visible to the operator and never be exceeded.
6. Raising and lowering sequences should only be initiated if adequate clearance is available.
7. The platform should be protected with edge guardrails, toeboards and provided with adequate means of access.
8. Emergency systems should only be used for that purpose, and not for operational reasons.
9. Unauthorised access into the work area should be prevented using ground barriers.
10. Contact with overhead power cables should be prevented, by preliminary site inspection and by not approaching closer than a given distance. This distance can be obtained in respect of the particular power lines, from the power supply company, where necessary.

Mobile elevated work platforms (MEWPS)

A wide variety of equipment falls into this category, ranging from small, mobile tower structures with self-elevating facilities, to large vehicle-mounted, hydraulically-operated platforms. Some of their uses may involve high risk situations — these have been identified by the HSE as where:

- There are protruding features which could catch or trap the carrier/platform;
- Nearby vehicles or mobile plant could foreseeably collide with the MEWP;
- The nature of the work may mean operators are more likely to lean out, or are handling work pieces which may move unexpectedly; and

■ Unexpected or rapid movement of the machine, or overturning, is possible.

The key points to be observed in their use are:

1. Operator controls should be at the platform level, with over-ride at ground level for emergencies.
2. There should be a levelling device fitted to the chassis to ensure verticality in use.
3. Supervision should prevent use of the equipment during adverse weather conditions.
4. Outriggers, where provided for increased stability, should be fully extended and locked into position before the equipment is used/raised, in accordance with the manufacturer's instructions. The wheels may also require locking.
5. Materials and/or persons should not be transferred to and from the platform while in the raised position.
6. Training is required for operators before they are allowed to use the equipment in field conditions unsupervised.
7. Operators and others on the platform who are wearing safety harnesses should secure them to the inside of the platform cage.
8. When fitted, scissor mechanisms require the provision of adequate fixed guards, so as to prevent trapping of the operator or others during raising or lowering. (Note that this does not apply where interlocked proximity sensors are fitted to cut power when resistance is felt between the scissor arms.)
9. The equipment requires regular inspection, servicing, maintenance and testing in accordance with the manufacturer's instructions.

Legal requirements

General requirements for safe access and equipment can be found in the Health and Safety at Work etc. Act 1974 and the more detailed requirements of the Management of Health and Safety at Work Regulations 1999, the Workplace (Health, Safety and Welfare) Regulations 1992 or the Construction (Health, Safety and Welfare) Regulations 1996 and the Provision and Use of Work Equipment Regulations 1998.

The most complete set of requirements is contained within the Construction (Health, Safety and Welfare) Regulations 1992. Work equipment such as scissor lifts is covered by the Provision and Use of Work Equipment Regulations 1998, and suspended access equipment is covered by the Lifting Operations and Lifting Equipment Regulations 1998.

Asbestos

Asbestos is the name given to a group of naturally-occurring mineral silicates, which are grouped because of their physical and chemical similarity and their consequent general properties. Asbestos is strong, inert, resilient and flexible — and therefore almost indestructible. It has been used in a wide range of products requiring heat resistance and insulation properties. In humans it has been claimed with varying degrees of certainty to be a factor in asbestosis, lung cancer, cancers of the stomach, intestines and larynx, and mesothelioma. There is no safe level of exposure to any form of asbestos; mostly, long periods of time pass before any effect upon an individual can be diagnosed.

Asbestos produces its effects more because of the size, strength, sharpness and jagged shape of the very small fibres it releases than as a result of its chemical constituents, although these do also have relevance. The health hazards arise when these small fibres become airborne and enter the body, and when they are swallowed. The body's natural defence mechanisms can reject large (visible) dust particles and fibres, but the small fibres reaching inner tissues are those that are both difficult to remove and the most damaging. They are particularly dangerous because they cannot be seen by the naked eye under normal conditions, and they are too small (less than 5 microns in length) to be trapped by conventional dust filter masks.

Asbestos found in construction work is normally encountered in the demolition or refurbishment processes, but even simple jobs such as drilling partitions or removing ceiling tiles can disturb it. It is important to be aware that asbestos is normally present in a mixture containing a low percentage of asbestos in combination with filling material. Therefore, neither the colour nor the fibrous look of a substance is a reliable guide. The only reliable identification of the presence of asbestos is by microscopic analysis in a laboratory. The only alternative to this is to adopt a default position of assuming that blue asbestos is present (which requires the strictest controls) and proceeding accordingly, but this will be so costly as to be unacceptable except in cases where it has already been identified at some points and the best course may be to treat a complete area as contaminated.

There has been much debate over whether it is better to remove asbestos wherever it is found as a matter of course, or to leave that which is in good condition in place, recorded and monitored. Arguments other than safety, health and environmental ones have a place in the risk debate, but the cost is an issue, as is the fact that all the fibres are never controlled when asbestos is removed and the background level of asbestos in air would rise.

Common forms of asbestos

Crocidolite — usually known as blue asbestos. The colour of asbestos can only be an approximate guide, because of colour washes and coatings following application, and also because the colour changes due to heat either in the manufacturing process or following application of a product or chemical action on it. This is the type of asbestos considered by some to be the most dangerous, linked with a high cancer risk and particularly with meso-thelioma — a specific form of lung cancer affecting the lining of the lungs and occasionally the stomach. The condition can take up to 25 years to develop following exposure, and as with other forms of asbestos-induced cancer even a minimal exposure may trigger the disease, which is invariably fatal.

Crocidolite has been banned in practice in the UK since 1969 but is still to be found in boiler lagging, piping, older insulation boards and sheeting. Asbestos lagging installed before 1940 has a high chance of containing a measurable percentage.

Amosite — brownish in colour unless subjected to high temperature, this form of asbestos is now considered as dangerous as crocidolite. In 1984 it was made subject to the same control limits.

Chrysotile — white asbestos and the most common form, found in cement sheets, older 'Artex' ceiling compounds, lagging and many other products, and with all other forms of asbestos now banned from use. Prolonged exposure causes asbestosis, which is a gradual and irreversible clogging of the lungs with insoluble asbestos fibres and the scar tissue produced by the body's defence mechanisms trying to isolate the fibres. Even short periods of exposure have been known to trigger lung cancer, especially in smokers.

Less common substances which fall within the legal definition of asbestos are fibrous actinolite, fibrous antho-phyllite and fibrous tremolite (and any mixture containing any of the above).

Work with asbestos — legal requirements

The information provided here is intended only as a summary and guide to the requirement of the Regulations, which must be studied together with the relevant Approved Codes to obtain an authoritative interpretation of the requirements.

Importation, supply and use (and reuse) of all types of asbestos has now been banned. The main direct legislation dealing with work with asbestos is the Control of Asbestos at Work Regulations 1987 (CAWR) as amended in 1992 and 1998. Anyone having a need to handle asbestos to any extent in the construction industry should obtain a copy of the current edition of the Approved Code of Practice (see page 183). Work on land contaminated by asbestos is not covered by a Code.

In simple terms, most work involving asbestos-containing material requires the contractor doing the work to hold a **licence** issued by the HSE under the Asbestos Licensing Regulations 1983 (as amended). Any mixture containing any of the above substances, regardless of the quantity or proportion, falls within the definition of 'asbestos' and therefore within the scope of the Regulations. The purpose of the mixture is not relevant, so that, for example, insulating board containing asbestos which is not insulating anything currently is covered.

Almost all of the huge range of asbestos insulation and coating products is covered — anything with any asbestos in it. There are some exceptions to the licensing requirements — articles made of rubber, plastic, resin or bitumen which also contain asbestos, such as vinyl floor tiles, roofing felts and electric cables, and other asbestos products which have no insulation purposes, such as gaskets and ropes.

The major common exception to the Licensing Regulations is for asbestos—cement mixture products such as roofing sheets, gutters and pipes. The defining point is the bulk density of the product, which to be classed as asbestos cement must exceed 1 tonne per cubic metre. Even though the risk from asbestos is less with these products because the fibres are firmly bound by the cement, the work is still covered by CAWR.

So the Licensing Regulations apply to work with asbestos insulation, coating or insulating board, and 'work' means removing, repairing or disturbing the material. Painting when the surface is in sound condition is not included in their scope, but sealing or painting damaged boards, insulation or coating is.

A licence is required for work with asbestos insulation, coating or insulating board except:

- Where any individual worker does not spend more than a total of 1 hour on the work in any period of 7 consecutive days, **and** the total time spent by all those working on that work does not exceed 2 hours (this exception allows minor repair work to be done without a licence, and the time refers only to the time spent removing, repairing or disturbing the asbestos, not the length of the whole job)
- Where the work is being done at the employer's own premises, **and** the employer has given at least 14 days notice of the work to the HSE **and** takes all the appropriate precautions
- Where the work is solely air monitoring, clearance inspections or the collection of samples for identification

Licences may be granted with conditions, and may refer to insulation, coating or board in any combination. All licences are issued by the HSE's Asbestos Licensing Unit, and can be refused to applicants with previous convictions for health and safety offences whether or not related to asbestos work, who have failed HSE inspections of their work previously, cannot demonstrate adequate competence or who have had two enforcement notices issued against them within a 2-year period. There is an appeal process against licence decisions, through the Department of the Environment, Transport and the Regions (DETR).

The techniques of asbestos stripping, building and testing of enclosures, employee protection and other important matters relevant to those employers working with licences are beyond the scope of this book.

Before starting any work with asbestos, CAWR requires a specific assessment to be made of likely exposure of employees to asbestos. This need not be repeated for repetitive work, but where there is significant variance between jobs, as in demolition, a new assessment will be required. There may also be a need for other assessments under the COSHH Regulations. The Approved Code of Practice for CAWR gives details of the scope of the assessment as:

- Description of the work and its expected duration

- Mention of the type of asbestos and the results of analysis
- Controls to be applied to prevent or control exposure
- The reason for the chosen work method
- Details of expected exposures, including whether they are likely to exceed the action level or control limit (see below) and the number of people affected
- Steps to be taken to reduce exposure to the lowest level reasonably practicable
- Steps to be taken to reduce release of asbestos into the environment and for the removal of waste
- Procedures for provision and use of personal protective equipment
- Procedures for dealing with emergencies where appropriate
- Any other relevant information

The assessment will need to be reviewed if circumstances change, such as fibre control or work methods.

Control limits and action levels

A control limit is that concentration of asbestos in the air, averaged over any continuous 4-hour or 10-minute period, to which employees must not be exposed unless they are wearing suitable respiratory protective equipment (RPE). Control limits are expressed in fibres per millilitre of air averaged over the period.

Action levels are cumulative exposures calculated over a longer time, any continuous 12-week period. The period chosen should be a 'worst case', not a 'best case' period. If any employee's exposure could or does exceed an action level, then the CAWR requirements for notification, designated areas and medical surveillance apply (see below). Action levels are expressed in fibre-hours per millilitre, calculated by multiplying the airborne exposure in fibres/ml by the time in hours for which the exposure lasts. The cumulative exposure is arrived at by adding all the individual exposures over the chosen 12-week period together.

The likely concentration of asbestos fibres to be encountered can be estimated and where necessary confirmed by measuring and monitoring. The latter is done by personal sampling followed by analysis. Monitoring results must be retained to supplement the health records, for the same length of time (see below).

Control limits and action levels for asbestos are given in

Table 15.1. If the assumption is made that the material is not chrysotile alone, then the more stringent limits and level must be used and there will not be a need to identify the type of asbestos.

Notification

CAWR requires notification of work where exposure to employees above the action level is likely to occur (Regulation 6). Licence holders only have to make a single one-off notification, others have to notify each time the work is planned. The enforcing authority depends on the main activity at the premises where the work is planned to be done.

Designated areas

The intention of marking asbestos areas (Regulation 14) is to make sure workers do not enter them unknowingly. A work area can be designated as an asbestos area (because the workers may exceed the action level) or a respirator zone (because either of the control limits may be exceeded), or both.

Medical surveillance

CAWR's Regulation 16 requires that anyone who is exposed above the action level must have been medically examined within the previous 2 years, and employers will need to check with the medical examiner or previous employer that the certificates are genuine. Employers must keep a health record for 40 years, containing prescribed information.

Work with asbestos cement

The general principles to be followed are:

- Reduce the risk of breathing fibres by avoiding the need to do the work where reasonably practicable

- Segregate the area, considering use of a physical barrier such as an enclosure
- Post appropriate warning notices
- Keep the material wet while working on it
- Avoid using abrasive power and pneumatic tools, choosing hand tools instead
- Where power tools are used, provide local exhaust ventilation and use at low speed
- Wear suitable PPE and RPE
- Organise essential high-risk tasks such as cutting and drilling at a central point
- Minimise dust disturbance by choosing appropriate cleaning methods (not sweeping)
- Provide training and information for employees

Asbestos cement sheets

Cleaning by dry scraping or wire brushing should never be allowed. Water jetting can produce slurry which contaminates the area, and the process can cause the sheets to break. Use of surface biocides followed by hand brushing from a safe working platform is the method of choice.

Removal of asbestos cement sheets causes problems because of their potential height and fragility, and weathering which produces fibrous surface dust. Basic principles to follow are:

- Do not allow anyone to walk or stand on sheets or their fixing bolts or purlins — they should be removed from underneath
- Take asbestos cement sheets off before any other demolition is done
- Do not break the sheets further
- Keep the material wet and lower it onto a clean and hard surface
- Remove waste and debris quickly to avoid further breakdown

Table 15.1: Control limits and action levels for asbestos			
Asbestos type	4-hour control limit (fibres/ml)	10-minute control limit (fibres/ml)	Action level (fibre-hours/ml)
Chrysotile (white asbestos) alone	0.3	0.9	72
Any other form, alone or in mixtures, including chrysotile mixtures with any other form of asbestos	0.2	0.6	48

- Do not dry sweep debris
- Follow safe waste disposal practice
- Consider background air sampling at the site perimeter to reassure neighbours as well as workers
- Conduct a final inspection to confirm removal of all debris and that a clean area has been left behind

Disposal of asbestos waste

Asbestos waste is subject to the Special Waste Regulations 1996, which require it to be consigned to a site authorised to accept it. Whether put into plastic sacks or another type of container, the requirement is for the container to be strong enough not to puncture and to contain the waste, be capable of being decontaminated before leaving the work area and be properly labelled and kept secure on site until sent for disposal. The label design is specified.

Large pieces or whole sheets should not be broken up, but transferred direct to covered trucks or skips, or wrapped in polythene sheeting before disposal. Smaller pieces can be collected and put into a suitable container. This should be double wrapped or bagged, and labelled to show it contains asbestos. Dust deposits are removed preferably using a type H vacuum cleaner. Outer surfaces of waste containers should be cleaned off before removal.

Personal protective equipment

Employers will decide as part of their initial assessment what protective equipment is required to complement the planned control measures. The assumption should be made that PPE will be necessary except where exposure is both trivial and infrequent. If asbestos fibres could land on clothing, then protective clothing will be required which is impervious to fibres. Dust jackets alone could be appropriate for infrequent exposures to low levels of fibre concentration.

The principles for clothing are that it must fit the wearer, prevent penetration by fibres, fit closely at the neck, wrists and ankles, have no pockets to trap dust and be easily decontaminated. Clothing that has been in contact with asbestos fibres should be taken off before leaving the work area, then sealed in a dust-tight bag and labelled if it is to be cleaned commercially; it should never be taken home. Separate storage will be required. Disposable protective equipment and respiratory protective equipment (RPE) should be treated as asbestos waste.

The test for the need for RPE is that it will be needed if, despite use of other control measures, exposures still could have the potential to exceed the control limits. 'Suitable' RPE is that which provides an adequate margin for safety in use. It must be matched to the job, the environment, the expected maximum exposure and to the wearer. In achieving all of this, it must reduce the concentration of inhaled asbestos fibres to a level which is below the control limit and below that where it is reasonably practicable to do so. The main parameters are the ability of the filtering medium to remove the contaminant over reasonable periods before cleaning, and the adequacy of the fit to the wearer so that no leaks occur.

Generally, a good fit between the RPE and the wearer's face is the key. For filtering facepieces, half and full face masks, this can only be achieved if the wearer is clean shaven. Fit testing as a specific exercise is not required by law in the UK as it is in some other jurisdictions. However, the initial selection of RPE should include this where practicable, under the advice of a supplier.

Children and third party safety

Other people can easily become casualties of construction besides site workers. Work affecting nearby roads must be planned to consider traffic controls and signs, the marking out of the work area to alert and protect both site workers and the public, the adequacy of footways, available light, vehicle movements and storage of materials — and other hazards which may become clear following the project risk assessment. Site workers should wear high visibility clothing to enable drivers of vehicles to see them clearly especially in poor weather conditions. The Code of Practice governing the New Roads and Street Works Act 1991 provides detailed guidance on signing, and the diversion of pedestrian footways.

Parts of site plant and equipment can easily intrude into public highways, sometimes with disastrous effect. Buckets and counterweights should not be allowed to swing into roadways or pedestrian areas, even with the presence of a banksman (who may be momentarily distracted).

Falling materials are major causes of injury and damage to third parties. Nearby residents and pedestrians must be protected by brickguards, netting or plastic sheeting. Covered skips and debris chutes improve the efficiency of waste removal as well as protection for those nearby. Scaffolding should not be overloaded or used for long-term storage of materials. Hoardings liable to be blown over (i.e. all of them) should be designed to take a wind loading.

Work in occupied premises should be subject to additional risk assessment to make sure that third party interests are observed and appropriate information is given to them.

Visitors to sites need to be controlled from the moment they arrive. Directing them to the site office for appropriate instructions and induction should be done by prominent signs, and visitors should be escorted at all times. This is especially important where projects are partially completed, as on housing developments. Access routes should be marked to separate visitors from hazards unfamiliar to them.

A perimeter fence with gates that can be locked at night should be the standard precaution against trespassers, but where the nature of the project or the cost does not allow such a fence, basic precautions must be taken to reduce the attraction of the working areas and to safeguard machinery, materials and substances which could be harmful. The major requirements include:

- Keeping containers of chemicals locked away
- Turning off gas cylinders and removing keyways
- Isolation of plant, removing keys from ignitions
- Removing or boarding first-run ladders from scaffolding
- Removing access from excavations and placing strong barriers or covers on them
- Checking that materials are not stored so they can topple easily or roll if climbed on by children

Where children are known to live nearby, it is often useful to arrange with teachers at local schools to show one of several films available on dangers of building sites. Contact with parent/teacher associations can help to spread awareness of construction dangers. A letter to parents of adjacent housing containing helpful advice, but not threats, is a tactic that has helped to limit the numbers of child trespassers on some sites.

Demolition

Research studies show that accidents during demolition work are more likely to be fatal than those in many other areas of construction work. The more significant causes of accidents which have high potential for serious injury are premature collapse of buildings and structures, and falls from working places and access routes.

A common feature of demolition accidents is that investigations show that there is usually a failure to plan the work sufficiently at the appropriate stage, which leaves operatives on-site to devise their own methods of doing the work without knowledge and information about the dangers that confront them. Failure to plan is, of course, not confined to demolition but it is difficult to think of many other situations where the consequences are visited so rapidly upon the employees. For this reason all construction work which includes an element of demolition falls within the scope of the Construction (Design and Management) Regulations 1994 (CDM).

Although there is no officially-sanctioned definition of what constitutes 'demolition' for the purposes of CDM, and the triggering of the need for method statements, an unofficial rule of thumb which can be followed with some confidence is that demolition proper involves the taking down of load-bearing structures and/or the production of a substantial quantity of demolished material — about 5 tonnes as a minimum. This rule attempts a practical definition based on the level of risk attached to the work. A stricter interpretation would mean that the removal of any part of a structure could be classed as demolition work, and a result of its being followed would mean there would be very few projects which did not fall within the scope of CDM — clearly not what was intended.

Planning for safety

As much information should be obtained about the work to be done, at the earliest possible time. The extent to which a client is willing or able to provide structural information will partly depend on whether it is actually available to him. The CDM Regulations require the client to provide factual information about the state of the structure to be demolished which he either has, or could find out by making reasonable enquiries. It should not be up to the contractor to discover a particular hazard of the structure or building to be demolished, although it is entirely possible that hazards may reveal themselves during work which could not reasonably have been anticipated. All parties to demolition work must remain alert to this possibility.

Demolition is normally carried out by specialist contractors with experience. Supervision and control of the actual work is required to be done under the supervision of a competent person — by people experienced in demolition — and this may not be achieved by a general contractor.

Information on storage and use of chemicals on a site due for demolition can usually be obtained from the owner, but in cases where the site has been vacant for some time or where previous ownership is unclear the services of a competent analyst may be needed. Architectural or archaeological items for retention must be defined before the preferred method of demolition is determined.

Demolition surveys

Using information supplied, prospective contractors should carry out a survey in sufficient detail to identify structural problems, and risks associated with flammable substances or substances hazardous to health. The precautions required to protect employees and members of the public from these risks, together with the preferred demolition procedure, should be set out in a method statement (see below).

The survey should:

- Take account of the whole site; access should be permitted for the completion of surveys and information made available in order to plan the intended method of demolition
- Identify adjoining properties which may be affected by the work — structurally, physically or chemically; premises which may be sensitive to the work (other than domestic premises) include hospitals, telephone exchanges and industrial premises with machines vulnerable to dust, noise or vibration
- Identify the need for any shoring work to adjacent properties or elements within the property to be demolished; weatherproofing requirements for the work will also be noted
- Identify the structural condition, as deterioration may impose restrictions on the demolition method

Preferably, the survey should be divided into structural and chemical aspects, the latter noting any residual contamination. Structural aspects of the survey should note variations in the type of construction within individual buildings and among buildings forming a complex due for demolition. The person carrying out the survey should be competent. An assessment of the original construction method including any temporary works required can be very helpful.

Preferred method of work

The basic ideal principle is that structures should be demolished in the reverse order to their erection. The method chosen should gradually reduce the height of the structure or building, or arrange its deliberate controlled collapse so that work can be completed at ground level.

Structural information obtained by the survey should be used to ensure that the intended method of work retains the stability of the parts of the structure or building which have not yet been demolished. The aim should be to adopt methods which make it unnecessary for work to be done at height. If this cannot be achieved then systems which limit the danger of such exposure should be employed. The use of balling machines, heavy duty grabs or pusher arms may avoid the need to work at heights. If these methods are possible, the contractor must be satisfied that sufficient area is available for the safe use of the equipment, and that the equipment is adequate for the job.

When work cannot be carried out safely from part of a permanent structure or building, working platforms can be used. These include scaffolds, towers and power-operated mobile work platforms. Where these measures are not practicable, safety nets or harnesses (properly anchored) may be used. The hierarchy of fall protection measures is set out clearly within the Construction (Health, Safety and Welfare) Regulations 1996.

Causing the structure or building to collapse by the use of wire ropes or explosives may reduce the need for working at heights, but suitable access and working platforms may still be needed during the initial stages.

A knowledge of structural engineering principles is necessary to avoid premature collapse, especially an understanding of the effect of pre-weakening by the removal, cutting or partial cutting of structural members. Assessment of risks posed by hazards such as these is the responsibility of the designer(s) and the planning supervisor for the project.

Method statements

In order to comply with section 2(2) of the Health and Safety at Work etc. Act 1974 in relation to the provision of a safe system of work, the production of a method statement is recognised as necessary for all demolition work. Because of the special demolition needs of each structure, an individual risk assessment must be made by the employer undertaking the work, in writing, to comply with the Management of Health and Safety at Work Regulations 1999.

The detailed method statement will be derived from the risk assessment carried out by the contractor. This must be drawn up before work starts, and communicated to all involved as part of the Health and Safety Plan. It should identify the work procedure, associated problems and their solutions, and should form a reference for site supervision.

Method statements should be easy to understand, agreed by and known to all levels of management and supervision, including those of subcontracting specialists.

The demolition method statement should include:

1. The sequence of events and method of demolition or dismantling (including drawings/diagrams) of the structure or building
2. Details of personnel access, working platforms and machinery requirements
3. Specific details of any pre-weakening of structures which are to be pulled down or demolished using explosives
4. Arrangements for the protection of personnel and public, and the exclusion of unauthorised people from the work area; details of areas outside the site boundaries which may need control during critical aspects of the work must be included
5. Details of the removal or isolation of electrical, gas and other services, including drains
6. Details of temporary services required
7. Arrangements for the disposal of waste
8. Necessary action required for environmental considerations (noise, dust, pollution of water, disposal/treatment of contaminated ground)
9. Details of controls covering substances hazardous to health and flammable substances
10. Arrangements for the control of site transport
11. Training requirements
12. Welfare arrangements appropriate to the work and the conditions expected
13. Identification of people with special responsibilities for the co-ordination and control of safety arrangements

Demolition techniques

Piecemeal demolition is done by hand, using hand-held tools, sometimes as a preliminary to other methods. Considerations include provision of a safe place of work and safe access/egress, and debris disposal. It can be completed or begun by machines such as balling machines, impact hammers or hydraulic pusher arms. Considerations for these include safe operation of the machines, clearances, capability of the equipment and protection of the operator.

Deliberate **controlled collapse** involves pre-weakening the structure or building as a preliminary, and completion by use of explosives or overturning with wire rope pulling. Considerations for the use of explosives include competence, storage, blast protection, firing programmes and misfire drill. Wire rope pulling requires a similar level of expertise, as well as selection of materials and clear areas for rope runs.

Demolition training

The importance of adequate training in demolition has been recognised by the introduction of the Construction Industry Training Board's Scheme for the Certification of Competence of Demolition Operatives. Training requirements are imposed by the Construction (Health, Safety and Welfare) Regulations 1996, the Health and Safety at Work etc. Act 1974, and specifically by the Management of Health and Safety at Work Regulations 1999. The selection of competent contractors to carry out the work, as required by the CDM Regulations, will lead to the choice of those with demonstrable qualification(s), knowledge and/or experience. Possession of certificated formal training together with experience will satisfy the requirement.

Legal requirements

Several Acts and Regulations control demolition work. The Health and Safety at Work etc. Act 1974 applies whenever demolition work is done, as do the Management of Health and Safety at Work Regulations 1999 and others introduced on 1 January 1993 such as the Manual Handling Operations Regulations. The importance of the CDM Regulations has already been mentioned in this section. The Construction (Health, Safety and Welfare) Regulations 1996 apply to all demolition work. Other Acts and Regulations which may be applicable include those covering COSHH, electricity, first aid, asbestos and lead. Laws covering the control of pollution and environmental protection are also likely to apply.

Electricity

The ratio of fatalities to injuries is higher for electrical accidents than for most other categories of injury. If an electrical accident occurs, the chances of a fatality are about one in 30 to 40. Despite the beliefs of some, including some electricians, the human body does not develop tolerance to electric shock.

The consequences of contact with electricity are: electric shock, where the injury results from the flow of electricity through the body's nerves, muscles and organs and causes abnormal function to occur (the heart stops, for example); electrical burns resulting from the heating effect of the current which burns body tissue; and electrical fires caused by overheating or arcing apparatus in contact with a fuel.

Causes of electrical failure

Failures and interruption of electrical supply are most commonly caused by:

- Damaged insulation
- Inadequate systems of work
- Inadequate overcurrent protection (fuses, circuit breakers)
- Inadequate earthing
- Carelessness and complacency
- Overheated apparatus
- Earth leakage currents
- Loose contacts and connectors
- Inadequate ratings of circuit components
- Unprotected connectors
- Poor maintenance and testing

Preventing electrical failures

These failures can be prevented by regular attention to the following points:

Earthing — providing a suitable electrode connection to earth through metal enclosure, conduit, frame, etc. Regular inspections and tests of systems should be carried out by a competent person. Portable appliance inspection and testing is part of this regime.

System of work — when working with electrical circuits and apparatus, switching and locking off the supply. Then, the supply and any apparatus should be checked personally by the worker to verify that it is 'dead'; permit-to-work systems should be used in high-risk situations previously identified. Working on live circuits and apparatus should only be permitted under circumstances which are strictly controlled and justified in each case. Rubber or other non-conducting protective equipment may be required. Barriers

and warning notices should be used to ensure that persons not involved in the work directly do not expose themselves to risk. The use of insulated tools and equipment will always be necessary.

Insulation — where work is required near uninsulated parts of circuits. In all circumstances, making the apparatus 'dead' must be considered as a primary aim, and rejected only if the demands of the work make this not practicable. A variety of permanent or temporary insulators may be used, such as cable sheathing and rubber mats.

Fuses — these are strips of metal placed in circuits which melt as a result of overheating in the circuit, effectively cutting off the supply. Different fuses 'melt' at predetermined current flows. A factor in the selection of fuses is that there is a variable and appreciable delay in their action, which may expose those at risk to uninterrupted current for unacceptable periods.

Circuit breakers — detect electromagnetically and automatically any excess current flow and cut off the supply to the circuit.

Residual current devices — detect earth faults and cut off the supply to the circuit. Their use should be considered a requirement on all temporary power circuits, because they will offer improved protection where, for example, connectors and cables 'downstream' have been damaged and not detected.

Competence — only properly trained and suitably experienced people should be employed to instal, maintain, test and examine electrical circuits and apparatus.

Electrical equipment

Electrical tools used should be selected and operated bearing in mind the following considerations:

Substitution — electrical tools and equipment may be replaced with pneumatic equipment — which have their own dangers.

Switching off circuits and apparatus — this must be readily and safely achievable.

Reducing the voltage — use of the lowest practicable voltage should be practised in every circuit. 110 volt circuits are now common in the industry, with lighting circuits often below that voltage. Where circumstances are proved to require the use of 240 volt circuits and above during the construction phase of work, a written risk assessment should be made specifying the steps taken to make the situation as safe as possible. Such steps would include protection of cables, earthing arrangements and authorisation.

Cable and socket protection − should be provided to protect against physical and environmental effects which could have adverse consequences on the integrity of circuits and apparatus, such as rain.

Plugs and sockets − must be of the correct type and specification, and meet Regulations and Codes of Practice requirements.

Maintenance and testing − should be carried out at regular and prescribed intervals by competent and experienced personnel. Recording of the results and values measured will provide a baseline figure for assessing any subsequent deterioration in performance or quality.

Explosive atmospheres − require the careful consideration and selection of equipment for service in dusty or flammable environments. The type of equipment to be used is specified by law.

Portable appliance testing

Regular visual inspection of electrical equipment will identify most faulty items, but not all faults can be seen. Because of this, testing at appropriate intervals is required, depending upon the treatment the equipment receives in use. The guidelines in Table 15.2 are given by the HSE on what constitutes 'appropriate' checks, inspections and testing for site electrical equipment.

Overhead and buried cables

Work near overhead and underground lines requires planning; incidents can nearly always be avoided if thought is given in advance to precautions that will prevent contact or proximity to live cables. Carrying metal ladders and scaffold tubing near overhead wires can provoke arcing − safe clearance distances must be obtained from electricity utilities. Unless cables can be re-routed or the current turned off, the best options, inadvertent contacts can be prevented by choosing traffic routes to avoid them, or building 'goalposts' or collapsible barriers beneath them on either side so that the barrier is hit before the cable. Suitable warning signs should also be provided.

Cables underground are typically struck by operators who were not aware of their likely presence, and so detailed

Table 15.2: HSE guidelines for inspecting and testing electrical equipment				
Equipment	Voltage	Check by user	Recorded visual inspection	Recorded inspection and test
All types	Less than 110 V with secondary winding centre tapped to earth = 55 V	Not required	Not required	Not generally required except yearly for hand lamps
Portable hand-held, lighting, leads, etc.	110 V with secondary winding centre tapped to earth = 55 V	Weekly	Monthly	Before first use on site, then every 3 months
Portable hand-held, lighting, leads, etc.	200−240 V mains supply through RCD	Daily/shift	Weekly	Before first use on site, then monthly
Heavy equipment including lifts, hoists, floodlights	200−240 V mains supply fuses or MCBs	Weekly	Monthly	Before first use on site, then 3-monthly
RCDs		Daily/shift	Weekly	Before first use on site, then monthly, except 3-monthly for fixed RCDs
Site office equipment	200−240 V	Monthly	Every 6 weeks	Before first use on site, then annually

drawings should be obtained wherever possible to show positions and allow them to be marked out. Where the precise position is not known, hand-digging only should be allowed until the cable's position is made clear.

Legal requirements

These are covered by the Electricity at Work Regulations 1989, which amplify the general duties placed on employers, employees and the self-employed by the Health and Safety at Work etc. Act 1974. Further requirements are contained in the Management of Health and Safety at Work Regulations 1999. Arrangements and provisions for the treatment of electric shock are contained in the Health and Safety (First Aid) Regulations 1981.

Excavations

The key to safety in excavation work is planning. For example, if the correct form of support has not been decided in advance, and the necessary support materials ordered in good time, the temptation to go ahead without them is almost too strong to resist. All excavations, however shallow, should be subject to risk assessment. The main hazards to be controlled are:

- Collapse of sides
- Falling materials from sides
- Falls into the excavation
- Undermining of structures in the area
- Discovery of underground services
- Access and egress to the excavation
- Fumes and lack of oxygen
- Occupational health considerations

Collapse of sides

People have been killed by excavation collapses — not only in deep trenches. Even those buried up to the waist can die, because of compression of the lungs from below and the resulting suffocation.

If space and time allow, battering or sloping back the sides of excavations to a safe angle of repose is a good alternative to provision of supporting material. One or the other is required except when working in rock, or when a geological report indicates that it is safe to do so. The guiding principle for support work is that those exposed within the trench or other excavation should always be protected. Sliding trench boxes provide good protection, but the open

ends are traps which must not be entered without additional protection. Excavation work must be supervised by a competent person who is aware of ground conditions and can give necessary instructions to those working in the excavation. Regular inspection of the work is also a requirement.

Falling materials

Spoil heaps should be kept a metre back from excavation edges, and vehicles such as trucks and dumpers should be kept well away from edges. Any making deliveries, such as tipping dumpers, should be restrained from over-running and falling in by anchoring timber stop blocks at the approach. Those within an excavation should wear safety helmets.

Falls into the excavation

Solid guardrails should protect the edges of excavations more than 2 metres deep, and anywhere the public can approach should be fenced off regardless of depth. Traffic routes should keep vehicles away from excavations to prevent overloading pressures to the soil in the area.

Undermining structures in the area

Footings and shallow foundations are easily undermined by excavation work involving even shallow trenches, so investigation should be made in advance to find out whether structural support is needed. Scaffolding that is undermined can be severely weakened, and suitable strengthening and bracing may be required to distribute the loads.

Discovery of underground services

A safe system of work is required to avoid the sudden discovery of services which, if interrupted, could be dangerous to the excavation worker and to those supplied by the service. Safe digging practices include hand-digging where necessary, rather than relying on 'fine movements' of an excavator bucket arm. Cable locators should always be used to augment information from plans, which is often relatively inaccurate.

Access and egress

Ladder or other fixed access to excavations is essential, especially for emergency escape. Ladders should be placed at least every 25 metres, and be fixed securely at the top with a good handhold available. It is also good practice to dig out behind the top rung at ground level, to enable a boot to tread fully on the top rung step-off.

Fumes and lack of oxygen

Even where not obvious from the type of work being done, such as a sewer connection, an excavation should be considered as a potential 'confined space' where the oxygen supply may be limited and fumes may collect. Where there is any possibility of this occurring, such as in deep excavations or those in boggy areas, the atmosphere should be tested by a competent person using a proprietary meter, and regular checks must be maintained to ensure that the position does not change.

Occupational health considerations

Those working in excavations should have tetanus injections kept up to date, or alternatively seek medical advice if dirt enters any small cuts and abrasions they may have. Leptospiral jaundice, or Weil's disease, can be passed to humans from rats indirectly through contaminated ground water. Workers exposed to this hazard should be advised to tell their doctor about their job, as this may help in the future diagnosis of illness. Those working in areas which may be contaminated should wear good protective clothing and equipment, including rubber boots and gloves. A high standard of personal hygiene, and the necessary facilities, should always be present.

Falls

The general principle is that fall prevention is far more effective than fall protection, which often involves personal protective equipment. Reliance on people to make the 'right' decision about wearing personal protective equipment has be shown by events to be unsatisfactory — they forget, decide not to wear it in view of the expected short exposure time, or do not wear or use it correctly. The first stage in fall prevention is during the design process, which influences the construction method. Early erection of stairways avoids the need to fence off open stairwells, and also provides safe access to the next level. Are so many floor slab penetrations necessary? Could the design be changed to allow a single common service duct?

The basic legal requirements for preventing falls by those working at heights are to be found in the summary of the Construction (Health, Safety and Welfare) Regulations 1996. They require a structured approach, which can be summarised as follows:

- Use a working platform fitted with guardrails and toeboards where practicable
- Only when this is not practicable, or when protection has been removed for a period, a method of arresting falls must be used
- Harnesses and lines do not prevent falls, but do provide protection for those falling

Openings and edges

Solid guardrails are a UK requirement where falls of more than 2 m could occur, in preference to other means of fall protection. They protect workers near the edge, and those who have to pass near the edge for access to other places. Protection from falling objects is also needed; high-rise structures allow falling objects to reach high speeds on the way to the ground. In many cases, an effective barrier can only be provided by sheeting below the guardrail down to floor level. In some cases it may be necessary to insist that tools and equipment are physically tethered to the worker by a lanyard to prevent them being dropped.

Where used, debris fans and netting must be placed within two floors of the work area to be effective, and must be placed below each open edge if they are to be relied upon.

A common cause of fatalities is the opening in a floor, covered by a piece of ply or metal sheet. People have been killed following picking up these covers for holes and walking forward into the void. All holes should be protected properly, either by guardrails if they are large enough, or by securing a cover over them that is prominently marked 'Danger — hole beneath'. All such covers should be secured to the floor in such way that they cannot be easily or accidentally removed. A fastening that requires a special tool to undo it is an effective precaution.

Another effective method of guarding floor slab penetrations is to leave the rebar mat in place for as long as possible, cutting it free only when work on passing services is about to begin.

Lift shaft openings must be clearly marked as such, and protected by toeboards or other means of preventing falls of material onto the heads of those working in the shaft.

Scaffolders

Fall protection for scaffolders during erection, alteration and dismantling scaffolding must be provided where they can fall 2 m or more. A useful publication by the National Access and Scaffolding Confederation provides a full description of an appropriate safe method of work. Fall arrest equipment will normally be relied upon, using a suitable anchor point when working over 4 m unless within the protection of a single guardrail and a platform three boards wide. During the raising and lowering of materials scaffolders should be tied off at all times unless protected as before, but this time behind a **double** guardrail. Climbing and working directly from the scaffold structure can only be done while tied off.

The fall arrest equipment used must include a full body harness with rear dorsal ring, and a 1.75 m fixed length lanyard with a shock absorber included in it and a scaffold hook with a 55 mm opening for use one-handed. Although full details of the system of work to be followed are given in the NASC guidance, the principle is that installation, alteration and dismantling must be done from the protection of the single guardrail. Tie-off points should be waist height or above to minimise the drop height following a fall.

Tie-off (anchorage) points on scaffold tubing are acceptable provided that the scaffold is tied in to a structure and that the tie-off is made on horizontal members only (transoms, ledgers or guardrails) with no joint in the same bay.

Rescue planning is advisable, and a copy of an emergency plan should be available and form part of toolbox talks. Emergency procedures will be based upon a risk assessment of the work and the conditions, including the number and experience of the scaffolders.

Maintenance

Maintenance can be defined as work carried out in order to keep or restore every facility (part of a workplace, building and contents) to an acceptable standard. It is not simply a matter of repair; some mechanical problems can be avoided if preventative action is taken in good time. Health and safety issues are important in maintenance because statistics show the maintenance worker to be at greater risk of accidents and injury. This is partly because these workers are exposed to more hazards than others, and partly because there are pressures of time and money upon

the completion of the maintenance task as quickly as possible. Less thought is given to the special needs for health and safety of the maintenance task than is given to routine tasks which are easier to identify, plan and control.

Maintenance standards are a matter for the organisation to determine; there must be a cost balance between intervention with normal operations by planned maintenance and the acceptance of losses because of breakdowns or other failures. From the health and safety aspect, however, defects requiring maintenance attention which have led or could lead to increased risk for the workforce should receive a high priority. Linking inspections with maintenance can be useful, so that work areas and equipment are checked regularly for present and possible future defects. Some plant items may be subject to statutory maintenance requirements, and the manufacturer's instructions in this respect should be complied with as well.

For the construction industry, maintenance of working plant and equipment is a priority. The need for maintenance of any piece of equipment should have been anticipated in its design. Lubrication and cleaning will still be required for machinery, but the tasks can be made safer by consideration of maintenance requirements at an early stage of design, with feedback from users.

A written plan for **preventative maintenance** is required, which documents the actions to be taken, how often this needs to be done, all health and safety matters associated, the training required (if any) before maintenance work can be done, and any special operational procedures such as permits to work and locking-off required.

Breakdown maintenance

The number of failures which require this will be reduced by planned maintenance, but many circumstances (such as severe weather conditions) can arise which require construction workers to carry out tasks beyond their normal work experience, and/or which are more than usually hazardous by their nature. Records of all breakdowns should be kept, to influence future planned maintenance policy revisions, safety training and design.

Injuries during maintenance of plant, equipment and premises are caused by one or more of the following factors:

- Lack of perception of risk by managers/supervisors, often because of a lack of necessary training

- Unsafe or no system of work devised, e.g. no permit-to-work system in operation, no facility to lock off machinery and electricity supply before work starts and until work has finished
- No co-ordination between workers, or communication with other supervisors or managers
- Lack of perception of risk by workers, potentially resulting in failure to wear protective clothing or equipment
- Inadequacy of design, installation, siting of plant and equipment
- Use of subcontractors who are inadequately briefed on health and safety aspects

Maintenance control to minimise hazards

The safe operation of maintenance systems requires steps to be taken to control the above factors. These steps can be divided into the following phases:

Planning — identification of the need for planned maintenance and arranging a schedule for this to meet any statutory requirements. A partial list of items for consideration includes air receivers and all pressure vessels, lifting equipment, electrical tools and machinery, fire and other emergency equipment, and structural items subject to wear, such as scaffold fittings.

Evaluation — the hazards associated with each maintenance task must be listed and the risks of each considered (frequency of the task and possible consequences of failure to carry it out correctly). The tasks can then be graded and the appropriate degree of management control applied to each.

Control — the control(s) for each task will take the above factors into account, and will include any necessary review of design and installation, training, introduction of written safe working procedures to minimise risk (which is known as a safe system of work), allocation of supervisory responsibilities, and necessary allocation of finances.

Monitoring — random checks, safety audits and inspections, and the analysis of any reported accidents for cause which might trigger a review of procedures, constitute necessary monitoring to ensure the control system is fully up to date. The introduction of new plant and equipment may have maintenance implications and should therefore be included in the monitoring process.

Legal requirements

Many pieces of health and safety legislation contain both general and specific requirements to maintain plant and equipment. The general duty is contained in section 2(2)(d) of the Health and Safety at Work etc. Act 1974, and more specific requirements are contained in the Management of Health and Safety at Work Regulations 1999. Other requirements are contained in the Workplace (Health, Safety and Welfare) Regulations 1992, the Personal Protective Equipment at Work Regulations 1992 and the Provision and Use of Work Equipment Regulations 1998.

Manual handling

It has long been recognised that the manual handling of loads in construction work contributes significantly to the number of injuries, with approximately a quarter of all reported accidents attributed to these activities. The majority of these injuries result in more than 3 days' absence from work, almost half of the total being sprains or strains of the lower back, with other types of injury including cuts, bruises, fractures and amputations. Many of the injuries are of a cumulative nature rather than being attributable to any single handling incident.

Until very recently, training of employees to lift concentrated on methods which would allow them to minimise the risks from moving and lifting the heavy loads expected of them. Films and videos explain how these techniques have helped them do the work without injury. Now, however, the aim is to reduce the opportunities for injury by reducing the amount of lifting and moving that the human body is required to do in the work environment. We should be asking not 'How do I lift this safely?' but 'Do I really need to lift this at all?'.

The casualties of the old approach are all around us — about 80% of the working population will suffer some form of back injury requiring them to take time off work at some point in their lives. Those who are injured are three times as likely to be injured again in the same way as those without a back injury. This situation needs to be addressed by informed employers and employees, seeking to identify and remove hazardous lifting and to find new ways of doing work that has traditionally involved human effort to move loads.

Injuries resulting from manual handling

Some of the more common types of injury resulting from manual handling are considered here. It is important to remember that the spine is undergoing a natural process of 'ageing' which will affect the integrity of the spine. For example, loss of 'spinal architecture' (the natural shape and curvatures of the spine) produces excessive pressure on the edges of the discs, wearing them out faster. These effects are not considered here.

Disc injuries

Ninety per cent of back troubles are attributable to disc lesions. The discs lie between the vertebrae, acting as shock absorbers and facilitating movement. They do not 'slip' in the conventional sense, as they are permanently fixed to the bone above and below. Roughly circular, the discs are made up of an outer rim of elastic fibres with an inner core containing a jelly-like fluid. When stood upright, forces are exerted directly through the whole length of the spine and in this position it can withstand considerable stress. However, when the spine is bent, most of the stress is exerted on only one part (usually at the part where the bending occurs). Also, due to the bending of the spine in one place, all the stress is exerted on one side of the intervertebral disc, thus pinching it between the vertebrae. This pinching effect may scar and wear the outer surface of the disc, so that at some time it becomes weak and eventually, under pressure, it ruptures.

Many authorities believe that all disc lesions are progressive, rather than sudden. It is also important to remember that the disc cores dry out with age, making them less flexible and functional, and more prone to injury. The disc contents are highly irritating to the surrounding parts of the body, causing an inflammatory response when they leak out.

Ligament and tendon injuries — Ligaments and tendons are connective tissues, and hold the back together. Ligaments are the gristly straps that bind the bones together while tendons attach muscles to other body parts, usually bones. Repetitive motion of the tendons may cause inflammation. Both can be pulled and torn, resulting in sprains. Any factor that produces tightness in ligaments and tendons predisposes the back to sprains (such as age effects and cold weather). Two main ligaments run all the way down the spine to support the vertebrae.

Muscular and nerve injuries

The muscles in the back form long, thick bands that run down each side of the spine. They are very strong and active, but tiring of these muscles can result in aches and pains, and can create stress on the discs. Postural deformities can result from damage to the muscles. Fibrositis (rheumatic pain) can also result. Nerves can become trapped between the elements of the spine causing severe pain and injury.

Hernias

A hernia is a protrusion of an internal organ through a gap in a wall of the cavity in which it is contained. For example, any compression of the abdominal contents may result in a loop of intestine being forced into one of the gaps or weak areas formed during the development of the body. When the body is bent forward, possibly during a lift, the abdominal cavity decreases in size causing a compression of internal components and increasing the risk of hernias.

Fractures, abrasions and cuts

These can result from dropping the objects that are being handled (possibly because of muscular fatigue), falling while carrying objects (perhaps as a result of poor housekeeping), from other inadequacies in the working environment such as poor lighting, or from the contents of the load.

Injury during manual handling

The injuries highlighted above can result from lifting, pushing, pulling and carrying an object during manual handling.

Lifting — Compressive forces on the spine, its ligaments and tendons can result in some of the injuries identified above. High compressive forces in the spine can result from lifting too much, poor posture and incorrect lifting technique. Prolonged compressive stress causes what is known as 'creep-effect' in the spine, squeezing and stiffening it. If the spine is twisted or bent sideways when lifting, the added tension in the ligaments and muscles when the spine is rotated considerably increases the total stress on the spine to a dangerous extent. The elastic disc fibres are also put under tension during repetitive twisting, and individual fibres damaged.

Pushing and pulling — Stresses are generally higher for pushing than pulling. Because the abdominal muscles are active as well as the back muscles, the reactive compressive force on the spine can be even higher than when lifting. Pushing also loads the shoulders and the ribcage is stiffened, making breathing more difficult.

Carrying — Carrying involves some static muscular work, which can be tiring for the muscles, the back, shoulder, arm and hand depending on how the load is supported. A weight held in front of the body induces more spinal stress than one carried on the back. Likewise, a given weight held in one hand is more likely to cause fatigue than if it was divided into equal amounts in each hand. As with pushing, carrying objects in front of the body or on the shoulders may restrict the ribcage. Thus, the way in which a load is carried makes a great difference to the fatiguing effects. This is why getting a good grip is important, as well as keeping the load close to the body, which places it closer to the body's centre of gravity.

The Manual Handling Operations Regulations 1992 establish a clear hierarchy of measures to reduce the risk of injury when performing manual handling tasks. To summarise, manual handling operations which present a risk must be avoided so far as is reasonably practicable; if these tasks cannot be avoided, then each such task, where there is a risk of physical injury, must be assessed. As a result of that assessment, the risk of injury must be reduced for each particular task identified so far as is reasonably practicable. Details of the assessment process are given in Chapter 6 Assessing the risks.

It is important to remember that what is required initially is not a full assessment of each of the tasks, but an appraisal of those manual handling operations which involve a risk that cannot be dismissed as trivial, to determine whether they can be avoided. Consideration of a series of questions will be useful in completing this stage of the exercise. These include:

Is there a risk of injury? An understanding of the types of potential injury will be supplemented with the past experiences of the employer, including accident/ill-health information relating to manual handling and the general numerical guidelines contained in official guidance.

Is it reasonably practicable to avoid moving the load? Some work is dependent on the manual handling of loads and cannot be avoided (as, for example, in most cases of glass handling). If it is reasonably practicable to avoid moving the load then the initial exercise is complete and further review will only be required if conditions change.

Is it reasonably practicable to automate or mechanise the operation? Introduction of these measures can create different risks (introduction of fork-lifts and material handlers creates a series of new risks, for example) which require consideration.

The aim of the full assessment is to evaluate the risk associated with a particular task and identify control measures which can be implemented to remove or reduce the risk (possibly mechanisation and/or training). For varied work (such as done in the construction industry) it will not be possible to assess every single instance of manual handling. In these circumstances, each type or category of manual handling operation should be identified and the associated risk assessed.

The assessment must be kept up to date. It needs reviewing whenever it may become invalid, such as when the working conditions or the personnel carrying out those operations have changed. Review will also be required if there is a significant change in the manual handling operation. There may, for example, be a change in the nature of the task, or the load, such as changing to different bulk sizes.

Before beginning an assessment, the views of management and employees can be of particular use in identifying manual handling problems. Their involvement in the assessment process should be encouraged, particularly in reporting problems associated with particular tasks.

Records of accidents and ill health are also valuable indicators of risk, together with absentee records, poor productivity and morale, and excessive material damage. There is now a considerable amount of literature identifying risks and solutions associated with manual handling operations, which will be a useful source of information.

Schedule 1 to the Manual Handling Operations Regulations specifies four interrelated factors which the assessment must take account of. The answers to the questions specified about the factors form the basis of an appropriate assessment. These are:

The tasks — Do they involve:

- Holding or manipulating loads at a distance from the trunk?
- Unsatisfactory bodily movement or posture, especially
  - □ twisting the trunk?

□   stooping?

□   reaching upwards?

■   Excessive movement of loads, especially

□   excessive lifting or lowering distances?

□   excessive carrying distances?

■   Excessive pushing or pulling of loads?

■   Risk of sudden movement of loads?

■   Frequent or prolonged physical effort?

■   Insufficient rest or recovery periods?

■   A rate of work imposed by a process?

The loads — are they:

■   Heavy?

■   Bulky or unwieldy?

■   Difficult to grasp?

■   Unstable, or with contents likely to shift?

■   Sharp, hot, or otherwise potentially damaging?

The working environment — are there:

■   Space constraints preventing good posture?

■   Uneven, slippery or unstable floors?

■   Variations in levels of floors or work surfaces?

■   Extremes of temperature or humidity?

■   Conditions causing ventilation problems or gusts of wind?

■   Poor lighting conditions?

Individual capability — does the job:

■   Require unusual strength, height, etc.?

■   Create a hazard to those who might reasonably be considered to be pregnant or to have a health problem?

■   Require special information or training for its safe performance?

Other factors

■   Is movement or posture hindered by PPE or clothing?

Reducing the risk of injury

In considering the most appropriate controls, an ergonomic approach to designing the manual handling operation will optimise the health, safety and productivity associated with the task. The task, the load, the working environment, individual capability and the inter-relationship between these factors are all important elements in deciding optimum controls designed to fit the operation to the individual

rather than the other way around. Techniques of risk reduction appropriate to the construction industry include:

Mechanical assistance — this involves the use of handling aids, of which there are many examples. One could be the use of a lever, which would reduce the force required to move a load. A hoist can support the weight of a load while a trolley can reduce the effort needed to move a load horizontally. Chutes are a convenient means of using gravity to move loads from one place to another.

Improvements in the task — changes in the layout of the task can reduce the risk of injury by, for example, improving the flow of materials or products. Improvements which will permit the use of the body more efficiently, especially if they permit the load to be held closer to the body, will also reduce the risk of injury. Improving the work routine by reducing the frequency or duration of handling tasks will also have a beneficial effect. Using teams of people, and personal protective equipment (PPE) such as gloves where appropriate, can also contribute to a reduced risk of injury.

Reducing the risk of injury from the load — the load may be made lighter by using smaller packages/containers, or specifying lower packaging weights. Additionally, the load may be made smaller, easier to manage, easier to grasp (for example, by the provision of handles), more stable and less damaging to hold (clean, free from sharp edges, etc.). On the other hand, the introduction of smaller and lighter loads may carry a penalty in the form of further bending and other repetitive movements to handle the load than were necessary before. This could result in no change in the overall risk.

Improvements in the working environment — this can be done by removing space constraints, improving the condition and nature of floors, improving housekeeping standards and ensuring that adequate lighting is provided.

Individual selection — clearly, the health, fitness and strength of an individual can affect ability to perform manual handling tasks. Health screening is an important selection tool. Knowledge and training have important roles to play in reducing the number of injuries resulting from manual handling operations. There is little point in enquiring about any previous back injuries on a job application form if no attempt is made to ensure that those who do admit to previous problems are not given work which is foreseeably likely to produce them.

Manual handling training

A formal training programme, and toolbox talks, should include mention of:

- Dangers of careless and unskilled handling methods
- Principles of levers and the laws of motion
- Functions of the spine and muscular system
- The effects of lifting, pushing, pulling and carrying, with emphasis on harmful posture
- Use of mechanical handling aids
- Selection of suitable clothing for lifting and necessary protective equipment
- Techniques of:
  - identifying slip/trip hazards
  - assessing the weight of loads and how much can be handled by the individual without assistance
  - bending the knees, keeping the load close to the body when lifting (but avoiding tension and knee-bending at too sharp an angle)
  - breathing, and avoiding twisting and sideways bending during exertion
  - using the legs to get close to the load, making best use of body and load weight
  - using the 'power grip' — the load is supported under the fingers bent at right angles to the palm and with the palmar ends of the fingers taking the load rather than the fingertips.

There are several guiding principles for the safe lifting of loads. These are:

- Take a secure grip
- Use the proper foot position
- Adopt a position with bent knees and comfortably straight back
- Keep arms close to the body
- Keep the chin tucked in
- Use body weight to advantage

All attending the training should have an opportunity to practise under supervision. Other factors which should be discussed with trainees include:

- Personal limitations (age, strength, fitness, girth)
- Nature of loads likely to be lifted (weight, size, rigidity)
- Position of loads
- Working conditions to minimise physical strain
- The requirements of the Manual Handling Operations Regulations 1992 and risk assessments

Mechanical handling

Mechanical handling techniques have improved efficiency and safety, but have introduced other sources of potential injury. Cranes and hoists, and powered trucks and fork-lifts are the primary means for mechanical handling on site. In all circumstances the safety of the equipment can be affected by the safety of operating conditions, site hazards and the operator.

Cranes

Basic safety principles for all mechanical equipment apply to cranes: the equipment should be of good construction, made from sound material, of adequate strength and free from obvious faults. All equipment should be tested and regularly examined to ensure its integrity. The equipment should always be properly used.

What can go wrong?

- **Overturning** can be caused by weak support, operating outside the machine's capabilities and by striking obstructions
- **Overloading** by exceeding the operating capacity or operating radii, or by failure of safety devices
- **Collision** with other cranes, overhead cables or structures
- **Failure** of support — placing over cellars and drains, outriggers not extended, made-up or not solid ground, or failure of structural components of the crane
- **Operator** errors from impaired/restricted visibility, poor eyesight, inadequate training
- **Loss of load** from failure of lifting equipment, lifting accessories or slinging procedure

Hazard elimination

Matters which require attention to ensure the safe operation of a crane include:

Identification and testing — every crane should be tested and a certificate should be issued by the seller and following each test. Each should be identified for reference purposes, and the safe working load clearly marked. This should never be exceeded, except under test conditions.

Maintenance — cranes should be inspected regularly, with any faults repaired immediately. Records of checks and inspections should be kept.

Safety measures — a number of safety measures should be incorporated for the safe operation of the crane. These include:

- Load Indicators — of two types:
  - □ load/radius indicator
  - □ automatic safe load indicator, providing audible and visual warning
- Controls — should be clearly identified and of the 'dead-man' type
- Overtravel switches — limit switches to prevent the hook or sheave block being wound up to the cable drum
- Access — safe access should be provided for the operator and for use during inspection and maintenance/emergency
- Operating position — should provide clear visibility of hook and load, with the controls easily reached
- Passengers — should not be carried without authorisation, and never on lifting tackle
- Lifting tackle — chains, slings, wire ropes, eyebolts and shackles should be tested/examined, should be free from damage and knots as appropriate, be clearly marked for identification and safe working load, and be properly used (no use at or near sharp edges, or at incorrect sling angles)

Operating area — all nearby hazards, including overhead cables and bared power supply conductors, should be identified and removed or re-routed where practicable. Solid support should be available and on fixed site installations the dimensions and strength of support required should be specified. The possibility of striking other cranes or structures should be examined.

Operator training — crane operators and slingers should be fit and strong enough for the work. Training should be provided for the safe operation of the particular equipment.

Hoists

Platform goods hoists and personnel hoists share a number of characteristics which in turn require the same general controls to be in place to assure safety.

What can go wrong?

- **Physical failure of the equipment** can be caused by inadequate ties or other support, operating outside the hoist's capabilities causing structural failure and by striking obstructions

- **Overloading** by exceeding the operating capacity, or by failure of safety devices
- **Collision** with overhead cables or structures, occasionally involving eccentrically-loaded platforms
- **Operator** errors from impaired/restricted visibility, poor eyesight, inadequate training and wrong operating position
- **Loss of load** from failure or bumping of the hoist platform
- **Falls from platforms** — standing on a moving platform, equipment failure whilst standing on a stationary platform, using the unprotected hoist platform as a working platform or means of access

Hazard elimination

Matters which require attention to ensure the safe operation of hoists include:

Identification and testing — every hoist should be tested and a certificate should be issued by the seller and following each test. Each should be identified for reference purposes, and the safe working load clearly marked. This should never be exceeded, except under test conditions.

Maintenance — hoists should be inspected regularly, with any faults repaired immediately. Records of checks and inspections should be kept.

Safety measures — a number of safety measures should be incorporated for the safe operation of hoists. These include:

- Information on load capacity for the loader and operator
- Overtravel switches or physical stops — to prevent the continued manual operation of platforms at the top and bottom of travel
- Access — safe access should be provided for the users, and for use during inspection and maintenance/ emergency. Users' access points should be fully boarded and fitted with edge guardrails and toeboards at the sides. Gates are required at each landing point, which must be kept closed when the platform is not at that level
- Operating position — should provide clear visibility of platform or cage, with the controls easily reached and marked where appropriate
- Passengers — must not be carried on goods hoists

Operating area — solid support should be placed under the ground level hoistway and approach areas.

Operator training — hoist operators should be fit and strong enough for the work. Physical effort is involved in operating cage hoists where the gates are operated from within the cage by the driver. Training should be provided for the safe operation of the particular equipment.

Powered rough terrain and other fork trucks

Trucks should be of good construction, free from defects and suitable for the purpose in terms of capacity, size and type. The type of power supply to be used should be checked, because the nature of the work area may require one kind of power source rather than another. In unventilated confined spaces, internal combustion engines will not be acceptable because of the toxic gases they produce.

Trucks should be maintained so as to prevent failure of vital parts, including brakes, steering and lifting components. Special facilities may be required for some tasks, such as mast replacement. Specific risk assessment should be made, which will take into account local conditions and availability of appropriate equipment. Any damage should be reported and corrected immediately. Overhead protective guards must be fitted for the protection of the operator. Trucks and their attachments should only be operated in accordance with the manufacturer's instructions.

What can go wrong?

- **Overturning** from manoeuvring with load elevated, driving at too high a speed, sudden braking, striking obstructions, use of forward mast tilt with load elevated, driving down a ramp with the load in front of the truck, turning on or crossing ramps at an angle, shifting loads, unsuitable road or support conditions
- **Overloading** by exceeding the maximum lifting capacity of the truck
- **Collision** with structural elements, pipes, stacks or with other vehicles
- **Floor failure** because of uneven or unsound floors, or by exceeding the load capacity of the floor. The designed load capacity of floors other than the ground level should always be checked before using trucks on them
- **Loss of load** can occur if devices are not fitted to stop loads slipping from forks

- **Explosions and fire** may arise from electrical shorting, leaking fuel pipes, dust accumulation (spontaneous combustion) and from hydrogen generation during battery charging. The truck itself can be the source of ignition if operated in flammable atmospheres
- **Passengers** should not be carried unless seats and other facilities are provided for them

Hazard elimination

Matters which require attention for the safe operation of all types of lift and rough terrain trucks include:

Operating area — storage and stacking areas should be properly laid out, with removal of blind corners. Passing places need to be provided where trucks and people are likely to pass each other in restricted spaces, and traffic routes need to be clearly defined with adequate visibility. Pedestrians should be excluded from operating areas. Suitable warning signs will be required to indicate priorities.

Lighting in incomplete structures should be adequate to facilitate access and stacking operations. Battery charging and LPG refuelling areas should be well-ventilated and lit with no smoking or naked lights permitted. Reversing lights and/or sound warnings should always be fitted where possible.

Training should be provided for operators in the safe operation of their equipment, followed by certification.

Legal requirements

Mechanical handling is covered by the general duties provisions of the Health and Safety at Work etc. Act 1974, and the more detailed requirements of the Management of Health and Safety at Work Regulations 1999. The Provision and Use of Work Equipment Regulations 1998 (PUWER 98) apply to all work equipment. Many kinds of mechanical handling equipment fall within the scope of the Lifting Operations and Lifting Equipment Regulations 1998 (LOLER), but others do not. In most cases, LOLER applies to work equipment which has as its principal function a use for lifting or lowering of the type associated with traditional lifting equipment, such as cranes, or accessories, such as chains. LOLER does not define the term 'lifting equipment'; however, similar standards of safety are always required by PUWER 98.

Noise

Noise enables us to communicate, and can create pleasure in the form of music and speech. However, exposure to excessive noise can damage hearing. Noise is usually defined as 'unwanted sound', but in strict terms noise and sound are the same. Noise at work can be measured using a sound level meter. Sound is transmitted as waves in the air, travelling between the source and the hearer. The frequency of the waves is the pitch of the sound, and the amount of energy in the sound wave is the amplitude.

How the ear works

Sound waves are collected by the outer ear and pass along the auditory canal for about 2.5 cm to the ear drum. Changes in sound pressure cause the ear drum to move in proportion to the sound's intensity. On the inner side of the ear drum is the middle ear which is completely enclosed in bone. Sound is transmitted across the middle ear by three linked bones, the ossicles, to the oval window of the cochlea, the organ of hearing which forms part of the inner ear. This is a spirally-wound tube, filled with fluid which vibrates in sympathy with the ossicles. Movement of the fluid causes stimulation of very small, sensitive cells with hairs protruding from them and rubbing upon a plate above them. The rubbing motion produces electrical impulses in the hair cells which are transmitted along the auditory nerve to the brain which then interprets the electrical impulses as perceived sound. Hair cells sited nearest the middle ear are stimulated by high frequency sounds, and those sited at the tip of the cochlea are excited by low frequency.

Noise-induced hearing loss occurs when the hair cells in a particular area become worn and no longer make contact with the plate above them. This process is not reversible, as the hair cells do not grow again once damaged.

Hearing damage

Excessive noise energy entering the system invokes a protection reflex, causing the flow of nerve impulses to be damped and as a result making the system less sensitive to low noise levels. This is known as threshold shift. From a single or short duration exposure, the resulting temporary threshold shift can affect hearing ability for some hours, but recovery then takes place. Repeated exposure can result in irreversible permanent threshold shift. The following damage can occur as a result of exposure to noise:

- Acute effects:
  - acute acoustic trauma from gunfire, explosions; usually reversible, affects ear drum, ossicles
  - temporary threshold shift from short exposures, affecting the cochlea
  - tinnitus (ringing in the ears) results from intense stimulation of the auditory nerves, usually wears off within 24 hours. Can be produced by drugs or as a result of illness
- Chronic effects:
  - permanent threshold shift from long duration exposure, affects the cochlea and is irreversible
  - noise-induced hearing loss from (typically) long duration exposure, affects ability to hear human speech, irreversible, compensatable. It involves reduced hearing capability at the frequency of the noises that have caused the losses
  - tinnitus, as the acute form, may become chronic without warning, often irreversibly

Presbycusis is the term for hearing losses in older people. These have been thought to be due to changes due to ageing in the middle ear ossicles, which causes a reduction in their ability to transmit higher frequency vibrations.

Noise measurement

The range of human hearing from the quietest detectable sound to engine noise at the pain threshold is enormous, involving a linear scale of more than 100 000 000 000 units. Measuring sound intensity on such a scale would be clumsy, and so a method of compressing it is used internationally. The sound intensity or pressure is expressed on a logarithmic scale and measured in bels, although as a bel is too large for most purposes, the unit of measurement is the decibel (dB). The decibel scale runs from 0 to 160 decibels (dB). A consequence of using the logarithm scale is that an increase of 3 dB represents a doubling of the noise level. If two machines are measured when running separately at 90 db each, the sound pressure level when they are both running together will not be 180 dB, but 93 dB. To establish noise levels on this scale, several different types of measurement are used.

Three weighting filter networks (A, B and C) are incorporated into sound level meters. They each adjust the reading given for different purposes, and the one most commonly used is the 'A' weighted dB. This filter recognises the fact that the human ear is less sensitive to low frequencies, and the circuit attenuates or reduces very low frequencies to mimic the response of the human ear, and attaches greater importance to the values obtained in the sensitive frequencies. Measurements taken using the A circuit are expressed as dB(A).

In construction work, noise levels vary continuously. A measurement taken at a single moment in time is unlikely to be representative of exposure throughout the work period, yet this needs to be known as the damage done to hearing is related to the total amount of noise energy to which the ear is exposed.

A measure called **LEq** — the continuous equivalent noise level — is used to indicate an average value over a period which represents the same noise energy as the total output of the fluctuating real levels. LEq can be obtained directly from a sound level meter having an integrating circuit, which captures noise information at frequent timed intervals and recalculates the average value over a standard period, usually 8 hours. It can also be calculated from a series of individual readings coupled with timings of the duration of each sound level, but this is laborious and relatively inaccurate.

Noise dose is a measure which expresses the amount of noise measured as a percentage, where 8 hours at a continuous noise level of 90 dB(A) is taken as 100%. If the work method and noise output is uniform, and the dose measured after 4 hours is 40%, then the likely 8-hour exposure will be less than 100%. However, if the dose reading after 2 hours is 60%, this will be an indication of an unacceptably high exposure.

LEP,d measures a worker's daily personal exposure to noise, expressed in dB(A). **LEP,w** is the measure of the worker's weekly average of the daily personal noise exposure, again expressed in dB(A).

Peak pressure is the highest pressure level reached by the sound wave, and assessments of this will be needed where there is exposure to impact or explosive noise. A meter capable of carrying out the measurement must be specially selected, because of the damping event of needle-based measuring which will consistently produce under-reading. A similar effect can be found in 'standard' electronic circuitry.

Controlling construction noise

This can be achieved by:

Engineering controls — purchasing equipment which has low vibration and noise characteristics, and achieving designed solutions to noise problems including using quieter processes and making mountings and couplings flexible. Rotating and reciprocating equipment should be operated as slowly as practicable.

Orientation and location — moving the noise source away from the work area, or turning the machine around. Much can be done to reduce site perimeter noise levels by careful positioning of noisy fixed plant, and by screening plant using sound baffles and earth banks.

Enclosure — surrounding the machine or other noise source with sound-absorbing material is of limited use unless total enclosure is achieved.

Use of silencers — these can suppress noise generated when air, gas or steam flow in pipes or are exhausted to atmosphere.

Lagging can be used on pipes carrying steam or hot fluids as an alternative to enclosure.

Damping can be achieved by filling proprietary damping pads, stiffening ribs or by using double skin construction techniques.

Absorption treatment in the form of wall applications or ceiling panels; these must be designed for acoustic purposes to have significant effect and are rarely encountered in the construction industry except as part of structures to be built or fitted out.

Isolation of workers in acoustically-quiet booths or control areas properly enclosed, coupled with scheduling of work periods to reduce dose. This will only be effective where there is little or no need for constant movement into or out of areas with high noise levels: even a short duration exposure to high sound pressure levels will exceed the permitted daily dose. Nevertheless, improved design of vehicle and plant cabs and (for example) batch plant and lifting equipment operating positions can do much to improve the health of operators.

Personal protection by the provision and wearing of ear muffs or plugs. This must be regarded as the last line of defence, and engineering controls should be considered in all cases. Areas where personal protective devices must be worn should be identified by signs, and adequate training should be given in the selection, fitting and use of the equipment, as well as the reasons for its use. Wrongly-used or poorly-fitting protection devices can lose much of their protective capability, for example when straps and other items lift the edges of ear muffs and allow sound energy to penetrate.

Hearing protection

Hearing protection should be chosen to reduce the noise level at the wearer's ear to below the recommended limit for unprotected exposure. Selection cannot be based upon A-weighted measurement alone, because effective protection will depend on the ability of the protective device to attenuate (reduce) the sound energy actually arriving at the head position. Sound is a combination of many frequencies unless it is a pure tone, and it can happen that a particular noise against which protection is required has a frequency component which is not well handled by the 'usual' protection equipment. Therefore, a more detailed picture of the sound spectrum in question should be made before selection, checking the results obtained by octave band analysis against the sound-absorbing (attenuation) data supplied by the manufacturers of the products under consideration.

There are two main forms of hearing protection — objects placed inside the ear canal to impede the passage of sound energy, and objects placed around the outer ear to restrict access of sound energy to any part of the inner, middle or outer ear. Neither of these forms of protection will prevent some sound energy from reaching the organ of hearing by means of bone conduction effects within the skull.

Ear plugs fit into the ear canal. They may be made from glass down, polyurethane foam or rubber, and most types are disposable. Reusable plugs suffer from the twin disadvantages that they may present hygiene problems unless great care is taken to clean them after use, and unless they are cast into the individual ear a good fit is unlikely to be achieved in every case. Even though some plugs are available in different sizes, the correct size can only be determined by a qualified person. One further difficulty is that in a reasonable proportion of people the ear canals are not the same size. Ear plugs lose their ability to attenuate sufficient sound energy at levels above 105 dB(A), and beyond this point their claimed attenuation figures at different frequencies should be compared with actual frequency measurements taken at the noise source.

Ear muffs consist of rigid cups which fit over the ears and are held in place by a head band. The cups generally have acoustic seals of polyurethane foam or a liquid-filled annular bag to obtain a tight fit. The fit achieved is a function of the cup design in relation to head shape and contours, the type of seal, and the tightness of the head band. The protective value of ear muffs can be seriously compromised if helmet straps, spectacles, goggle straps or anything else intrudes under or past the annular seals. Again hygiene is important and suitable cleaning facilities and uncontaminated accommodation is required to keep them in good condition.

Legal requirements

Specific legislation on noise in all places of work, including construction, is contained in the Noise at Work Regulations 1989. The Personal Protective Equipment at Work Regulations 1992 require PPE to be selected according to criteria established in the risk assessment. Equipment provided must conform to EC standards.

Although exposure to noise at work has been recognised as damaging to health for a number of years, until relatively recently there has been little specific legal control. The Agriculture (Tractor Cabs) Regulations 1974, the Offshore Installations (Operational Safety, Health and Welfare) Regulations 1976 and the Woodworking Machines Regulations 1974 have or had specific controls included in them, and there was a general use of the Code of Practice issued in 1972 by the Department of Employment to determine compliance with the general duties imposed by the Health and Safety at Work etc. Act 1974.

The Noise at Work Regulations became effective on 1 January 1990. They require the protection of all persons at work, with a broad general requirement for employers to reduce the risk to employees arising from noise exposure as far as is reasonably practicable (Regulation 6). The Regulations also affect the self-employed and trainees.

The Regulations introduced a control framework by requiring assessment of the extent of the problem, by carrying out noise surveys to identify work areas and employees at risk; control by engineering measures, isolation or segregation of affected employees; supply and use of protective equipment and administrative means; and monitoring by reassessment at intervals to ensure that the controls used remain effective.

Regulation 6 imposes a general duty to reduce the risk of hearing damage to the lowest level that is reasonably practicable. Three **noise action levels** are defined, which determine the course of action an employer has to take if employees are exposed to noise at or above the levels. These are:

- First action level — daily personal noise exposure (LEP,d) of 85 dB(A)
- Second action level — daily personal noise exposure of 90 dB(A)
- Peak action level — peak sound pressure level of 200 Pascals (Pa) or more. This action level is important in circumstances where workers are subjected to small numbers of loud impulse noises during an otherwise relatively quiet day, for example during piling operations or when working with cartridge-operated fixing tools. Anyone firing more than about 30 cartridges per shift or working day should be considered to be at risk

Where the daily noise exposure exceeds the first action level, employers must:

- Carry out noise assessments (Regulation 4), using only competent persons to carry out the assessments (Regulation 5) and keep records of these until new ones are made
- Provide adequate information, instruction and training for the employees — about the risks to hearing, the steps to be taken to minimise the risks, how employees can obtain hearing protectors if they are exposed to levels between 85 and 90 dB(A) LEP,d and their obligations under the Regulations (Regulation 11)
- Ensure that hearing protectors (complying with the requirements as to suitability) are provided to those employees who ask for them (Regulation 8(1)) and that the protectors are maintained or repaired as required (Regulation 10(1)(b))
- Ensure so far as is practicable that all equipment provided under the Regulations is used (Regulation 10(a)) — apart from the hearing protectors provided on request as noted above
- Where the noise exposure exceeds the second action level or the peak action level, employers must in **addition** to the requirements detailed above:
  - □ take steps to reduce noise exposure so far as is reasonably practicable by means other than provision of hearing protection (Regulation 7). This is an important requirement, introducing good occupational health and hygiene practice to

ensure hazards are designed out from a process as opposed to employee involvement only to reduce the risks
  - □ establish hearing protection zones, marking them with notices so far as is reasonably practicable (Regulation 9)
  - □ supply hearing protection (complying with the requirements as to suitability) to those exposed (Regulation 8(2)) and ensure they are worn by them (Regulation 10(1)(a)), so far as is practicable
  - □ ensure that all those entering marked hearing protection zones use hearing protection, so far as is reasonably practicable
- Duties are imposed on employees by the Regulations to ensure their effectiveness. The requirements (Regulation 10(2)) are:
  - □ on being exposed to noise at or above the first action level, they must use any noise control equipment other than hearing protection which the employer may provide and report any defects discovered to the employer
  - □ if exposed to the second action level or peak action level, they must in addition wear the hearing protection supplied by the employer

The Regulations also extend duties placed on designers, importers, suppliers and manufacturers under section 6 of the Health and Safety at Work etc. Act to provide information on the noise likely to be generated, should an article (which can mean a machine or other noise-producing device) produce noise levels reaching any of the three action levels (Regulation 12). In practice, this means supplying data from noise tests conducted on the machine, in addition to other information regarding safe use, installation, etc, already required by section 6.

Occupational health basics

Occupational health anticipates and prevents health problems that are caused by the work which people do. In some circumstances the work may aggravate a pre-existing medical condition, and stopping this is also the role of occupational health. Health hazards often reveal their effects on the body only after the passage of time; many have cumulative effects, and in some cases the way this happens is still not fully understood. Because the effects are often not immediately apparent, it can be difficult to understand and persuade others that there is a need for caution and control. Good occupational hygiene practice encompasses the following ideas:

- Recognition of the hazards or potential hazards
- Quantification of the extent of the hazard, usually by measuring physical/chemical factors and their duration, and relating them to known or required standards
- Assessment of risk in the actual conditions of use, storage, transport and disposal
- Control of exposure to the hazard, through design, engineering, working systems, the use of personal protective equipment and biological monitoring
- Monitoring change in the hazard by means of audits or other measurement techniques, including periodic re-evaluation of work conditions and systems

Health hazards in construction

Health hazards can be divided into four broad categories: physical, chemical, biological and ergonomic. Examples of the categories are:

- Physical — heat, noise, radiant energy, electric shock
- Chemical — exposure to toxic materials such as dusts, fumes and gases
- Biological — infection, such as tetanus, hepatitis and Legionnaire's disease
- Ergonomic — work conditions and stress

Physical hazards are the most obvious, while chemical, biological and ergonomic hazards are often more subtle in their effects, which can also take longer to build up. To understand something of the chemical category it is necessary to have an understanding of the basic terms involved.

Toxicity of substances

Toxicity is the ability of a substance to produce injury once it reaches a site in or on the body. The degree of harmful effect which a substance can have depends not only on its inherent harmful properties but also on the **route** and the **speed** of entry into the body. Substances may cause health hazards from a single exposure, even for a short time (**acute effect**) or after prolonged or repeated exposure (**chronic effect**). The substance may affect the body at the point of contact, when it is known as a **local agent**, or at some other point, when it is described as a **systemic agent**. **Absorption** only occurs when a material has gained access to the bloodstream and may consequently be carried to all parts of the body.

Toxicity factors

The effect a substance will have on the body cannot always be predicted with accuracy, or explained solely on the basis of physical and chemical laws. The influence of the following factors combine to produce the effective dose:

- Quantity or concentration of the substance
- The duration of exposure
- The physical state of the material, e.g. particle size
- Its affinity for human tissue
- Its solubility in human tissue fluids
- The sensitivity to attack of human tissue or organs

Substances which are toxic can have a toxic effect on the body after only one single, short exposure. In other circumstances, repeated exposure to small concentrations may give rise to an effect. A toxic effect related to an immediate response after a single exposure is called an **acute effect**. Effects which result after prolonged (hours or days or much longer) are known as **chronic effects**. 'Chronic' implies repeated doses or exposures at low levels; they generally have delayed effects and are often due to unrecognised conditions which are therefore permitted to persist.

The body fights back

The body's response against the invasion of substances likely to cause damage can be divided into external or **superficial** defences and internal or **cellular** defences. These defence mechanisms inter-relate, in the sense that the defence is conducted on a number of levels at once, and not in a stage-by-stage pattern. The cellular defences are beyond the scope of this book.

The superficial mechanisms work by the action of cell structures, such as organs and functioning systems. The body's largest organ, the skin, provides a useful barrier against many foreign organisms and chemicals (but not against all of them). Its effect is limited by its physical characteristics. Openings in the skin, including sweat pores, hair follicles and cuts, can allow entry, and the skin itself may be permeable to some chemicals such as toluene. The skin can withstand limited physical damage because of its elasticity and toughness, but its adaptation to cope with modern substances is usually viewed by its owner as unhelpful — dermatitis, with thickening and inflammation, is painful and prominent.

Defences against inhalation of substances harmful to the body begin in the respiratory tract, where a series of reflexes activate the coughing and sneezing mechanisms to expel forcibly the triggering substance. Many substances and micro-organisms are successfully trapped by nasal hairs and the mucus lining the passages of the respiratory system. The passages are also well supplied with fine hairs which sweep rhythmically towards the outside and pass along larger particles. These hairs form the ciliary escalator. The respiratory system narrows as it enters the lungs, where the ciliary escalator assumes more and more importance as the effective defence. In the deep lung areas, only small particles are able to enter the alveoli (where gas exchange with the red blood cells takes place), and cellular defence predominates there.

For ingestion of substances entering the mouth and gastrointestinal tract, saliva in the mouth and acid in the stomach provide useful defences to substances which are not excessively acid or alkaline, or present in great quantity. The wall of the gut presents an effective barrier against many insoluble materials. Vomiting and diarrhoea are additional reflex mechanisms which act to remove substances or quantities which the body is not equipped to deal with without damage to itself. Thus, there are a number of primitive defences, useful at an earlier evolutionary stage to prevent man unwittingly damaging himself, which are now available to protect against a newer range of problems as well as the old.

Eyes and ears are potential entry routes for substances and micro-organisms. The eyes prevent entry of harmful material by way of the eyelids, eye lashes, conjunctiva (the thin specialised outer skin coating of the eyeball), and by bacteria-destroying tears. The ears are protected by the outer shell or pinna, and the ear drum is a physical barrier at the entrance to the sensitive mechanical parts and the organ of hearing. Waxy secretions protect the ear drum, and trap larger particles.

It can be seen that there are three main ways that harmful substances can enter the body — absorption through the skin (organic solvents are good examples), ingestion through the mouth and inhalation through the lungs — easily the commonest way.

How much is too much?

Control depends upon measurement, followed by comparison with published standards.

The **maximum exposure limit (MEL)** is the maximum concentration of an airborne substance, averaged over a stated reference period (often 8 hours) to which employees may be exposed by inhalation under any circumstances. Some substances have been assigned short-term MELs with very short reference periods, such as 15 minutes. These substances give rise to acute effects and therefore these limits should never be exceeded.

The **occupational exposure standard (OES)** is the concentration of an airborne substance, averaged over a reference period, at which there is no current evidence that repeated (day after day) exposure by inhalation will be injurious to the health of employees. OESs should not be exceeded, but where this occurs effective steps should be taken as soon as practicable to reduce the exposure. These values are given in units of parts per million (ppm) and milligrams per cubic metre ($mg\,m^{-3}$). They are given for two periods: long-term exposure (8-hour time-weighted average (TWA)) and short-term exposure (15-minute TWA). Some substances are designated 'SK' which indicates that they can be absorbed through the skin.

The current standards can be found in HSE Guidance Note EH40 which is published annually and contains the current statutory list of maximum exposure limits and approved occupational exposure standards.

The exposure of employees to substances hazardous to health is to be prevented or, where this is not reasonably practicable, adequately controlled as a fundamental requirement of the COSHH Regulations 1999. EH40 lists the limits to be used in determining the adequacy of the control of exposure by inhalation. There is separate legislation for lead and asbestos, and these substances are not covered in detail in EH40.

The lists contained in Tables 1 and 2 of EH40 are legally binding; COSHH 1999 imposes requirements by reference to these lists, which are the only official record of substances assigned MELs and OESs. It should be remembered that all occupational exposure limits refer to healthy adults working at normal rates over normal work periods and patterns. In practice, it is advisable to work well below the standards set in EH40, and to bear in mind the desirable goal of progressive risk reduction over time.

Dermatitis

'Dermatitis' means 'inflammation of the skin'. It is probably the most common form of occupational ill-health in the construction industry; some estimates claim that as many of 40% of workers in the industry suffer or have suffered from it. In the course of work the skin can be attacked by a large number of chemicals that can irritate it, and the response is most often 'non-infective' dermatitis. The effect on the skin depends upon the concentration of the chemical and the duration of the exposure, and most (but not all) those exposed will be affected. Solvents, acids and alkalis are examples of dermatitic agents; they act by attacking the greasy surface of the skin and removing it so that the outer skin layers are exposed to further attack. The severity gradually worsens until action is taken; usually the condition can be avoided by good personal hygiene (not washing the hands in petrol or other hydrocarbon solvents, for example) or the use of barrier creams and protective clothing.

A smaller number of chemicals, known as 'sensitisers', invoke a response in the blood which produces dermatitis on continuing exposure. People who are thus sensitised become abnormally sensitive to the chemical and can develop severe reactions to its presence. Often, their only recourse is to exclude themselves from further work with or near the chemical substance. Cement, some wood types and resins are examples of sensitisers found in the industry.

Vibration

Exposure to vibration from work with power tools held by hand, or from other kinds of vibrating equipment, can cause reduced circulation of the blood and damage to the nerves and muscles of the arms. Collectively, the conditions that can result from vibration exposure are known as hand–arm vibration syndrome (HAVS). The best-known form of HAVS is vibration white finger (VWF), which is a prescribed (reportable) industrial disease prevalent amongst those who work with older tools such as chainsaws and jack-hammers. In recent years the makers of vibrating tools have become aware of the risks to users (and the likelihood of claims for compensation) arising from use of unmodified equipment, and have put considerable efforts into researching and developing methods of reducing or removing the vibration.

Symptoms of HAVS include tingling in the fingers or localised numbness, often accompanied by a whitening of the area due to reduced bloodflow beneath the skin. There are now established techniques of assessing risks, and of developing strategies to control risks by altering the nature of the work as well as changing or modifying the offending

equipment. There is also a role for health surveillance of victims and potential sufferers from HAVS. The reader interested in learning more about this complex subject is recommended to study the suggested references.

Personal protective equipment

Personal protective equipment, or PPE, has a serious general limitation — it does not eliminate a hazard at source. If the PPE fails and the failure is not detected, the risk increases greatly. Where used, equipment must be appropriately selected and its use and condition monitored. Workers required to use it must be trained. For a PPE scheme to be effective, three elements must be considered:

- Nature of the hazard — details are required before adequate selection can be made, such as the type of contaminant and its concentration
- Performance data for the PPE — the manufacturer's information will be required concerning the ability of the PPE to protect against a particular hazard
- The acceptable level of exposure to the hazard — for some hazards the only acceptable exposure level is zero. One example is the protection of the eyes against flying particles. Occupational exposure limits can be used to 'gauge acceptable' exposure, bearing in mind their limitations

Factors affecting use

There are three inter-related topics to consider before adequate choice of PPE can be made:

- The workplace — what sorts of hazards remain to be controlled? How big are the risks which remain? What is an acceptable level of exposure or contamination? What equipment or processes are involved? What movement of objects or people will be required?
- The work environment — what are the physical constraints? They can include temperature, humidity, ventilation, size and movement requirements for people and plant
- The PPE wearer — points to consider include:
  - □ training — users (and supervisors) must know why the PPE is necessary, any limitations it has, the correct use, how to achieve a good fit and the necessary maintenance and storage for the equipment
  - □ fit — a good fit for the individual wearer is required

to ensure full protection. Some PPE is available only in a limited range of sizes and designs

- acceptability — how long will the PPE have to be worn by individuals? Giving some choice of the equipment to the wearer without compromising on protection standards will improve the chances of its correct use
- wearing pattern — are there any adverse health or safety consequences which need to be anticipated? For example, any need for frequent removal of PPE which may be dictated by the nature of the work may affect the choice of design or type of PPE
- interference — regard for the practicability of the item of equipment is needed in the work environment. Some eye protection interferes with peripheral vision, other types cannot easily be used with respirators. Correct selection can alleviate the problem, but full consideration must be given to the overall protection needs when selecting individual items, so that combined items of equipment may be employed. For example, the 'Airstream' helmet gives respiratory protection, and has fitted eye protection incorporated into the design
- management commitment — the sine qua non of any safety programme, required especially in relation to PPE because it constitutes the last defence against hazards. Failure to comply with instructions concerning the wearing of it raises issues of industrial relations and corporate policy

Types of PPE

The types of PPE have different functions, including hearing protection, eye protection, respiratory protection, protection of the skin, and general protection in the form of protective clothing, and safety helmets, harnesses and lifelines. Hearing protection was discussed in the previous section.

Eye protection

Assessment of potential hazards to the eyes and the extent of the risks should be made in order to select equipment effectively. There are three types of eye protection commonly available:

Safety spectacles/glasses — which provide protection against low-energy projectiles such as metal swarf, but do not assist against dusts, are easily displaced and have no protective effect against high-energy impacts.

Safety goggles to protect against high-energy projectiles and dusts. They are also available as protection against chemical and metal splashes with additional treatment. Disadvantages include a tendency to mist up inside (despite much design effort by manufacturers), lenses which scratch easily, limited vision for the wearer, lack of protection for the whole face, and high unit cost. Filters will be required for use against non-ionising radiation.

Face shields — offering high-energy projectile protection, also full-face protection and a range of special tints and filters to handle various types of radiation. Field of vision may be restricted. The high initial cost of equipment is a disadvantage, although some visors allow easy and cheap replacement of shields. Weight can be a disadvantage, but this is compensated by relative freedom from misting up.

Respiratory protective equipment (RPE)

There are two broad categories: respirators, which purify the air by drawing it through a filter to remove contaminants, and breathing apparatus, which supplies clean air to the wearer from an uncontaminated external source. Most equipment will not provide total protection; a small amount of contaminant entry into the breathing zone is inevitable.

Four main types of **respirator** are available:

- Filtering half-mask — a facepiece covers nose and mouth which is made of a filtering medium which removes the contaminant. Generally used for up to an 8-hour shift and then discarded
- Half-mask respirator — which has a rubber or plastic facepiece covering the nose and mouth, and which carries one or more replaceable filter cartridges
- Full-face respirator — covering the eyes, nose and mouth, and having replaceable filter canisters
- Powered respirator — supplies clean, filtered air to a range of facepieces (including full, half and quarter masks, hoods and helmets) via a battery-operated motor fan unit

Respirators do not provide **any** protection in oxygen-deficient atmospheres.

There are three main types of **breathing apparatus** which

provide continuous airflow (in all cases the delivered air must be of respirable quality):

- Fresh air hose apparatus — which supplies clean air from an uncontaminated source, pumped in by the breathing action of the wearer, by bellows or an electric fan
- Compressed airline apparatus (CABA) — using flexible hosing delivering air to the wearer from a compressed air source. Filters in the airline are required to remove oil mist and other contaminants. Positive pressure continuous flow full masks, half masks, hoods, helmets and visors are used
- Self-contained breathing apparatus (SCBA) — in which air is delivered to the wearer from a cylinder via a demand valve into a full-face mask. The complete unit is usually worn by the operative, although cylinders can be remote and connected by a hose

To make proper selection of any type of RPE, an indication is needed of its likely efficiency, when used correctly, in relation to the hazard guarded against. The technical term used in respiratory protection standards to define the equipment's capability is the **nominal protection factor**, or **NPF**. For each class of equipment, it is the total inward leakage requirement set for that class, and therefore does not vary between products meeting the class standards. Manufacturers will supply information on the NPF for their product range — it is the simple ratio between the contaminant outside the respirator (in the ambient air) and the acceptable amount inside the facepiece. Examples of typical NPFs are:

Disposable filtering half-mask respirator	4.5 to 50
Positive pressure powered respirator	20 to 2000
Ventilated visor and helmet	10 to 500
Self-contained breathing apparatus (SCBA)	2000
Mouthpiece SCBA	10 000

The NPF can be used to help decide which type of RPE will be required. What is needed is to ensure that the concentration of a contaminant inside the facepiece is as far below the OEL as can reasonably be achieved. The required protection factor to be provided by the RPE will be expressed as:

$$\frac{\text{measured ambient concentration}}{\text{occupational exposure limit}}$$

This value must be less than the NPF for the respirator type under consideration. Other limitations may apply, includ-

ing restrictions on the maximum ambient concentration of the contaminant based on its chemical toxicity, and its physical properties such as the lower explosive limit.

In 1997, the revised BS 4275 presented a new approach to the efficiency issue, aimed at dispensing with the theoretical calculations of NPF. Instead, the term **assigned protection factor (APF)** was introduced to describe the level of protection that can reasonably be achieved in the workplace, given that workers have been appropriately trained and the RPE is properly fitted and working. Used in the same way as NPF, the APF provides a higher margin of safety. A full discussion of the use of the APF is beyond the scope of this book. Guidance on best practice and the setting up of an RPE programme can be found in the Standard, to which the reader is referred.

Protective clothing

This is intended to provide body protection against a range of hazards, including heat and cold, radiation, impact damage and abrasions, water damage and chemical attack. Items relevant to the construction industry include:

Head protection — provided by two types of protectors: the safety helmet and the scalp protector (known also as the 'bump cap') which is usually brimless. Their function is to provide protection against sun and rain, and against impact damage to the head. The ability of the scalp protector to protect against impacts is very limited, and its use is mainly to protect against bruising and bumps in confined spaces. It is not suitable for use as a substitute for a conventional safety helmet. Safety helmets have a useful life of about 3 years, which can be shortened by prolonged exposure to ultraviolet light and by repeated minor or major impact damage. They should not be painted or decorated other than by their manufacturer.

Protective outer garments — normally made of PVC material and often of high-visibility material to alert approaching traffic. PVC clothing can be uncomfortable to wear because of condensation, and vents are present in good designs. Alternatively, non-PVC fabric can be used which allows water vapour to escape, but garments made from this material are significantly more expensive.

Gloves — must be carefully selected, taking account of use requirements such as comfort, degree of dexterity required, temperature protection offered and ability to grip in all conditions likely to be encountered, against considerations of cost and the hazards likely to be encountered by the wearer. For example, they can also become entangled in machinery. The main types of material and their features are shown below.

Glove material	Good for
Leather	Abrasion protection, heat resistance
PVC	Abrasion protection, water and limited chemical resistance
Rubber	Degreasing, paint spraying
Cloth/nylon, latex coated	Hand grip
Latex	Electrical insulation work
Chain mail	Cut protection

Footwear — designed to provide protection for the feet, especially for the toes, if material should drop or fall onto them. It should also protect against penetration from beneath the sole of the foot, be reasonably waterproof, provide a good grip, and be designed with reference to comfort. Steel toecaps are inflexible, and it is important to purchase the right size of footwear.

Skin protection

Where protective clothing is not a practicable solution to a hazard, barrier creams may be used together with a hygiene routine before and after work periods. There are three types of barrier cream commonly found: water miscible, water repellent and special applications.

Failure by construction workers to protect their skins results in a small but increasing number of cases of skin cancer each year, including the potentially fatal malignant melanoma.

Safety harnesses

These are not replacements for effective fall prevention practices. Only where the use of platforms, nets or other access and personal suspension equipment is impracticable is their use permissible. The functions of belts and harnesses are to limit the height of any fall, and to assist in rescues from confined spaces. In addition to comfort and freedom of movement, selection of this equipment must take into account the need to provide protection to the enclosed body against energy transfer in the event of a fall. Because of this, harnesses are preferable to belts except for a very limited number of applications where belts are required because of the movement needs of the work.

Harness attachments to strong fixing points must be able to withstand the snatch load of any fall. A basic principle is to attach the securing lanyard to a fixing point as high as possible over the area of the work, so as to limit the fall distance. Similarly, a short lanyard should be provided. Equipment which has been involved in fall arresting should be thoroughly examined before further use, according to the manufacturer's instructions. In many cases replacement of the item is required.

The CE mark

One of the definitions of 'suitable' PPE is that it must conform to legal requirements. Workwear and other PPE which meets the relevant strict rules for products must be sold carrying the mark 'CE'. This stands for *Communauté Européene*, and the letters show the item complies with necessary design and manufacturing standards based on several levels and classes of protection. The same rules apply throughout the European Union, so that PPE which complies and bears the CE mark can be sold in any member state, as part of the harmonisation of standards. There are three categories or levels of 'conformance' within the CE mark system — basically equipment is designated as being suitable for protection against low risks, medium risks and high risks. In some cases PPE in the higher protection categories is subject to examination by an approved body before the mark can be displayed, and the category details will be displayed alongside the mark.

Workwear in the first **simple** category will protect against, for example, dilute detergents. Gloves capable of handling items at less than 50° celsius will also be in this category, and equipment capable of protecting against minor impacts. The manufacturer of products in this category can assess his own product and award himself the CE mark to be shown on the equipment, but must keep a technical file.

In the second **intermediate** category, the manufacturer must submit a technical file for review, which holds details of the product's design, and claims made for it, and details of the quality system used by the manufacturer. The product itself is liable to be tested to ensure it can do what is claimed for it.

The **high risk**, or 'complex design' third category mostly covers hazards where death is a potential outcome of exposure. Equipment which can protect against limited chemical attack, ionising radiation protection and respiratory equipment protecting against asbestos are examples of products in this group. Manufacturers must comply with all the requirements for the second category, and also submit to an EC examination annually or have a recognised quality system in place, such as ISO 9000.

Legal requirements

The general duties sections of the Health and Safety at Work etc. Act 1974 require a safe place of work to be provided together with safe systems of work, and these may involve use of PPE. General requirements for provision, use, maintenance and storage of PPE are contained in the Personal Protective Equipment at Work Regulations 1992. Specific Acts, Regulations and Orders also contain requirements for PPE, and should be consulted for particular applications.

Radiation

Energy which is transmitted, emitted or absorbed as particles or in wave form is called radiation. Radiation is emitted by a variety of sources and appliances used in the construction industry, and it is also a natural feature of the environment. Transmission of radiation is the way in which radios, radar and microwaves work. The human body absorbs radiation readily from a wide variety of sources, mostly with adverse effects. All types of electromagnetic radiation are similar in that they travel at the speed of light. Visible light is itself a form of radiation, having component wavelengths which fall between the infrared and the ultraviolet portions of the spectrum. Essentially, there are two forms of radiation, ionising and non-ionising, which can be further subdivided.

Ionisation and radiation

All matter is made up of elements, which consist of similar atoms. These, the basic building blocks of nature, are made up of a nucleus containing protons and orbiting electrons:

■ Protons have a mass and a positive charge
■ Electrons have a negligible mass and a negative charge
■ Neutrons have a mass but no charge

If the number of electrons in an atom at a point in time is not equal to the number of protons, the atom has a net positive or negative charge, and becomes ionised. Ionising radiation is that which can produce ions by interacting with matter, including human cells, which leads to functional changes in body tissues. The energy of the radiation dislodges electrons from the cell's atoms, producing ion pairs (see below), chemical free radicals and oxidation products. As body tissues are different in composition and form as well as in function, their response to ionisation is different. Some cells can repair radiation damage, others cannot. The cell's sensitivity to radiation is directly proportional to its reproductive capability.

Ionising radiation

Ionising radiations found in industry are alpha, beta and gamma, and X-rays. Alpha and beta particles are emitted from radioactive material at high speed and energy. Radioactive material is unstable, and changes its atomic arrangement so as to emit a steady but slowly diminishing stream of energy. Alpha particles are helium nuclei with two positive charges (protons), and thus are comparatively large and attractive to electrons. They have short ranges in dense materials, and can only just penetrate the skin. However, ingestion or inhalation of a source of alpha particles can place it close to vulnerable tissue, so essential organs can be destroyed. Beta particles are fast-moving electrons, smaller in mass than alpha particles but have longer range, so they can damage the body from outside it. They have greater penetrating power, but are less ionising.

Gamma rays have great penetrating power and are the result of excess energy leaving a disintegrating nucleus. Gamma radiation passing through a normal atom will sometimes force the loss of an electron, leaving the atom positively charged — an ion. This and the expelled electron are called an ion pair. Gamma rays are very similar in their effects to X-rays, which are produced by sudden acceleration or deceleration of a charged particle, usually when high-speed electrons strike a suitable target under controlled conditions. The electrical potential required to accelerate electrons to speeds where X-ray production will occur is a minimum of 15 000 volts. Equipment operating at voltages below this will not, therefore, be a source of X-rays. Conversely, there is a possibility of this form of radiation hazard being present at higher voltages. X-rays and gamma rays have high energy, and high penetration power through fairly dense material. In low density substances, including air, they may have long ranges.

Common sources of ionising radiation in the construction industry are X-ray machines and isotopes used for non-destructive testing (NDT). They can also be found in laboratory work and in communications equipment.

Non-ionising radiation

Generally, non-ionising radiations do not cause the ionisation of matter. Radiation of this type includes that in the electromagnetic spectrum between ultraviolet and radio waves, and also artificially-produced laser beams.

Ultraviolet (UV) radiation comes from the sun, and is also generated by equipment such as welding torches. Much of the natural ultraviolet in the atmosphere is filtered out by the ozone layer. Sufficient penetrates to cause sunburn and even blindness. Its effect is thermal and photochemical, producing burns and skin thickening, and eventually skin cancers. Electric arcs and ultraviolet lamps can produce a photochemical effect by absorption on the conjunctiva of the eyes, resulting in 'arc-eye' and cataract formation. In construction work, education is required to explain to workers the risks they run from continued exposure of the skin to UV radiation, which can lead to irreversible skin effects and the production of skin cancers, including the potentially lethal malignant melanoma.

Infrared radiation is easily converted into heat, and exposure results in a thermal effect such as skin burning, and loss of body fluids. The eyes can be damaged in the cornea and lens which may become opaque (cataract). Retinal damage may also occur if the radiation is focused, as in laser radiation. This is a concentrated beam of radiation, having principally thermal damage effects on the body.

Radio-frequency radiation is emitted by microwave transmitters including ovens and radar installations. The body tries to cool exposed parts by blood circulation. Organs where this is not effective are at risk, as for infrared radiation. These include the eyes and reproductive organs. Where the heat of the absorbed microwave energy cannot be dispersed, the temperature will rise unless controlled by blood flow and sweating to produce heat loss by evaporation, convection and radiation. Induction heating of metals can cause burns when touched.

Controls for ionising radiation

The intensity of radiation depends upon the strength of the source, the distance from it and the presence and type of shielding. Intensity will also depend on the type of radiation emitted by the source. Radiation intensity is subject to the inverse square law — it is inversely proportional to the square of the distance from the source to the target. The dose received will also depend on the duration of the exposure. These factors must be taken into account when devising the controls.

Elimination of exposure is the priority, to be achieved by restricting use and access, use of shielded enclosures and written procedures to cover:

- Use, operation, handling, transport, storage and disposal of known sources
- Identification of potential radiation sources
- Training of operators
- Identification of operating areas
- Monitoring of radiation levels around shielding
- Monitoring of personal exposure of individuals, by dosimeters
- Medical examinations for workers at prescribed intervals
- Hygiene practice in working areas
- Wearing of disposable protective clothing during work periods
- Clean-up practice
- Limiting of work periods when possible exposure could occur

Controls for non-ionising radiation

Protection against ultraviolet radiation is relatively simple; sunbathers have long known that anything opaque will absorb ultraviolet light. That emitted from construction processes can be isolated by shielding and partitions, although plastic materials differ in their absorption abilities. Users of emitting equipment, such as welders, can protect themselves by the use of goggles and protective clothing — the latter to avoid 'sunburn'. Assistants often fail to appreciate the extent of their own exposure, and require similar protection.

Exposure of construction workers to ultraviolet radiation is a recognised health hazard. Education is required to improve awareness of the risk of skin damage; the number of cases of malignant melanoma is rising among Northern European workers whose lighter skins cannot provide adequate protection unaided.

Visible light can, of course, be detected by the eye which

has two protective control mechanisms of its own, the eyelids and the iris. These are normally sufficient, as the eyelid has a reaction time of 150 milliseconds. There are numerous sources of high-intensity light which could produce damage or damaging distraction, and sustained glare may also cause eye fatigue and headaches. Basic precautions include confinement of high-intensity sources, matt finishes to nearby paintwork, and provision of optically-correct protective glasses for workers in snow, sand or near large bodies of water.

Problems from infrared radiation derive from thermal effects and include skin burning, sweating and loss of body salts leading to cramps, exhaustion and heat stroke. Clothing and gloves will protect the skin, but the hazard should be recognised so that effects can be minimised without recourse to personal protective equipment.

Controls for laser operations depend on prevention of the beam from striking persons directly or by reflection. Effects will depend on the power output of the laser, but even the smallest is a potential hazard if the beam is permitted to strike the body and especially the eye. Low output lasers are normally used in the construction industry as levelling devices. Workers with lasers should know the potential for harm of the equipment they work with, and should be trained to use it and authorised to do so. If the beam cannot be totally enclosed in a firing tube, eye protection should be worn which is suitable for the class of laser being operated (classification is based on wavelength and intensity). A specific risk assessment should be made before the use of laser equipment is authorised.

Equipment which produces microwave radiation can usually be shielded to protect the users. If size and function prohibits this, restrictions on entry and working near an energised microwave device will be needed. Metals, tools, flammable and explosive materials should not be left in the electromagnetic field generated by microwave equipment. Appropriate warning devices should be part of the controls for each such appliance. Commercially-available kitchen equipment is subject to power restrictions and controls over the standard of seals to doors, but regular inspection and maintenance by manufacturers is required to ensure that it does not deteriorate with use and over time.

Control of exposure to radiation

The following general principles must be observed:

1. Radiation should only be introduced to the work place if there is a positive benefit.
2. Safety information must be obtained from suppliers about the type(s) of radiation emitted or likely to be emitted by their equipment.
3. There is a requirement under the Management of Health and Safety at Work Regulations 1999 for written assessments of risks to be made, noting the control measures in force. All those affected, including employees of other employers, the general public and the self-employed must be considered, and risks to them evaluated. They are then to be given necessary information about the risks and the controls.
4. All sources of radiation must be clearly identified and marked. Where CDM applies, the future presence of sources of radiation must be identified in the project safety plan.
5. Protective equipment must be supplied and worn so as to protect routes of entry of radiation into the body. Protective equipment so provided must be suitable and appropriate, as required by relevant Regulations. It must be checked and maintained regularly.
6. Safety procedures must be reviewed regularly.
7. A radiation protection adviser should be appointed, with specific responsibilities to monitor and advise on use, precautions, controls and exposure.
9. Emergency plans must cover a potential radiation emergency, as well as providing a control strategy for other emergencies which may threaten existing controls for radiation protection.
10. Written authorisation by permit should be used to account for all purchase, use, storage, transport and disposal of radioactive substances.
11. Workers should be classified by training and exposure period. Those potentially exposed to ionising radiation should be classed as 'persons especially at risk' under the Management of Health and Safety at Work Regulations 1999.

Legal requirements

'General duties' sections of the Health and Safety at Work etc. Act 1974 apply to exposure to all forms of radiation, with its potential for harm. Specifically, the Ionising Radiations Regulations 1985 are the main control measure, together with duties concerning risk assessments and information provision contained in the Management of Health and Safety at Work Regulations 1999.

Roofing work

Roofing work is a high-risk activity. The most common causes of injury are:

- **Falls while gaining access**, such as a ladder slipping while weight is being transferred from the step-off rung to the roof. Resting an unsecured ladder against a gutter is not a safe practice; the ladder should be secured by tying to an anchor bolt while a person foots it, or one of several stabilising devices can be fitted to the ladder top to allow it to rest against a vertical surface without slipping.

- **Falls from exposed edges** — leading edges are notoriously difficult to protect as their extent and location are often changing rapidly. Advance planning may allow the use of proprietary guarded staging based on runners on purlins. Otherwise, there is often little alternative to the use of harnesses and running lines in order to allow movement over a reasonable area. This work should be risk assessed beforehand, and a method statement produced detailing the safe system of work to be followed.

- **Falls through fragile roofing material**, rooflights and other openings. Roof sheets made of asbestos, plastic or even steel can become fragile with age even if they were not originally. Their supports may have deteriorated. The use of crawling boards and roof ladders is usually required, on both flat and pitched roofs. Openings where rooflights are to be fitted are traps for the unwary, and should be fitted with temporary covers. Roof valleys may appear to offer safe access, but it is possible to fall through adjacent fragile material on losing balance.

- **Falls while walking** on covered or open purlins and on ridges. It is never safe to walk on purlins or lines of bolts.

- **Being struck** by materials falling, being blown from roofs or being thrown down. Debris chutes which are enclosed to retain materials and fed into covered skips are the best solution, unless the material can be safely lowered in covered skips or other containers. Carrying flat materials up ladders onto high roofs, or breaking boarding and sheeting packs at roof level, can offer unsuspected opportunities for hang gliding.

Only work of very short duration (certainly less than a day) should be carried out on pitched roofs accessed only from ladders. Pitched and other roof edges will normally require protection in the form of guardrails and toeboards fitting.

Steel erection and decking

Fall prevention is far more effective than fall protection — the former removes the hazard by eliminating or reducing the need to work at height. Decisions about the work methods and the erection sequence to be used have to be taken during the planning phase, as liaison will normally be required with designers and manufacturers.

All possible bolting up and connecting should be done at low, preferably ground, level. Work that has to be done at height is best carried out by using other means such as mobile towers or MEWPs for access, provided that the ground conditions are suitable.

The hazards to be considered and controlled during the steel erection process include falls of people and material, the potential for structural collapse through temporary instability, the manual handling and slinging of the steelwork, and the means of lifting it into place.

Steel erectors should work from independent platforms, but if it is necessary to work directly on the beams they should be straddled or crabbed. In either case, a full safety harness tied off to the steelwork or a running line is considered essential.

Similar considerations apply to laying decking. The potential for falls is high during all stages of the laying process: falls can occur from leading and fixed edges, and from decked areas when carrying sheets, especially in windy conditions. The effective solution is to plan erection sequences in detail well in advance, so that packs of sheets can be split at ground level or a railed area of previously-laid decking, and running lines for harness attachment are installed in advance. Where practicable, decking sheets should be laid from MEWPs or tower scaffolds, but where this cannot be done then guardrails at openings and fixed edges are mandatory, with harness protection for those working at leading edges.

A particular area of risk is the laying of the first sheets, which should be covered specifically in method statements. Standing on stacks of decking sheets is not considered to be safe; generally another safe place of work must be provided.

Transport on site

Accidents involving site transport frequently result in fatalities. If site traffic is not managed, using planned layouts, banksmen and instructions to vehicle drivers and pedestrians alike then everyone will be left to make their own choices about critical factors which will involve the safety of others. A traffic management plan should be included in every project safety plan.

Starting with the main access and egress points to the site, there needs to be good visibility and room for manoeuvering wherever possible. The most important point is to keep pedestrians and vehicles physically separated. This can be done by identifying and marking separate entry and exit points, with rails or barriers separating roadways from walkways. The point should be emphasised during induction training; it is easy for discipline to slacken if the rules are not enforced.

On site, a one-way system is always preferable to a 'free-for-all', with turning areas marked so that the need to reverse is minimised. Banksmen should always be used to assist drivers in reversing, and they need to be trained in the traffic management plan and signalling. High-visibility clothing should be worn by everyone using pedestrian footways that may cross vehicle routes.

On larger projects a speed limit should be set, and prominent signs displayed. Traffic routes may need extra lighting. Their physical condition should be improved wherever possible, with no potholes, extreme or adverse camber angles, overhead cables or tight bends. Any remaining hazards to traffic should be signed and drivers briefed about them. Where vehicles need to pass close to scaffolding, excavations or overhead cables, warning signs should be posted.

Maintenance of vehicles on site should only be done where appropriate facilities are present. Jacking vehicles using handy supports rather than jacks or axle stands can result in instability, as can reaching under unpropped vehicle bodies. Refuelling stations should be bunded to contain fuel spills. Where maintenance is done in parking areas, these should be level and clear of obstruction. Parked vehicles should have ignition keys removed, brakes applied and wheels may need to be chocked.

Many accidents have happened because people have fallen from moving vehicles when hitching a lift. There should be a firm policy that no riding takes place on vehicles unless there is a fixed seat provided for the passenger(s). Riding on top of loads that may shift on bends should be forbidden. Drivers should be required to check their loads for security before moving off, and their vehicles for mud that may drop off once the vehicle has passed through the wheel-washing facility at or near to the exit point.

Site-based vehicles and plant

Rollover protection for site-based vehicles should be mandatory, together with seat belts for drivers and passengers wherever possible. Speed limits on site must apply to site-based plant and equipment as well as to visitors.

All mobile driven equipment should be in the charge of a certificated driver, and evidence should be produced to establish the currency of certificates before driving is allowed. A licence to drive a car or other road vehicle should not be accepted as evidence of competence to drive a site dumper — the techniques involved are very different. During induction training, drivers should be reminded that they must not allow anyone else to drive their equipment, even for a short distance. Many injuries have been caused by people trying to 'help out' in this way. Special conditions, including insurance and licensing, may apply to plant such as fork lifts and dumpers, that have to be driven on public roads. Roads and other areas that may seem to be private may in fact be classed legally as public highways and construction workers driving on them will be bound by the same requirements as other people.

References

Access equipment

HSG150	Health and safety in construction
CIS 10	Tower scaffolds — Construction Information Sheet (free)
CIS 49	General access scaffolds and ladders — Construction Information Sheet (free)
PASMA	(Prefabricated Aluminium Scaffolding Manufacturer's Association — Operator's Code of Practice
IPAF	(International Powered Access Federation) — Operator's Safety Guide
IRATA	(Industrial Rope Access Trade Association) — Guidelines on the use of rope access methods for industrial use
SAEMA	(Suspended Access Equipment Manufacturers Association) — User's guide to temporary suspended access

Asbestos

Control of Asbestos at Work Regulations 1987, SI 1987 No 2115, HMSO as amended by CAW (Amendment) Regulations 1992 and 1998

Asbestos (Licensing) Regulations 1983, SI 1983 No 1649, HMSO

Special Waste Regulations 1996, SI 1996 No 972, HMSO

Control of Asbestos at Work: Approved Code of Practice, L27 (3rd edn)

L28	Work with asbestos insulatioan, asbestos coating and asbestos insulating board: Approved Code of Practice (3rd edn)
L11	Guide to the Asbestos (Licensing) Regulations 1983 as amended
HSG189/1	Controlled asbestos stripping techniques for work requiring a licence
HSG189/2	Working with asbestos cement
INDG264	Selecting respiratory equipment for work with asbestos
EH10	Asbestos: exposure limits and measurement of airborne dust concentrations

Demolition

HSR 25	Memorandum of Guidance on the Electricity at Work Regulations
HSG 47	Avoiding danger from underground services
HSG 85	Electricity at work: safe working practices
HSG 107	Maintaining portable and transportable electrical equipment
HSG 141	Electrical safety on construction sites

Excavations

HSG 185	Health and safety in excavation

Falls

HSG 32	Safety in falsework for in situ beams and slabs
SG4:00	Use of fall arrest equipment whilst erecting, altering and dismantling scaffolding, National Access & Scaffolding Confederation

Maintenance

L22	Safe use of work equipment
L113	Safe use of lifting equipment
INDG 229	Using work equipment safely (leaflet)
INDG 271	Buying new equipment (leaflet)

Manual handling

L23	Guidance on Regulations, Manual Handling Operations Regulations 1992
HSG149	Backs for the future
CIS37(rev)	Handling heavy building blocks (free)
HSG115	Manual handling: solutions you can handle

Mechanical handling

HSG 6	Safety in work with lift trucks
HSG 150	Health and safety in construction

Noise

INDG 127	Noise in construction
L108	Guidance on the Noise at Work Regulations 1989
HSG 138	Sound solutions

Occupational health basics

HS(G)88	Hand—arm vibration
INDG 175L	Hand—arm vibration: advice for employers
INDG 126L	Hand—arm vibration: advice on vibration white finger for employees and the self-employed
HSG 170	Vibration solutions
HSG 61	Surveillance of people exposed to health risks at work

Personal protective equipment

L25	Personal protective equipment at work
INDG 288	Selection of suitable respiratory protective equipment for work with asbestos
HSG 53	The selection, use and maintenance of respiratory protective equipment
BS 4275:1997	Recommendations for selection, use and maintenance of RPE

Roofing work

HSG 33	Health and safety in roofwork
HSG 150	Health and safety in construction
INDG 284	Working on roofs

Steel erection and decking

GS28/2	Safe erection of structures — site management and procedures
GS28/3	Safe erection of structures — working places and access
INDG 284	Working on roofs
HSG 33	Health and safety in roofwork

Transport on site

Rollover Protection Structures for Construction Plant (EEC Requirements) Regulations 1988

HSG 144	Safe use of vehicles on construction sites

16 Quick Reference Guide

1. Access scaffolding
2. Asbestos-containing materials
3. Bitumen boilers
4. Cartridge tools
5. Chainsaws
6. Clearing sites
7. Cofferdams
8. Confined spaces
9. Demolition
10. Disc cutters
11. Disposal of waste materials
12. Driving vehicles
13. Dumpers
14. Electrical work to 415 volts
15. Erection of structures
16. Excavations
17. Falsework
18. Fire on site
19. Fork lift trucks
20. Gas welding and cutting
21. Hand tools
22. Joinery workshops
23. Ladders and stepladders
24. Lasers
25. Lifting equipment, general

26. Materials hoists
27. Mobile cranes
28. Mobile elevating work platforms
29. Mobile towers
30. Portable electrical equipment
31. Powered tools — woodworking
32. Public protection
33. Pressure testing
34. Roadworks
35. Sewage connections
36. Site transport
37. Skips
38. Steam & water cleaners
39. Storage of materials on site
40. Storage and use of LPG
41. Storage and use of HFL
42. Temporary electrical supplies
43. Working alone
44. Work on fragile roofs
45. Work with flat glass
46. Working at heights, general
47. Working in occupied premises
48. Work near underground services
49. Work under power lines
50. Work over water

1 ACCESS SCAFFOLDING

Significant hazards

- Falls of persons
- Falling materials
- Collapse of structure

Regulatory requirements include

- Construction (Health, Safety and Welfare) Regulations 1996
- Provision and Use of Work Equipment Regulations 1998 and ACoP (L22)
- British Standard BS 5973:1993 — Code of Practice for access and working scaffolds

Pre-planning issues

- Ensure that only authorised personnel will erect, modify or dismantle scaffolding. For structures over 5 m in height CITB certification of erectors should be specified and checked.
- Design drawings should be produced for load-bearing scaffolds and non-standard structures, including those affected by abnormal wind loading.
- Erection practice should be specified to follow BS 5973:1993 and the 1996 Construction Regulations, and where erection has been carried out by a specialist contractor a handover certificate should be asked for.
- Refer lighting arrangements to the local authority for approval, for structures on the public highway or right of way.
- Consider the use of debris guards, netting and fans for high-rise scaffolds and those close to public areas.
- Consider harness requirements for scaffolders.

Basic physical control measures

- Guardrails and toeboards removed for access must be replaced after access has been gained. Unused ladder gaps must be closed off. Ties removed for any purpose must be replaced, or alternative ties fitted at once.
- Restrict traffic movements around and access to scaffold bases.
- Monitor excavations near to scaffold bases to ensure the stability of the structure is not affected.
- Harnesses will be required for scaffold erectors and dismantlers who could fall more than 2 m.

Management tasks

- All scaffolding should be inspected on handover to or from other contractors. After alteration or adverse weather conditions, scaffolding should be inspected by site management competent to do so.
- Arrangements must be made to inspect scaffolding with working platforms over 2 m in height every 7 days, and before use, and for the results of the inspections to be recorded. All such inspections must be made by a competent person. In addition, scaffolds should be checked regularly to ensure their correct use and that unauthorised adaptations and alterations have not been made.

Training needs

- Persons erecting scaffolding must be adequately trained. This will normally be verified by production of a Scaffolder Certificate issued by the CITB. Training for site management may be required to enable them to inspect scaffolds, as above.

Circumstances requiring special physical controls

- Use of slung or suspended scaffolding
- When scaffolds are to carry loads in excess of normal requirements
- Erection of scaffolds next to overhead power lines, railway property or over water
- (See also information on mobile towers, mobile elevated working platforms and ladders)

Selected reference

HSE Guidance Booklet: HS(G)150 — Health and safety in construction

2 ASBESTOS-CONTAINING MATERIALS

Significant hazards

- ⁃ Inhalation of asbestos fibres
- ⁃ Inhalation of other dusts

Regulatory requirements include

- Control of Asbestos at Work Regulations 1987 and Amendment Regulations 1998
- Asbestos (Licensing) Regulations 1983 and Amendment Regulations 1998
- HSE Approved Code of Practice L27 — Control of Asbestos at Work
- HSE Approved Code of Practice L28 — Work with asbestos insulation, asbestos coating and asbestos insulating board

Pre-planning issues

Make written enquiries of prospective clients concerning the presence of asbestos if there is any likelihood that asbestos or material containing asbestos may be encountered. Failing to make enquiries may result in exposing people to asbestos — and possible prosecution as a result. It is essential that before any work takes place on any material suspected to contain asbestos, a sample is taken for analysis to establish the asbestos content and type. Samples are to be taken by a trained and competent person. After analysis the HSE ACoPs are to be used to compare the expected fibre release with the exposure standards. Only work of a minor nature as detailed in the Approved Code of Practice can be done without a licence issued by the Health and Safety Executive. The HSE area office must be informed 14 days before commencement of major work. All work involving asbestos should be covered by a written method statement prepared before the work starts.

Basic physical control measures

Access to work areas will be strictly controlled as appropriate. Asbestos products removed will be double-bagged or placed in sealed containers for disposal at a licensed tip. During removal, asbestos will be dampened or sealed to prevent fibre emission. PPE will be worn as required by specific assessment of the work, and the minimum standard will be impervious overalls with hood in addition to respiratory protection.

Management tasks

Except for work of the simplest kind, work will need to be carried out by specialist contractors. Management must ensure a method statement is available before work begins, that it is followed, and that PPE is used. Management should be aware that even sampling suspect insulation can place workers at risk, and provide appropriate protective equipment.

Training needs

Contractors should be required to produce a copy of their licence and proof of the competence training of operatives. All those working with asbestos must be trained in the hazards associated with the removal techniques, use of PPE and hygiene requirements.

Circumstances requiring special physical controls

Modern experience is that all forms of asbestos are equally hazardous, although for different reasons. Previously, attention was focused on blue asbestos (crocidolite), followed by amosite, or brown asbestos. The latter is often a component of tiles. Chrysotile, white asbestos, not only produces asbestosis following even quite modest exposures, but also cancer of the lung. Analysis determines the fibre type or types, which indicates the requirements for protection although this work will be restricted to HSE licence holders.

Selected references

HSE Guidance booklets: HSG 189/1 — Controlled asbestos stripping techniques
HSG 189/2 — Work with asbestos cement
HSE free leaflets: INDG223 — Managing asbestos in workplace buildings.
INDG288 — Selecting respiratory equipment for use with asbestos

3 BITUMEN BOILERS

Significant hazards

- Fire
- Molten bitumen — burn injuries
- Manual handling

Regulatory requirements include

- Construction (Health, Safety and Welfare) Regulations 1996
- Provision and Use of Work Equipment Regulations 1998 and ACoP (L22)
- Manual Handling Operations Regulations 1992
- Highly Flammable Liquids and Liquefied Petroleum Gases Regulations 1972
- Personal Protective Equipment Regulations 1992

Pre-planning issues

- Check the proposed operating area of the boiler in advance for suitability. Factors include access, surface load-bearing capacity, flammability in the area, ease of loading raw bitumen to the area, ease of removal and carrying of hot material.
- Plan the location so as to avoid or reduce the risk of fire.

Basic physical control measures

- Boilers must only be sited on firm, level standing, and must never be left unattended or moved when lit. A minimum of 3 m space must be left between boilers and any LPG cylinder in the area, including those supplying the boiler.
- Boilers must not be overfilled. Containers used to carry hot bitumen should be fitted with lids to minimise splashing.
- Gas cylinder hose connections must be secured using metal clips at both ends.
- At least one dry powder fire extinguisher must be readily available for use by bitumen boilers.
- A safe access route is required between the boiler and the point of use of the bitumen.
- Operatives handling bitumen must wear appropriate PPE, which will always include industrial leather gloves.

Management tasks

- A PPE assessment must be made for those working with bitumen boilers by their employer; consideration will be given to the need for eye protection in addition to gloves, also knee pads and other protective clothing or devices for those applying hot bitumen.
- Regular inspections should be made by supervisory staff to check temperature and boiler levels.
- A fire check must be made at the end of each work period, both in the area of the boiler and where the bitumen is laid.

Training needs

- Those working with bitumen and boilers must be trained in the action to take in the event of fire, and in first-aid treatment and procedure for dealing with bitumen burns.

Circumstances requiring special physical controls

- Use of bitumen boilers other than at ground level
- Use of bitumen boilers in an enclosed area

Selected reference

HSE Guidance Booklet: HS(G)168 — Fire safety in construction work

4 CARTRIDGE TOOLS

Significant hazards

- Flying splinters
- Flying fixing pins and nails
- Misfired cartridges
- Misuse of unfired cartridges
- Noise

Regulatory requirements include

- Construction (Health, Safety and Welfare) Regulations 1996
- Provision and Use of Work Equipment Regulations 1998 and ACoP (L22)
- Personal Protective Equipment Regulations 1992
- British Standard BS4078:Part 1:1987 — Powder actuated fixing systems — Code of Practice for safe use
- Noise at Work Regulations 1989

Pre-planning issues

- Strict control of the use of tools and ammunition is needed to prevent injury to users and others in the vicinity. Secure storage is needed for cartridges, and the general principle to follow is to treat the tools and the cartridges as firearms. Users should be asked to justify the need for high-power cartridges.
- Only tools with captive plungers should be permitted — these minimise the risk of a free-flying nail and/or plunger, but do not remove the risk completely.

Basic physical control measures

- Eye and hearing protection are mandatory for those who use these tools. The use of these tools should be limited, together with the presence of other people in the working area; this minimises the risk from ricochets and projectiles passing through materials.
- Select and issue cartridges to suit the materials being fixed.
- Misfire drills must be followed strictly, with all misfired cartridges being retained for examination by the supplier or manufacturer. It is important that misfired cartridges are removed according to the specific instructions of the manufacturer — there may be a special tool for doing so.

Management tasks

It is important to ensure that anyone proposing to use cartridge tools is trained and competent to do so. The management role is to ensure that the tools are not used inappropriately — in explosive or flammable atmospheres, for example. Supervisory staff will need to make specific arrangements for the issue, storage and use of the tools and their cartridges, including the safe locked storage of cartridges and the return of misfires, and the selection of cartridges suitable for the task. These arrangements should be detailed in a method statement.

Training needs

Those carrying out training must be competent to do so. Training must include hazards, precautions and storage arrangements, also the manufacturer's misfire drill for the particular make of tool to be used. Operators should possess a current certificate of training issued by the manufacturer or by a recognised body.

Circumstances requiring special physical controls

- When others need to work close to the firing area
- When the material to be worked on requires the use of full-strength cartridges
- When the material to be worked on is unusually thin or fragile

Selected reference

Manufacturer's literature, specific to the tool to be used.

5 CHAINSAWS

Significant hazards

- Contact with moving parts of equipment
- Contact with ejected cut material
- Vibration
- Noise

Regulatory requirements include

- Provision and Use of Work Equipment Regulations 1998 and ACoP (L22)
- Personal Protective Equipment Regulations 1992
- British Standard BS EN 608:1995 — Portable chainsaws, safety
- Noise at Work Regulations 1989

Pre-planning issues

Restrict the use of chainsaws to ground clearance activities. Chainsaws should be subject to a planned maintenance programme. Where chainsaws are to be used, appropriate PPE should be on site before the use is allowed to begin. Arrange for an appropriate first-aid kit and a qualified first-aider to be present before use is authorised.

Basic physical control measures

- Eye and hearing protection are mandatory for those who use chainsaws. Other requirements include safety helmet, face visor, ear defenders, safety boots, gloves with padded backs, and tight-fitting loose-woven nylon leggings and jacket. The latter are usually made from ballistic nylon.
- A clear work area is required to ensure that other people are not at risk from the activity.
- Older chainsaws are unlikely to provide sufficient protection against vibration because of their design. Vibration white finger is a common occupational complaint, and specific enquiries should be made concerning this when selecting a suitable saw for use.

Management tasks

- It is important to ensure that anyone proposing to use a chainsaw is trained and competent to do so. The management role is to ensure that the tools are not used inappropriately — in the cutting of joists, shuttering or other building timber, for example. Chainsaw operators should not be allowed to work alone.
- Subcontractors using chainsaws should be required to produce proof of training and competence.
- Site management should verify that the necessary PPE and first-aid facilities are present before authorising the work to begin.

Training needs

Selection of operatives may be required, to ensure that only those with experience of the work and who are physically fit are allowed to use chainsaws. Training should include the hazards and control measures, techniques of safe use, appropriate storage arrangements, and the manufacturer's maintenance and daily inspection requirements. Operators should possess a current certificate of training issued by the manufacturer or by a recognised body.

Circumstances requiring special physical controls

- Use of chainsaws other than at ground level
- Use of chainsaws for any other purpose than ground clearance
- Use of chainsaws for tree felling

Selected references

- Manufacturer's literature, specific to the saw to be used
- HSE Guidance Note GS48 — Training and standards of competence for people working with chainsaws
- HSE Agricultural Safety leaflet AS20 — Safety with chainsaws

6   CLEARING SITES

Significant hazards

- Chemical pollution of land, water and wildlife
- Exposure to hazardous substances
- Contact with moving plant and machinery
- Noise

Regulatory requirements include

- Provision and Use of Work Equipment Regulations 1998 and ACoP (L22)
- Personal Protective Equipment Regulations 1992
- Control of Substances Hazardous to Health Regulations 1999
- Control of Pesticides Regulations 1986 and ACoP L9 (Rev)
- Noise at Work Regulations 1989

Pre-planning issues

Make a prior assessment of the need to use chemicals and burning, and their probable effects on humans, animals and the environment. COSHH assessments for any hazardous substances to be used or produced must be made and copies held on site before their use or production. Make arrangements for the safe storage of chemicals such as pesticides to be used. Environmental impact studies may be necessary before clearance begins.

Basic physical control measures

PPE must be provided and worn as required following COSHH assessments. Spraying of chemicals must take account of environmental conditions, including wind speed, and should not be done while other workers or the general public are in the area. Application rates of chemicals must not be exceeded. After any pesticide application, thorough cleaning of equipment is required, and safe disposal of unused solutions. Consider the effects of noise and other pollution, including possible chemical overspray, before the work begins.

Management tasks

Frequency of chemical application and the mixing of solutions must be strictly controlled. Only authorised and certificated persons should be allowed to use plant and equipment in connection with site clearance. A site survey should have revealed any additional hazards which require control, including the presence of voids, tanks, disused equipment and buried objects.

Training needs

Spray operators must be trained and certificated to the appropriate NTPTC (National Training Proficiency Test Council) level. Relevant certification is required for plant and equipment operators.

Circumstances requiring special physical controls

- Introduction of new or different chemicals
- Non-routine work, including outbreaks of pests or vermin
- Work on known contaminated land
- (See also the notes on demolition and disposal of waste)

Selected references

HSE Agricultural Safety leaflets AS27 and 28 on pesticide use

7 COFFERDAMS

Significant hazards

- Unintentional collapse of cofferdam
- People and objects falling into the cofferdam
- Flooding of the cofferdam
- Presence of hazardous atmospheres
- Electric shock (water contact)
- Manual handling activities associated with piling work and equipment movement

Regulatory requirements include

- Provision and Use of Work Equipment Regulations 1998 and ACoP (L22)
- Personal Protective Equipment Regulations 1992
- Manual Handling Operations Regulations 1992
- Construction (Health, Safety and Welfare) Regulations 1996
- British Standards BS 8004:1986 — Foundations Code of Practice
             BS 5930:1981 — Code of Practice for Site Investigations

Pre-planning issues

Carry out a full site investigation before planning the work. Competent persons must be appointed to design temporary support works, and support equipment will be needed as determined by the design and investigation, with de-watering equipment appropriate to the needs.

Basic physical control measures

- Guardrails will be required where sheeting does not extend 1 m above ground level. Ladder or other means of access to the cofferdam will be needed. Plant must not be operated close to the edges of the cofferdam, and stop boards may be needed to restrict vehicle and plant movement. Where poor ventilation is identified as a possibility, the atmosphere must be continually monitored by a competent person.
- Safety helmets and other PPE must be provided as required by risk assessment.
- If water is present, the electrical supply should be limited to a maximum of 65 volts.

Management tasks

COSHH assessments will be required of substances likely to be found or produced during the work. Other risk assessments to be produced include PPE and manual handling, and all of these must be completed before work begins. Personnel selected for work in cofferdams must be physically fit, and experienced unless under direct supervision. All cofferdam work should be done under the supervision of a competent person. Support equipment must be inspected before use, and the cofferdam itself at the start of every shift. Results of inspections must be recorded. Air monitoring and rescue equipment must be provided if ventilation is poor and/or the presence of toxic gases or oxygen lack is suspected. Cofferdam work may be classified as work in a confined space, to which more detailed requirements apply.

Training needs

Supervisors must have received training in COSHH appreciation, principles of confined space entry, general site safety, and theory and practice of excavation work. Operatives must have received similar training in cofferdam work and the use of support equipment, and also in the use of air monitoring devices and the PPE provided.

Circumstances requiring special physical controls

- If the presence of contaminated soil or toxic gases is confirmed
- Deep excavation precautions appropriate to confined space entry may apply

Selected reference

HSE Guidance Booklet: HS(G)150 — Health and safety in construction

8 CONFINED SPACES

Significant hazards

- Contact with toxic or asphyxiant gases
- Lack of oxygen
- Explosion
- Fire
- Excessive heat
- Drowning

Regulatory requirements include

- Personal Protective Equipment Regulations 1992
- Construction (Health, Safety and Welfare) Regulations 1996
- Confined Spaces Regulations 1997 and ACoP L101 — Safe working in confined spaces

Pre-planning issues

- Eliminate the need for confined space entry or the use of hazardous materials by design, or by selection of alternative methods of work or materials. Common construction industry 'confined space situations' include (but are not restricted to): sewer work, paint or solvent spraying in confined areas, tunnelling, and work in compressed air environments.
- Assess: available ventilation and potential local exhaust ventilation requirements, potential hazardous gases and atmospheres, process byproducts, need for improved hygiene and welfare facilities.

Basic physical control measures

- A documented entry system is required, preferably in the form of a permit-to-work.
- Detection equipment must be present and used before entry to check on levels of oxygen and the presence of explosive or toxic substances. The work area must be monitored continuously for these factors during the presence of anyone in the confined space.
- Adequate ventilation must be present or arranged, with breathing apparatus or airlines provided where this is not possible. Where no breathing apparatus is assessed as being required, emergency BA and rescue harnesses must be provided.
- Necessary rescue equipment includes lifting equipment, safety lines and harnesses, and resuscitation facilities.
- A communication system must be established with those inside the confined space.
- Precautions for the safe use of any plant or heavier-than-air gases must be established before entry.
- Necessary PPE and hygiene facilities must be provided for those entering sewers.
- No confined space must have its atmosphere 'sweetened' with oxygen.

Management tasks

- The management role is to decide on the nature of the confined space and to put a safe system into operation, including risk assessment. General location, the potential for flooding and the need for isolations must also be included in the assessment.

Training needs

- Full training and certification is required for all those entering and managing confined spaces.
- A rescue surface party must also be trained, which must include first-aid and operation of testing equipment.

Circumstances requiring special physical controls

- If the presence of contaminated soil or toxic gases is confirmed
- Deep excavation precautions appropriate to confined space entry may apply

Selected reference

HSE Guidance IND(G)258 — Safe working in confined spaces

9 DEMOLITION

Significant hazards

- Unintentional collapse of structure or part of it
- People and objects falling from heights
- Inhalation of dust and toxic substances from the structure
- Exposure to excessive noise
- Striking overhead or underground services
- Manual handling activities associated with demolition and material movement

Regulatory requirements include

- Provision and Use of Work Equipment Regulations 1998 and ACoP (L22)
- Personal Protective Equipment Regulations 1992
- Manual Handling Operations Regulations 1992
- Construction (Health, Safety and Welfare) Regulations 1996
- Environmental Protection Act 1990
- British Standard BS 6187:1982 — Code of Practice for demolition

Pre-planning issues

A full site investigation is required before planning the work. Competent persons must be appointed to design temporary support works, and support equipment will be needed as determined by the investigation. The pre-start survey establishes hazards, adjacent affected premises, structural stability, related affected services, presence of lead or asbestos, foreseen contaminants from previous use(s) of structures. Ensure that all relevant COSHH assessments are present before work begins.

Basic physical control measures

Underground services must be located and marked where applicable, using plans, etc. Overhead power lines should be disconnected, but otherwise fenced/signed. Water, gas and electricity supplies must be disconnected. Fall prevention and protection techniques will be required in accordance with the Construction Regulations hierarchy of control. PPE required by assessments must be provided and worn. Dust should be controlled by a combination of damping down, housekeeping, work methods and local exhaust ventilation. Defined noise protection zones, barriers and work rotation combine to tackle noise above the action levels. Site boundary noise output may be the subject of local authority controls.

Management tasks

COSHH assessments will be required of substances likely to be found or produced during the work. Other risk assessments to be produced include PPE and manual handling, and all of these must be completed before work begins. The work sequence must be defined and planned. Suspect substances or contamination must be investigated before work is allowed to continue. Safe access and egress to the work areas must be maintained. Monitoring must be continuous, to ensure that all voids and open edges are covered and/or fenced off as appropriate. A fire and emergency procedure should be published and practised. Demolition must be planned and carried out only under the supervision of a competent person.

Training needs

Supervisors must have received training in COSHH appreciation, principles of demolition and general site safety. Operatives must be trained in the operation of the machinery and equipment used, and in demolition techniques. They will be briefed on the results of relevant assessments, including noise exposure. This also applies to specialist subcontractors.

Circumstances requiring special physical controls

- Presence of asbestos or lead compounds, or other hazardous contaminants
- Demolition of complex structures

Selected reference

HSE Guidance Booklet HS(G)150 — Health and safety in construction

10   DISC CUTTERS

Significant hazards

- Contact with moving parts of equipment
- Contact with ejected cut material, including broken disc
- Flying particles (eye injuries)
- Inhalation of dust
- Vibration
- Exposure to hazardous noise levels

Regulatory requirements include

- Provision and Use of Work Equipment Regulations 1998 and ACoP (L22)
- Personal Protective Equipment Regulations 1992
- Noise at Work Regulations 1989

Pre-planning issues

Ensure that workers who have been trained to change discs correctly are available at the workplace. The correct discs for the type of machine, spindle speed and material to be cut must be ordered and supplied. Abrasive disc cutters should be subject to planned maintenance programmes. Special permit-to-work clearance will be needed before using these machines in potentially explosive or flammable atmospheres.

Basic physical control measures

- An assessment of PPE requirements is mandatory for those who use disc cutters. This will include safety helmet, suitable eye protection, hearing protection, gloves and safety footwear appropriate for the work and the machine. Operators must not wear loose clothing and ties.
- Disc cutters must only be used when standing on a firm, level base. Third parties must be kept away from areas where dust or sparks are directed. Equipment and discs must be visually checked by the machine operator for any damage before use. Discs can fracture easily when incorrectly stored, undue pressure is used or they are incorrectly mounted onto the machine. Training and visual inspection can minimise these risks.
- Older disc cutters are unlikely to provide sufficient protection against vibration because of their design. Vibration white finger is a common occupational complaint, and specific enquiries should be made concerning this when selecting a suitable saw for use. Operators should be made aware of symptoms to look out for.

Management tasks

Suitable storage facilities must be available to store spare discs in dry conditions, without materials rested on them. All equipment and spare wheels must be checked for visible signs of damage before use. A list of operatives authorised and trained to use disc cutters should be maintained.

Training needs

- All personnel changing discs and using disc cutting equipment should be trained and appointed in writing by their employer. This is no longer a specific legal requirement, following the revocation of the Abrasive Wheels Regulations 1970, but the practice should be continued in view of the potential need to prove compliance with the more general requirements within PUWER 98.
- Selection may be required of operatives who have experience of the work and are physically fit.

Circumstances requiring special physical controls

- Filling the tank of a petrol-operated disc cutter
- Use of a disc cutter on tanks and drums

Selected reference

HSE Guidance Booklet: HS(G)150 — Health and safety in construction

11 DISPOSAL OF WASTE MATERIALS

Significant hazards

- General building and demolition waste — contact and environmental hazard(s)
- ◂ Asbestos and asbestos-containing products
- Flammable materials, flashpoint >21° Celsius
- Lead and lead compounds
- Organic halogen compounds
- Acids and alkalis
- Inorganic metallic and non-metallic compounds

Regulatory requirements include

- Control of Substances Hazardous to Health Regulations 1999
- Manual Handling Operations Regulations 1992
- Environmental Protection Act 1990
- Environmental Protection (Duty of Care) Regulations 1991
- Controlled Waste (Registration of Carriers & Seizure of Vehicles) Regulations 1991

Pre-planning issues

- Site safety plans and contract documentation should include the waste disposal procedures to be followed for foreseeable items of controlled waste. These are likely to include any of the materials listed above as hazards, together with any broken or surplus materials or substances.
- The collection and disposal of waste materials must be contracted only to authorised and registered contractors/carriers, who should be required to produce proof of this before being awarded a contract.

Basic physical control measures

Ensure that skips and containers are clearly marked to indicate restrictions on the disposal of particular kinds of waste.

Management tasks

Site management must ensure that building and other controlled waste is placed in suitable containers, so that transfer notes can be completed accurately as to their contents. Monitor the disposal of waste into containers/skips to ensure that unauthorised disposal is prevented. Contact the local authority's waste disposal officer when in doubt. Waste material must be passed only to registered carriers, completing transfer notes before its removal from site and keeping record copies on site.

Training needs

Contractors and personnel should be briefed on the requirements for waste disposal. Site management need training in the basic requirements of the above Regulations. (See also the selected References below.)

Circumstances requiring special physical controls

- Disposal of asbestos, lead compounds and radioactive waste
- Disposal of CFCs
- Disposal of fluorescent lamps
- Disposal of equipment containing PCBs

Selected references

Department of Environment Code of Practice: Waste Management — the Duty of Care
Department of Environment Circular 6/96 — Special Waste Regulations 1996

12 DRIVING VEHICLES

Significant hazards

- Road traffic accidents/incidents
- Fire/explosion (carriage of hazardous materials)
- Occupants struck by loose articles
- Property damage

Regulatory requirements include

- Road traffic legislation and the Highway Code
- Carriage of Dangerous Goods by Road Regulations 1996
- Construction (Health, Safety and Welfare) Regulations 1996
- Provision and Use of Work Equipment Regulations 1998 and ACoP (L22)

Pre-planning issues

- Check the system of maintenance to ensure that the manufacturer's instructions and service intervals are being complied with. The advance planning of transportation and handling of materials is important to ensure that drivers and their vehicles are capable and competent to carry out the task.
- On site, traffic routes must be planned and established. Specific traffic rules may be necessary and should be included in the site's safety plan.

Basic physical control measures

- A system must be in place to ensure that all accidents involving damage to vehicles, property or third parties will be reported immediately to management. Not all road vehicle accidents are classed as Road Traffic Act accidents and therefore (currently) non-reportable to the Health and Safety Executive, and best practice demands that this distinction should be ignored in any case.
- Up to 5 kg of LPG can be carried, provided that the driver is trained and provided with information (TREMCARD or similar), the vehicle is adequately ventilated, and a fire extinguisher is provided. If more than 500 kg of LPG are carried, two orange hazard labels.
- Materials and loads in vehicles must be evenly distributed and adequately secured.
- It is especially important to make sure that no-one is allowed to sit on a load whilst in transit, and each passenger in the vehicle must be provided with a constructed seat.

Management tasks

There must be a system to ensure that only authorised, licensed and insured persons drive their employer's vehicles. Management's role extends to verifying that the loads are within the limitations of the vehicle and driver detailed. The driver must be briefed on hazardous materials to be carried.

Training needs

Drivers must hold a current licence for the type of vehicle they will drive, and must be instructed in the features, controls and operation of any vehicle with which they are not familiar, before driving. They must also be instructed and informed about any hazardous material they may be required to carry in an employer's vehicle, the dangers that can arise, and the action to take in an emergency.

Circumstances requiring special physical controls

Carriage of:
- Hazardous or special waste
- Any hazardous substance in containers of 200 litres or more
- Any hazardous substances or flammable solids with controlled temperature conditions

Selected references

- HSE Guidance Booklet: HS(G)161 — Guidance for road vehicle operators and others involved in the carriage of dangerous goods by road
- HSE Guidance IND(G)78(L) — Transport of LPG cylinders by road
- Department of Transport Code of Practice — Safety of loads on vehicles

13 DUMPERS

Significant hazards

- Vehicle overturning
- Vehicle falling into excavations
- Noise emissions
- Persons falling from vehicle
- Unplanned movement of vehicle
- Persons in vicinity struck by vehicle

Regulatory requirements include

- Construction (Health, Safety and Welfare) Regulations 1996
- Provision and Use of Work Equipment Regulations 1998 and ACoP (L22)
- Supply of Machinery (Safety) Regulations 1992 and Amendment 1994

Pre-planning issues

- A planned maintenance programme should apply to dumpers. Where dumpers will operate in enclosed areas, such as within building structures, or confined spaces, adequate ventilation must be ensured. Public and operatives' access to vehicle routes should be limited where practicable. Selection of dumpers is required; roll-over protection (ROPS) is required where there is a risk of the dumper rolling or toppling over.

Basic physical control measures

- Dumpers must be checked by drivers before use and secured afterwards. Relevant daily checks include brake testing.
- When loading, tipping or parked, the handbrake must be applied. Even loading of the dumper skip is important, with no projecting materials. Drivers must wear safety helmets, and dismount during loading. Dumpers must not be left unattended with engines running, and the skips are to be kept clean to facilitate unloading of free-flowing materials.
- Where the driver's vision could be limited, or the operating area is congested, a banksman should be used to guide the driver and warn others.
- Passengers must not be carried unless an additional seat is provided.
- Dumpers must not be driven at excessive speed, but only in accordance with site conditions and any rules in the safety plan.
- Extra care must be taken when working on slopes, especially when crossing the gradient.

Management tasks

- Drivers of dumpers must be competent and authorised to drive the type of dumper, as handling characteristics vary greatly. The site management role includes creating and enforcing speed restrictions, monitoring dumper use on sloping ground, and foreseeing the need for stop boards at excavation sides and other unloading points where falls could occur.
- Hearing and head protection must be available for drivers.

Training needs

- Driver training and certification to CITB standard is required; driving by unauthorised persons should not be permitted under any circumstances. Possession of a car or other type of driving licence is not proof of competence.

Circumstances requiring special physical controls

- Use of dumpers on public roads (licensing and tax requirements)

Selected references

- HSE Guidance — Working with small dumpers
- HSE Guidance Booklet: HS(G)150 — Health and safety in construction

14 ELECTRICAL WORK TO 415 VOLTS

Significant hazards

- Electrocution and other contact effects
- Fire
- Falls, etc., consequent to electrical contact

Regulatory requirements include

- Provision and Use of Work Equipment Regulations 1998 and ACoP (L22)
- Supply of Machinery (Safety) Regulations 1992 and Amendment 1994
- Management of Health and Safety at Work Regulations 1999
- Electricity at Work Regulations 1989
- IEE Wiring Regulations, 16th edn and guidance
- British Standard BS 6423:1983 — Code of Practice for maintenance of switchgear and control gear up to 1 kV

Pre-planning issues

- Wherever possible avoid live work. Where this cannot be avoided, follow the assessment procedure in Figure 1 of HS(G)85 and devise a written safe system of work. Provide adequate PPE at the workplace.
- A written method statement is advisable, including layout plans.

Basic control measures

- All circuits to be worked on must be treated as live until verified as dead. There must be no exceptions to this requirement; experience of employees is irrelevant.
- Control the access to live conductors, with appropriate signs also in place. Make written information and instructions available for work on complex circuits, including control, metering and parallel circuits. Clear access of 1 m, gloves and matting (BS 697 and BS 291) are required for live working. Electrical test equipment to be insulated and fused to GS38 requirements and in date for calibration. Electricity supply seals must not be broken. Ensure written authority is given for final connections

Management tasks

- Live work only to be done by authorised and competent electricians under direct supervision of nominated supervisors. Do not allow unaccompanied work on live connections above 125 volts unless specifically authorised, and where good communications are in place. Adequate PPE, first-aid and trained first-aiders must be available at the workplace where live work is done.

Training needs

- Competence of all persons carrying out electrical work to be verified by inspection of current certificates of training/ experience. Training required in safe work practices (HS(G)85), and in any written safe work system, also treatment of electric shock and burns.

Circumstances requiring special controls

- Work on circuits involving voltages in excess of 415 volts
- Work on equipment which may be energised by third parties

Selected references

- HSE Guidance GS38 — Test equipment for use by electricians
- HSE Guidance Booklet: HS(R) 25 — Memorandum of guidance, Electricity at Work Regulations
                               HS(G) 85 — Electricity at Work Regulations, safe working practices

15 ERECTION OF STRUCTURES

Significant hazards

- Falls from heights of people and materials
- Structural instability during the erection process

Regulatory requirements include

- Construction (Health, Safety and Welfare) Regulations 1996
- Construction (Design and Management) Regulations 1994 and ACoP

Pre-planning issues

Design of permanent and temporary structures must take safe erection methods into account. Design issues include stability at all stages including loadings, connection methods and access and transport, handling, storing of components. Methods of support must be identified at this stage, allowing for anticipated loading conditions. Connections must be detailed to minimise time spent at heights. Built-in climbing aids and features should be included in the design. Weights of bundles and sub-assemblies should be calculated to allow for appropriate crane capacities. Planned delivery sequence may be required. Co-ordination and liaison must be planned before work starts.

Basic physical control measures

Fall protection systems should only apply when all possible means of fall prevention have been taken. Simple but significant standards should be enforced, including not using ladders as working platforms but only as temporary access between levels, mandatory requirement for safety harnesses and the provision of appropriate anchor or tie-off points. Tail-ropes should be used to control suspended loads. Personal fall protection is needed in conjunction with nets and other devices, or where these cannot be used, in protection against vertical free falls of more than 2 m. Lanyards should limit the drop to 1 m, or be fitted with a shock-absorbing device. Any equipment subjected to impact loading must be exchanged. The anchor point for lanyard attachment should be above the wearer's harness attachment. Suitable anchorage points will support at least twice the potential impact load and be located as high and vertical to the load as possible. Nets should not be hung more than 8 m below the work level, with sufficient clearance to avoid contact with surfaces or structures below. Debris nets should be placed on top of personnel safety nets but not compromise their design, construction or performance. Frequent inspection is required of all these systems.

Management tasks

Detailed method statement(s) will be required to ensure appropriate organisation of the work. Responsibilities must be clearly defined within the method statement, including verification of all changes that may be needed as the work progresses. Supervision by those trained and experienced in the type and size of structure to be erected is essential. Detailed drawings must be present before erection begins. Weather conditions must be monitored for change. Exclusion of access for those not involved must be achieved. The appropriate allocation of storage and assembly areas is a matter for management. Emergency first aid facilities must be provided.

Training needs

Erectors and their supervisors must be appropriately trained in relation to the type of structure being erected. Those wearing or using personal protective equipment require training before doing so.

Circumstances requiring special physical controls

- Use of tandem lifting
- Erection of proprietary structural systems. Special requirements must be incorporated into the planning process and the manufacturer's erection sequence should be followed in detail

Selected reference

HSE Guidance Note: GS28 — Safe erection of structures (out of print)

16 EXCAVATIONS

Significant hazards

- Collapse of sides
- Striking existing services
- Persons, plant and materials falling into excavations
- Flooding of excavations
- Presence of contaminated atmospheres
- Presence of contaminated soil

Regulatory requirements include

- Construction (Health, Safety and Welfare) Regulations 1996
- Provision and Use of Work Equipment Regulations 1998 and ACoP (L22)
- Control of Substances Hazardous to Health Regulations 1999
- Management of Safety and Health at Work Regulations 1999
- British Standard BS 6031:1981 — Earthworks

Pre-planning issues

Achieve compliance with BS 6031 and R97. Sufficient numbers of trained staff and adequate supervision must be present on site before works start. Sufficient and suitable plant and equipment for trench support must be on site before work starts. Suitable monitoring equipment and personnel trained in its use must be present before known exposure to toxic substances or lack of oxygen starts. Information on ground conditions and location of existing services must be obtained before work starts.

Basic control measures

The sides of excavations likely to collapse must be supported. Where flooding exists, pumps, cofferdams and caissons will be needed. Substantial barriers are required where the depth exceeds 2 m, with stop barriers to prevent vehicle entry. Stack spoil and materials back from edges. Ladders must be provided for safe access and egress. Trace buried services before work begins, using cable detectors and supply authority drawings. Signs and barriers will be needed to give warning of the work. Atmospheric monitoring may be required.

Management tasks

A safe system of work must be provided, taking account of weather and traffic conditions, and existing structures. Suitable PPE will be needed. Inspect excavations before each shift and record details, also after any destabilising event or fall of material. Selection may be required of capable, fit and experienced workers unless under direct supervision. COSHH assessment will be required for any hazardous substances likely to be found or produced during the work.

Training needs

Supervisors must be trained in the theory and practice of excavation work, also COSHH appreciation. Workers must have received training in excavation support procedures appropriate to the work method selected, and the use of cable detection devices.

Circumstances requiring special controls

- Excavations deeper than 2 m not battered to a safe angle
- Excavations with reduced ventilation or potential confined spaces
- Excavations in soil known to be contaminated
- Where explosive or toxic gases may be present, including methane

Selected reference

- CIRIA publication R97 — Trenching practice

17 FALSEWORK

Significant hazards

- Collapse of falsework
- Falls during erection and striking
- Being struck by falling materials

Regulatory requirements include

- Construction (Health, Safety and Welfare) Regulations 1996
- Provision and Use of Work Equipment Regulations 1998 and ACoP (L22)
- British Standard BS 5975:1996 — Code of Practice for Falsework

Pre-planning issues

- Fully-detailed design drawings must be produced before work starts. Support material and equipment must be selected and checked against specifications to ensure fitness for purpose.
- Plan the use of access equipment so as to allow safe access in both preparation and construction phases.
- Check the load-bearing properties of the support base surface in relation to the loading proposed.

Basic control measures

- PPE requirements include safety helmets, and those working in or near wet concrete must wear PPE as required by COSHH assessment(s).
- Route site traffic so as to avoid the area where falsework is being erected as far as practicable, and where not practicable allow for traffic passage in the calculations.
- Support equipment must be checked as it is selected to ensure it is in good condition and the correct type and capacity.
- Check that supports are positioned vertically, and as indicated on drawings. Even a small out of plumb angle substantially reduces load-bearing capacity of props.
- Clear all areas beneath falsework of personnel before unloading or loading of the structure. The design must allow for all anticipated live loading.
- Where safe access and egress cannot be maintained because of the nature of the work, make adequate arrangements for fall prevention by design, or fall protection by the provision of safe working platforms, safety nets, fall arrest equipment and safety harnesses.

Management tasks

- The role of site management is to ensure that all support equipment is in place and checked before loads are applied to it. All the necessary drawings must be available on site before the work begins. Continual checks are required to assure public and worker safety. Temporary support works must be co-ordinated with other trades and contractors who may be affected.

Training needs

- Everyone involved in the erection or dismantling of falsework must receive appropriate instruction and training, including both workers and supervisory staff.

Circumstances requiring special controls

- Large-scale works, where a Falsework Co-ordinator is to be appointed

Selected references

- HSE Guidance — Working with small dumpers
- HSE Guidance Booklets: HS(G) 32 — Safety in falsework
  HS(G)150 — Health and safety in construction

18   FIRE ON SITE

Significant hazards

- Hot work — welding, cutting
- Smoking near flammable materials
- Use of LPG plant — heaters and cookers
- Electrical faults
- Arson

Regulatory requirements include

- Construction (Health, Safety and Welfare) Regulations 1996
- Fire Precautions Act 1971 as amended by the Fire Safety and Safety of Places of Sport Act 1987
- Fire Certificates (Special Premises) Regulations 1976
- BEC/Loss Prevention Council Code of Practice: Fire prevention on construction sites
- BS4363:1991 — Specification for distribution assemblies for electricity supplies to construction

Pre-planning issues

Site planning and safety rules must include fire detection provisions, supply and maintenance of fire fighting equipment, control of hot work, emergency procedures in the event of fire, control of smoking on site as needed and prevention of the build-up of flammable materials in work areas and in waste skips. Adequate means of escape and access for emergency vehicles must be allowed for during all stages of construction.

Basic control measures

Fire emergency exit routes must be established, adequately signed and kept free of obstruction. Security measures must be taken as practicable, to restrict access to the site work areas, especially out of working hours. Any smoking restrictions must be enforced, adequately notified to contractors and signed, where flammable materials are, or are likely to be, present. Hot work and use of naked flame appliances must be controlled as necessary, including the use of permit-to-work systems as necessary. Temporary electrical systems will comply with legal standards and any changes made necessary by contract conditions or practical requirements must only be carried out by a competent electrician.

Management tasks

An emergency fire and evacuation procedure must be produced for every contract; the procedure will be continuously reviewed and updated as required. Emergency evacuation drills must be devised and practised. Exit routes must be clear of obstructions. Hot work must be strictly controlled, and work areas inspected 30 minutes after completion of the work to check for possible fire hazards. All work areas and site buildings must be inspected on completion of works for potential fire hazards. Quantities of highly flammable liquids and LPG within work areas must be restricted and suitable storage facilities provided. Records must be maintained of routine fire inspections and the maintenance and testing of fire fighting equipment.

Training needs

All site workers should be trained in fire and evacuation procedures during induction training. Those using highly flammable materials or carrying out hot work must be trained in appropriate fire prevention measures. Site management must be aware of the requirements of relevant Standards, Codes and Regulations.

Circumstances requiring special controls

(All construction sites should be evaluated for fire risk as part of best practice safety management, but the Fire Precautions (Workplace) Regulations 1997 as amended do not apply)

Selected references

- HSE Guidance EH9 — Spraying of highly flammable liquids
            HS(G)140 — Safe use and handling of flammable liquids
            HSG 168 — Fire safety in construction work

19 FORK-LIFT TRUCKS

Significant hazards

- Fall of load from forks
- Overturning of vehicle
- Noise and pollution emissions
- Persons falling from vehicle
- Unplanned movement of vehicle
- Persons in vicinity struck by vehicle
- Unplanned lowering of forks (mechanical failure)

Regulatory requirements include

- Construction (Health, Safety and Welfare) Regulations 1996
- Lifting Operations and Lifting Equipment Regulations 1998 and AcoP
- Provision and Use of Work Equipment Regulations 1998 and AcoP
- Supply of Machinery (Safety) Regulations 1992 and Amendment 1994

Pre-planning issues

- Proposed operating areas should be checked in advance for suitability and stability.
- A planned maintenance programme should apply to fork-lift trucks. Where trucks will operate in enclosed areas, such as within building structures, or confined spaces, adequate ventilation must be ensured. Public and operatives' access to vehicle routes should be limited where practicable. Selection of trucks is required; roll-over protection (ROPS) is required where there is a risk of the truck rolling or toppling over. Type of truck should be checked to ensure suitability for the loads to be moved and ground conditions.

Basic physical control measures

- Fork-lift trucks are not to be overloaded beyond manufacturers' recommendations.
- Daily driver checks must include brake testing.
- Fork-lift trucks must not be left unattended with engines running or forks raised.
- Where the driver's vision could be limited, or the operating area is congested, a banksman should be used to guide the driver and warn others.
- Passengers must not be carried unless an additional seat is provided.
- Fork-lift trucks must not be driven at excessive speed, but only in accordance with site conditions and any rules in the safety plan. At blind corners, signs and audio-visual warnings should be considered; in workshops and stores, warning signs will be displayed, and operating areas and overhead obstructions painted to highlight hazards.
- Extra care must be taken when working on slopes, especially when crossing the gradient.

Management tasks

- Fork-lift truck drivers must be competent and authorised to drive the type of truck. Vehicles must be checked by drivers before use and be secured afterwards. The site management role includes creating and enforcing speed restrictions, monitoring fork-lift truck use generally and especially on sloping ground, and foreseeing the need for site traffic management.

Training needs

- Specific driver training is required, in accordance with COP26; driving by unauthorised persons should not be permitted under any circumstances. Possession of a car or other type of driving licence is not proof of competence. This also applies to subcontractors and the self-employed.

Circumstances requiring special physical controls

- Use of vehicles on public roads (licensing & tax requirements)

Selected references

- HSE Guidance: COP26 — Rider-operated lift trucks:operator training
- HSE Guidance Booklet: HS(G)6 — Lift trucks: safety in working

20 GAS WELDING AND CUTTING

Significant hazards

- Fire
- Cylinder explosion
- Inhalation of toxic fumes
- Asphyxiation
- Injury to eyes
- Burns

Regulatory requirements include

- Workplace (Health, Safety and Welfare) Regulations 1992
- Management of Health and Safety at Work Regulations 1999
- Construction (Health, Safety and Welfare) Regulations 1996
- Provision and Use of Work Equipment Regulations 1998 and AcoP
- Supply of Machinery (Safety) Regulations 1992 and Amendment Regulations 1994
- Control of Substances Hazardous to Health Regulations 1999
- Highly Flammable Liquids and Liquefied Petroleum Gases Regulations 1972
- Relevant British Standards on PPE and equipment

Pre-planning issues

- Pre-start planning will take into account: location of welding and cutting work, fire safety/prevention needs, materials to be worked, available ventilation, other works in progress in the area, local site rules and statutory requirements. A permit-to-work system may apply.
- Written arrangements should be made for storage of gas cylinders. All work must be planned in advance, taking account of the above.

Basic physical control measures

- Suitable screens and fire blankets must be readily available to protect flammables and persons from sparks and heat. Where ventilation is poor, local exhaust ventilation should be provided.
- Minimum numbers of gas cylinders should be kept at the workplace. Cylinders must be stored upright and secured. Flashback arrestors must be fitted to all gas regulator sets. The length of hoses should be restricted to a maximum of 5 m.
- PPE, such as gloves, boots, overalls, aprons, eye protection (and respiratory protection where applicable) must be provided and worn.

Management tasks

- All necessary fire prevention equipment, and fire extinguishers, must be in place prior to the commencement of work. Management's role is to check specifically that all required precautions are taken. No work should take place in areas where the presence of flammable or explosive substances is suspected without specific assessment, or the area and any containers to be cut are certified as 'Gas Free'. Fire sentries must be posted where there is an assessed significant risk of fire or heat transfer to adjacent areas. A check of the work area should be carried out by management on completion of the shift or the work for any possibility of latent fire hazards, including smouldering.

Training needs

- Operatives carrying out this work will have received extensive trade and safety training.
- Supervisors and managers should be trained in the basics of fire prevention and welding safety.

Circumstances requiring special physical controls

- Arc welding
- Welding/cutting where flammable or explosive vapours are suspected (e.g. tanks)
- Transport of gas cylinders by road

Selected references

- HSE Guidance Booklet: HSG 168 — Fire safety in construction work
- HSE Guidance Note: EH55 — Control of exposure to fume from welding and allied processes

21 HAND TOOLS

Significant hazards

- Injury to eyes
- Injury to other body parts

Regulatory requirements include

- Provision and Use of Work Equipment Regulations 1998 and AcoP L22

Pre-planning issues

Tools provided by the employer must be assessed to ensure that they are suitable for the purpose and the environment in which they are to be used. Tools are to be in good working condition prior to issue to employees.

Basic physical control measures

Eye protection must be provided and used whenever work involves using cold chisels, drills, grinders or other tools where there is a risk of flying particles or pieces of the tool breaking off. Open-bladed knives, screwdrivers and other sharp tools are to be carried and used so as not to cause injury to the user or others. Non-ferrous (spark-free) tools must be used in flammable atmospheres. Insulated tools must be used where there is a possibility of live electrical work.

Management tasks

Management will monitor the condition of hand tools which can deteriorate with use, to ensure they are sharpened or replaced as necessary, and to ensure the correct tools are being used properly. Specific visual checks should be made as follows:

- Chisels for mushroom heads
- Hammer and file handles for deterioration and exposed tangs
- Open-ended spanners for splayed jaws

Training needs

Workers must be instructed in the correct method of use and in maintenance requirements at induction if not already included as part of craft training. All employees must be aware that they have a duty to inform their employer of any defects or deficiencies in hand tools provided for use at work.

Circumstances requiring special physical controls

None

Selected reference

- HSE Guidance Booklet: HSG 150 — Health and safety in construction

22 JOINERY WORKSHOPS

Significant hazards

- Inhalation of softwood dust
- Inhalation of hardwood dust
- Noise
- Vibration
- Rotating moving machinery
- Inhalation of solvents
- Inhalation of wood treatment materials

Regulatory requirements include

- Construction (Health, Safety and Welfare) Regulations 1996
- Lifting Operations and Lifting Equipment Regulations 1998 and AcoP
- Provision and Use of Work Equipment Regulations 1998 and AcoP L22
- Control of Substances Hazardous to Health Regulations 1999
- Supply of Machinery (Safety) Regulations 1992 and Amendment 1994
- Noise at Work Regulations 1989
- British Standards BS6854:1987 — Code of Practice for safeguarding wood-working machines
      BS EN 847-1 — Tools for woodworking; safety requirements — milling tools and circular saw blades

Pre-planning issues

- Machine layout should be planned to provide safe access and clear working area for machinists. COSHH assessments must be available for all hazardous materials used, including wood dusts. Noise assessments should be available for all machines and processes where action levels are exceeded.
- Exhaust ventilation systems and PPE must be supplied where appropriate. This will be related to the assessments, and used as required by them. First-aid kits and a qualified first-aider must be available whenever work is done using power machinery. Working alone on this equipment must not be permitted.
- Specifications of proposed new equipment should be examined, particularly in relation to noise output and CE compliance.

Basic physical control measures

Engineering methods of noise control should be used in preference to hearing protection for reduction of exposure to noise. Noise hazard zones must be established and marked where applicable. Local exhaust ventilation must be installed to reduce exposure levels of wood dust, glues, paints and wood treatment chemicals and must be inspected regularly, as required by the COSHH Regulations. Dust extraction equipment must be fitted on all machines which produce dust. Machine guards must be provided as required by the Standards listed, and always used when machines are in motion.

Management tasks

- Only trained and competent persons should be allowed to operate woodworking machines.
- Good housekeeping must be maintained to prevent build-up of flammable materials.
- Management should verify the presence of adequate first-aid facilities before any work with woodworking machinery. Systems of work must be monitored to ensure correct use.
- Adequate fire-fighting equipment must always be available, and emergency procedures for fire evacuation must be in place.

Training needs

Only trained and experienced operatives who are physically fit should be selected and appointed to operate woodworking machines. Training of operatives should include information from COSHH and noise assessments, training on safe use and the precautions to be followed on each type of machine. Machine training should include demonstration and work under supervision. Management will also be trained in these aspects, plus safety management in joinery workshops and the requirements of the COSHH and Noise Regulations.

Circumstances requiring special physical controls

None

Selected references

- HSE Guidance L114 — Safe use of woodworking machinery
- HSE Woodworking Information Sheets (produced by Woodworking NIG)

23   LADDERS AND STEPLADDERS

Significant hazards

- Falls from ladder, stepladder
- Objects dropped by ladder user
- Persons being struck by falling objects, including ladder/stepladder

Regulatory requirements include

- Provision and Use of Work Equipment Regulations 1998 and ACoP L22
- Construction (Health, Safety and Welfare) Regulations 1996
- British Standards: BS 1129:1990 — Timber ladders, steps etc.
                        BS 2037:1994 — Aluminium ladders, steps and trestles

Pre-planning issues

- Ladders and stepladders must be checked to ensure correct length, type and condition before use.
- Ladders and stepladders should be subject to a planned maintenance programme.
- Ladder work should be restricted to that which can be carried out using one hand only, and stepladder work to that which can be carried out ensuring the stability of the stepladder is maintained.

Basic physical control measures

- The ground base for ladder and stepladder use must be firm and level.
- The ladder must be of sufficient length to extend a sufficient height above the step-off point when used for access.
- The correct angle of rest for a ladder is 75°, or a base to height ratio of 1:4.
- Ladders must be secured against slipping, by securing at the top or at the bottom. Ladders may only be footed as a sole precaution against movement if less than 3 m high.
- Stepladders must be placed at right angles to the work and the user's hips should be below the top platform to ensure stability and prevent side loading. Over-reaching must be avoided in all cases.

Management tasks

- Supervisors must check ladders before use to ensure they are sound. The use of ladders should be monitored regularly, to ensure that operatives are not over-reaching, or using two hands to work. Damaged ladders must be broken up or removed from the workplace immediately. Painted ladders must not be accepted for use — the paint can mask cracks and other faults. Stepladders must be used fully open, with adjustment cords or other similar provisions taut or locked in place.
- Managers must check method statements supplied by subcontractors and others, e.g. window cleaners, to ensure that ladders will be used correctly and that safe access will be available.

Training needs

All operatives must be trained in the safe use of ladders and the hazards which are to be avoided. This training will normally be carried out at induction.

Circumstances requiring special physical controls

- If two hands are required to carry objects up a ladder, or if two hands are required to work from a ladder, **alternative access equipment is required.**

Selected references

- HSE Guidance Booklet: HSG 150 — Health and safety in construction
- HSE Construction Information Sheet 49 — General access scaffolds and ladders

24 LASERS

Significant hazards

- Injury to eyes
- Damage to skin

Regulatory requirements include

- Provision and Use of Work Equipment Regulations 1998 and ACoP
- Supply of Machinery (Safety) Regulations 1992 and Amendment 1994
- British Standard 7192: Specification for radiation safety of laser products
- British Standard BS EN 60825:1992 Radiation safety of laser products, equipment classification, requirements and users' guide

Pre-planning issues

- Equipment used on site must be Class 1 or Class 2 equipment only (the appointment of a Laser Safety Officer is required where higher-powered equipment is used).

Basic physical control measures

- Laser beams must not be directed towards personnel or vehicles.
- Laser equipment must be switched off when not in use.

Management tasks

- The management role is to ensure that those using laser beams are trained, and that physical precautions are observed.

Training needs

- Workers must be trained in the precautions and safe use of lasers. In particular, the laser beam should be terminated at the end of its path where reasonably practicable, and workers should not look deliberately into the laser beam.

Circumstances requiring special physical controls

- Use of higher-powered lasers

Selected references

None

25 LIFTING EQUIPMENT, GENERAL

Significant hazards

- Unintentional release of the load
- Unplanned movement of the load
- Damage to equipment and structures
- Crush injuries to personnel and others nearby

Regulatory requirements include

- Provision and Use of Work Equipment Regulations 1998 and ACoP L22.
- Lifting Equipment and Lifting Operations Regulations 1998 and ACoP L113
- British Standards:
    BS 6166: Part 3: 1988 — Guide to selection and use of lifting slings for multi purposes
    BS 6210: 1983 — Code of Practice for the safe use of wire rope slings for general purposes

Pre-planning issues

- Arrangements must be made for carrying out statutory inspections and keeping of records.
- Before selecting lifting equipment, the above Standards must be considered together with the weight, size, shape and centre of gravity of the load.
- Lifting equipment should be subject to a planned maintenance programme.

Basic physical control measures

- All items of lifting equipment must be individually identifiable, and stored so as to prevent physical damage or deterioration.
- Safe working loads of lifting equipment must be established before use.
- Packing will be used to protect slings from sharp edges on the load. All items of lifting equipment must be visually examined for signs of damage before use.
- Swinging of loads must be checked by ensuring that the eyes of strops are directly below the appliance hook, and that tail ropes are fitted to control larger loads.

Management tasks

- Only lifting equipment which is in date for statutory examination will be used. Manufacturers' instructions must be checked to ensure that methods of sling attachment and slinging arrangements generally are correct.

Training needs

- All those involved in the slinging of loads and the use of lifting equipment should be required to be trained to CITB or equivalent standard.
- Supervisors must be trained in the supervision of lifting operations.

Circumstances requiring special physical controls

- Any lifting where the manufacturer's or other recognised slinging method cannot be used.

Selected references

None

26 MATERIALS HOISTS

Significant hazards

- Falls of materials from hoist
- Persons being struck by moving hoist
- Falls of persons into open hoist shaft

Regulatory requirements include

- Construction (Health, Safety and Welfare) Regulations 1996
- Lifting Operations and Lifting Equipment Regulations 1998 and ACoP
- Provision and Use of Work Equipment Regulations 1998 and ACoP
- Supply of Machinery (Safety) Regulations 1992 and Amendment 1994
- British Standards: BS4465:1989 — Electric hoists (temporary installations)
  BS3125:1959 — Hoists for materials
  BS7212:1989 — Code of Practice for construction hoists

Pre-planning issues

- Hoists must be installed, tested and maintained in accordance with BS7212:1989. When planning the provision of a hoist the location, fixing points, expected loads and statutory inspections must be checked.

Basic physical control measures

- The SWL of the hoist must be clearly marked on the hoist platform. All gates must be signed 'Gate must be shut at all times'.
- An enclosure 2 m high must be placed around the base of the hoist. An enclosure or gate must be fitted at all levels where a person could be struck by a moving hoist. A device must be fitted to prevent over-run, to ensure that the platform does not over-run the masthead.
- The hoist should be controlled from one position only, and an effective means of signalling is required if the operator's view is obstructed.

Management tasks

- Wheelbarrows and loose materials must not be carried unless the platform is enclosed or materials are secured to prevent falling. Hoist inspections should be carried out weekly, and the results recorded.
- Management will inspect frequently to ensure that all hoist and landing gates are kept closed at all times. No person is to ride on the platform of a material hoist.

Training needs

- Hoist drivers must be authorised, trained and competent. Persons making inspections and signing for weekly inspections must be trained and competent to do so.

Circumstances requiring special physical controls

- Hoists required for carriage of passengers as well as materials

Selected references

- HSE Guidance Note: PM63 — Inclined hoists in building and construction
- HSE Guidance booklet: HSG 150 — Health and safety in construction

27 MOBILE CRANES

Significant hazards

- Unplanned release or dropping of load, damage to people, equipment or property
- Overturning of crane
- Persons crushed between load and fixture or vehicle, or fixtures and moving parts of crane
- Striking by falling objects
- Striking or arcing from overhead power cable

Regulatory requirements include

- Lifting Operations and Lifting Equipment Regulations 1998 and ACoP L113
- Provision and Use of Work Equipment Regulations 1998 and ACoP L22
- British Standards: BS7121: Part 1: 1989 — Safe use of cranes
  BS1757: 1986 — Power driven mobile cranes
  CP3010: Safe use of cranes (part superseded)

Pre-planning issues

All lifting operations must be under the control of an appointed person; where contracted, the contract should require that the work 'will be carried out in accordance with BS7121 Part 1 under the control of a competent person nominated by the contractor'. Planning should consider the weight of the load, radius of lift, overhead clearance, ground conditions, positioning of the crane and lifting equipment required. Relevant test certificates or copies must be kept available for inspection, and be present before the work starts. Sufficient trained banksmen and slingers must be available, and must be readily identifiable on site by, for example, the distinctive colour of fluorescent jackets or safety helmets.

Basic physical control measures

Cranes must be positioned on firm ground with stabilisers extended and wooden blocks beneath stabiliser pads regardless of the apparent ground conditions. No person should enter the crane operating area without permission, and loads must not be swung over personnel, site huts, buildings or public areas. Lifting equipment must be selected considering the weight and stability of the load. Guide ropes should be used on large loads to guide and steady the lift. All personnel involved with lifting operations must wear safety helmets, gloves and safety footwear. Telephone/radio communication or the system of hand signals specified in BS7121 should be used to communicate between the driver and the slingers and banksmen. The table of safe working loads for various radii must be clearly visible to the driver. Fitted audible alarm signals must be functioning correctly at all times when the crane is in use. Safety helmets must be worn by all those within the swinging radius of the crane.

Management tasks

Manufacturer's information on weight, centre of gravity and slinging arrangements for the load should be obtained in advance where practicable. Lifting operations must be under the control of an appointed person. Banksmen should be used when the driver's vision of the load is obstructed. Wind conditions should be monitored and work stopped if the stability of the load is affected. The area within the arc of operations should be cleared of personnel, and workers should not be allowed to stand beneath a suspended load. All the equipment used must be in date for servicing and statutory inspection.

Training needs

Supervisors, crane drivers and slingers must be trained in the requirements of BS7121. The appointed persons should be trained and competent in the theory and practice of crane lifting operations and equipment used.

Circumstances requiring special physical controls

- Tandem lifts
- Occasions when it is proposed that a mobile crane will be based on a road or pavement, when underground services and pipe runs may affect stability
- Work near or under overhead power lines

Selected references

- None

28   MOBILE ELEVATING WORK PLATFORMS

Significant hazards

- Falls of persons or materials from platform
- Unintentional lowering of platform
- Striking against overhead obstructions
- Platform overturning
- Vehicles or plant striking platform
- Trapping by scissor action of platform mechanism

Regulatory requirements include

- Construction (Health, Safety and Welfare) Regulations 1996
- Lifting Operations and Lifting Equipment Regulations 1998 and ACoP L113
- Provision and Use of Work Equipment Regulations 1998 and ACoP L22
- Supply of Machinery (Safety) Regulations 1992 and Amendment 1994
- British Standard BS 7171: 1989 — Specification for mobile elevating working platforms
         BS EN 1570: 1999 — Safety requirements for lifting tables

Pre-planning issues

- Control of traffic and pedestrians must be planned.
- Platform capacity must be checked to ensure sufficient height and SWL for the work undertaken, before use. This equipment should be subject to a planned maintenance programme. Where hired, proof of thorough examination will be required.

Basic physical control measures

- The area of the work must be fenced off and signed.
- Platforms must not be operated outside the limits set by the manufacturer. The operating area and surface is to be firm and level. Stabilisers must be extended before the platform is raised. Platforms must not be left unattended in the raised position. Platforms must not be moved until they are clear of loose material.
- The use of safety harnesses by all those working from MEWPs should be made mandatory; they should be tied off at an appropriate point within the platform.

Management tasks

- Platforms require regular maintenance, which must be arranged at appropriate intervals.
- Managers are responsible for ensuring that only trained and authorised personnel operate the platforms.

Training needs

- All operators and other workers must be specifically trained in the safe use of these platforms.

Circumstances requiring special physical controls

None

Selected reference

- International Powered Access Federation — MEWP Operator's Guide

29 MOBILE TOWERS

Significant hazards

- Falls of persons
- Falls of material
- Collapse of tower
- Overturning of tower

Regulatory requirements include

- Construction (Health, Safety and Welfare) Regulations 1996
- Provision and Use of Work Equipment Regulations 1998 and ACoP L22
- British Standards: BS 5973: 1993 — Access and working scaffolds
  BS 1139: Part 3: 1994 — Prefabricated mobile access and working towers

Pre-planning issues

Tower scaffolding should be subject to planned maintenance. Only authorised personnel should erect, modify or dismantle scaffolding. Towers should not be specified for use in the vicinity of overhead power lines. Specification for use of tower scaffolds must take into account the site ground conditions expected, height restrictions and obstructions. Work must be tendered for taking into account the relevant Standards.

Basic physical control measures

Towers will be erected by trained personnel in accordance with relevant standards and manufacturer's instructions. Ladder access should be internal and fixed to the narrowest side. Maximum height to base ratios, as specified by the manufacturer, will not be exceeded. As an approximate guide 3.5:1 for inside use and 3:1 for external use without ties should not be exceeded. Ties must be used in exposed or windy conditions. All tower platforms must be fully boarded and fitted with toeboards and guardrails. Wheels must be braked or locked when the tower is in use. Personnel and materials will be removed before a tower is moved. Manufacturer's advice on maximum loadings must be observed.

Management tasks

All working platforms should be inspected on handover to or from other contractors. After alteration or adverse weather conditions, mobile towers should be inspected by management. Towers over 2 m in height must be inspected every 7 days and the results recorded. All scaffold inspections must be carried out by a competent person. Scaffolds should be checked regularly to ensure their correct use and that unauthorised adaptations have not been made.

Training needs

All those erecting scaffolding must be competent. Where training has been provided by manufacturers or hirers, some proof of training should be obtained. Equally, inspections of scaffolding must be carried out only by those who are trained and competent to do so.

Circumstances requiring special physical controls

- If base to height ratios above are to be exceeded
- Use of tower scaffolds in the vicinity of overhead power lines
- If platform loadings will exceed manufacturer's advised limits
- If erected towers must be pushed by hand a significant distance

Selected reference

- HSE Guidance booklet: HSG 150 — Health and safety in construction

30   PORTABLE ELECTRICAL EQUIPMENT

Significant hazards

- Electrocution
- Fire
- Damage to equipment
- Vibration

Regulatory requirements include

- Provision and Use of Work Equipment Regulations 1998 and ACoP (L22)
- Electricity at Work Regulations 1989

Pre-planning issues

All portable electrical equipment must be identified individually, and is subject to planned maintenance and inspection. Equipment supplied to site must be fit for its intended purpose with regard to voltage and power, as well as environmental conditions.

Basic physical control measures

- All equipment found to be defective must be switched off, taken out of the working area, marked appropriately and reported as out of order. Visual inspection of equipment is required before use.
- Leads and cables must be routed so as to minimise the likelihood of damage and trip hazards. Where festoon leads are permitted, damaged lamps must be replaced promptly, and only moulded socket holders used.
- Generally, only equipment operating at 110 volts or less should be permitted on site; in exceptional cases, higher voltages must be authorised in writing by site management prior to first use. Higher voltage hazards must be controlled by ensuring leads are as short as practicable and that the circuits are protected by residual current devices.

Management tasks

- Subcontractors should be made aware of site policy concerning use of electrical equipment.
- Trained first-aiders should be available at all times when electrical equipment is being used.
- Safety management responsibilities include definition of site temporary electrical requirements and loads, including the arrangements for design, testing and installation of circuits and their protection by fuses and by residual current devices or their equivalent.
- Management's role is also to ensure that only trained and competent persons test, repair and maintain portable electrical equipment.

Training needs

- Training is required in the precautions and safe use of portable electrical equipment. Site first-aiders should receive training in electric shock treatment (CPR).

Circumstances requiring special physical controls

- Work in damp or wet conditions, including sewers, tunnels and quarries, involving the use of portable electrical equipment
- Battery charging
- Use of electrical equipment in flammable atmospheres

Selected references

- HSE Guidance: HS(R)25 — Memorandum of guidance to the Electricity at Work Regulations
- HSE Guidance Booklet: HSG 107 — Maintenance of portable and transportable electrical equipment
- British Standard EN 60903:1993 (replaces BS 607, which is 'still current') — Code of Practice for distribution of electricity on construction sites

31   POWERED TOOLS — WOODWORKING

Significant hazards

- Inhalation of softwood dust
- Inhalation of hardwood dust
- Noise
- Vibration
- Rotating and moving machinery

Regulatory requirements include

- Provision and Use of Work Equipment Regulations 1998 and ACoP L22
- Supply of Machinery (Safety) Regulations 1992 and Amendment 1994
- Noise at Work Regulations 1989
- Control of Substances Hazardous to Health Regulations 1999
- British Standard BS6854: 1987 — Code of Practice for safeguarding woodworking machines

Pre-planning issues

Will include COSHH and noise assessments, suitability of power tools, suitability and installation of power supply. PPE appropriate for the work must be supplied. First-aid kit and a qualified first-aider must be available when woodworking machines are being used.

Basic physical control measures

- Hearing, eye, head and foot protective equipment and respiratory protective equipment as required by COSHH assessments must be supplied and used. Machines must be checked before use to ensure guards are correctly fitted and work properly. Only 110 volt equipment is to be used where practicable, with leads positioned to prevent trip hazards and damage to loads.
- Extensive use of machines will require the provision of an area to be set aside, where a noise control zone and dust controls can be introduced.

Management tasks

- Only trained and competent persons may operate woodworking machines. The system of work must be monitored to ensure the safety of operators, and others who may be affected by dust or noise.
- Management must verify the presence of adequate first-aid facilities before any work with woodworking machines.

Training needs

- Only trained and experienced operatives who are physically fit will be selected. Training of operatives will include information from noise and COSHH assessment, training on the safe use and the precautions to be followed on each type of machine. Machine training must include demonstration and work under supervision. Management must also be trained in these aspects, also safety management and the requirements of the COSHH and Noise Regulations.
- Only those who are specifically appointed to do so may use woodworking machines.

Circumstances requiring special physical controls

- Use of woodworking machines for purposes beyond their designed use

Selected references

- HSE Guidance Booklet: L114 — Safe use of woodworking machines
- HSE Woodworking Information Sheets — (Woodworking NIG)

32 PUBLIC PROTECTION

Significant hazards

- "Children and visitors entering the site during working hours
- Children and other trespassers entering site during non-working hours
- Pedestrians and road users affected by site traffic
- Neighbours directly or indirectly affected by site work
- Noise
- Environmental nuisance or damage

Regulatory requirements include

- Health and Safety at Work etc. Act 1974 Sections 3—5
- Management of Health and Safety at Work Regulations 1999
- Control of Substances Hazardous to Health Regulations 1999

Pre-planning issues

Identify hazards and risk levels at pre-construction phase to decide on necessary control measures, then include details of controls in pre-tender safety plan. Construction phase safety plan should contain full details of measures to be taken and checked on regularly. Written risk assessments must be available. Planning permission may impose conditions such as noise and working hours restrictions, also delivery hours. Environmental section of safety plan must be reviewed for completeness. Access and security arrangements including fencing need attention at an early stage.

Basic physical control measures

Control of access to work areas as appropriate, including physical barriers and signage. Cover holes and uneven ground, check pedestrian areas for loose cables and materials. Lighting will be required for night or late access. Common site routes must be checked regularly for obstructions. Immobilise all plant when not in use. Barriers to open excavations should be at least 1 m away from edges; these should not be left open at night where possible. Materials should not be stored on pavements.

Management tasks

Co-operation with other employers sharing work areas must be achieved, especially concerning responsibilities, controlling access and emergency procedures. Authorisation for visitors and workers may be required. Access requirements for residents, the disabled and children should be discussed during planning and induction of contractors. Likely noise output of equipment should be estimated for control purposes. Visits to schools and letters to parents should be considered as a means of reducing the risk of children entering the site out of working hours.

Training needs

Management and site workers are likely to require additional training during site induction to cover site-specific precautions and work rules designed to protect third parties.

Circumstances requiring special physical controls

- Anticipated presence of dust or mud
- Presence of children in local community or nearby schools
- Use or creation of substances hazardous to health during the work process, as identified by prior assessment
- Presence of scaffolding or work overhead — evaluate for need for fans and netting

Selected references

- HSE Guidance: Construction Information Sheet No 24 — Chemical cleaners
- HSE Guidance Booklet: HSG 151 — Protecting the public — your next move

33   PRESSURE TESTING

Significant hazards

- Explosion
- Injury from ejected materials and fittings
- Leakage of substances hazardous to health

Regulatory requirements include

- Provision and Use of Work Equipment Regulations 1998 and ACoP L22
- The Pressure Systems Safety Regulations 2000
- HSE Approved Code of Practice COP37 — Safety of pressure systems
- Control of Substances Hazardous to Health Regulations 1999

Pre-planning issues

A formal testing procedure must be prepared before testing commences. Leak testing must precede pressure testing. Testing should be planned at the lowest pressure acceptable to the specifier and the specification, and should be avoided unless there are sound technical reasons for carrying out the tests.

Basic physical control measures

Access to the test equipment and the adjacent area must be restricted. Pressure test gauges need to be correctly dated to ensure correct calibration. Fire protection and first-aid equipment must be available before testing begins. PPE such as goggles and gloves must be available. Vulnerable items such as meters and pressure switches should be removed from the test area. Access equipment must be provided as necessary to enable easy access to the controls to be gained.

Management tasks

Only authorised, experienced and competent persons should carry out pressure testing. Work must be controlled, preferably by a permit-to-work system. The manager or supervisor must carry out an inspection of the system before starting the test, to ensure that isolations, blanks and other specified safety devices are in place before any testing begins.

Training needs

All workers must be briefed on the hazards, precautions and emergency procedures before testing starts. If a permit system is used, they must also be trained in the permit procedure and actions required by the system.

Circumstances requiring special physical controls

- If the pressurised substance is deemed to be hazardous (a COSHH substance)

Selected reference

- HSE Guidance Note: GS4 (Rev) — Safety in pressure testing

34 ROADWORKS

Significant hazards

- Workers struck by moving vehicles
- Workers struck by surface plant
- Traffic control risks to the public
- Traffic struck by surface plant
- Noise and other emissions from vehicles
- Surface plant contact with existing services
- Surface plant contact with overhead services
- Health hazards from the work

Regulatory requirements include

- Highways Act 1980
- New Roads and Streetworks Act 1991
- Construction (Safety, Health and Welfare) Regulations 1996
- Noise at Work Regulations 1989
- British Standard 5228 — Noise control on construction and open sites
           BS EN 500 series: Mobile road construction equipment — safety

Pre-planning issues

Traffic control will be planned in liaison with local authority and police. First-aid, welfare and emergency requirements will be assessed at the planning stage. Plans must be obtained to locate any existing hazards from services. Safe pedestrian routes will be planned. COSHH, manual handling, PPE and noise assessments must be reviewed, brought to site and the information passed to workers.

Basic physical control measures

PPE must be provided as indicated by local assessment including hearing, eye, foot and head protection. All those on site, including visitors, will wear high-visibility clothing. Traffic control signs, lighting and safety zones will be set out in accordance with the Approved Code of Practice. All plant and vehicles will be fitted with amber flashing beacons. Minimum widths for moving traffic will be maintained. Sufficient lighting must be provided for night work. Additional signs and warning lights for use in adverse weather conditions must be available on site.

Management tasks

Continual monitoring of work progress and traffic control systems is required. Certification of drivers must be checked. Any changes to planned arrangements are to be authorised in writing. Precautions must be reviewed with other contractors. Traffic and pedestrian controls must be checked at the end of each work period and working day.

Training needs

Plant operator training and certification to CITB standard is required. Supervisors must be trained in traffic management. Supervisors and operatives required to be qualified to Street Works (Qualification of Supervisors and Operatives) Regulations 1992.

Circumstances requiring special physical controls

- Motorway maintenance

Selected references

- Health and Safety Executive Booklet: HSG 151 — Protecting the public, your next move
- Traffic Signs Manual, Chapter 8
- DoT publication — Safety at Street Works and Roadworks, ACoP

35 SEWAGE CONNECTIONS

Significant hazards

- Leptospirosis (Weil's disease)
- Presence of toxic gases
- Presence of flammable gases
- Lack of oxygen
- Oxygen enrichment

Regulatory requirements include

- Construction (Health, Safety and Welfare) Regulations 1996
- Confined Space Regulations 1997 and ACoP L101
- Control of Substances Hazardous to Health Regulations 1999
- Personal Protective Equipment Regulations 1992
- Management of Health and Safety at Work Regulations 1999

Pre-planning issues

Planning must take into account the potential hazards, assessments for excavations and available natural ventilation. Where ventilation is poor or the excavation is deep, specific method statements will be required. Sufficient numbers of trained operatives and competent supervisory staff must be available on site before work starts. Monitoring equipment and personnel trained in its use will be required. Relevant COSHH assessments must be present before work starts. Contingency plans for possible flooding of the excavation must be made. Adequate welfare provision must be available on site.

Basic physical control measures

Monitoring equipment capable of detecting flammable gases, toxic or explosive atmosphere and lack of oxygen must be used (Exotox or similar). Sewage pipes must be ventilated by removing manholes above and below the work area prior to the start of any work, ensuring suitable fences and signs are present. PPE including waterproof gloves, foot protection, overalls, head, eye and hearing protection must be used (where relevant) by those working on connections. Hygiene and first-aid facilities must be available. Excavations must be covered or suitably fenced when not being worked on. A suitable means of access/egress must be provided. A communications system must be established between the ground level and the working area.

Management tasks

Leptospirosis medical instruction cards must be issued to all workers. The atmosphere will be monitored before entry and continuously during work. A top man, who is trained in emergency procedures, must be present outside at all times. All workers must be trained in emergency procedures. No work will continue if the monitoring alarm operates.

Training needs

Workers must be trained in procedures to combat foreseen hazards, use of monitoring equipment, personal hygiene and emergency procedures. Management must be equally trained, with special reference to use of rescue equipment and confined space entry routines.

Circumstances requiring special physical controls

- Working within sewage works
- Entering live sewers

Selected reference

- National Joint Health and Safety Committee for the Water Services — Guidance 2 — Safe working in sewers and at sewage works

36   SITE TRANSPORT

Significant hazards

- Contact with power lines
- Overturning of vehicles
- Collision with pedestrians
- Collision with structures
- Collision with other vehicles
- Causing collapse of or falling into excavations

Regulatory requirements include

- Provision and Use of Work Equipment Regulations 1998 and ACoP L22
- Control of Substances Hazardous to Health 1999
- Noise at Work Regulations 1989
- Construction (Health, Safety and Welfare) Regulations 1996
- Road Traffic Acts
- Road Vehicles (Construction and Use) Regulations 1986
- Controlled Waste (Registration of Carriers and Seizure of Vehicles) Regulations 1991

Pre-planning issues

The site must be surveyed and vehicle routes planned to avoid danger to pedestrians, contact with buildings, structures or overhead power lines, and to be clear of excavations. Vehicles should be subject to a planned maintenance programme.

Basic physical control measures

Suitable fencing must be provided around excavations. Barriers and notices must be erected at overhead power lines as required by Guidance Note GS6. Vehicle operating areas and traffic routes must be clearly signed and fenced off where practicable. Stop boards will be fitted to excavations where tipping is to take place. Drivers must ensure loads are placed evenly in vehicles and that the vehicle is not overloaded. Drivers must not remain in or on vehicles being loaded or unloaded. Tipping vehicles will not move with the body raised.

Management tasks

Vehicles used for removal of site waste must be registered with the local authority. Those intended for use on the public highway must comply with current licensing requirements. Signs requesting visiting vehicle drivers to report to the site office must be displayed. Speed restrictions must be displayed and enforced as necessary. Where audible warnings are not fitted to vehicles, banksmen must be used to give warning to others when vehicles are reversing.

Training needs

Visiting drivers must be briefed on site rules and hazards. All employed and contractors' drivers will hold a current driving licence, and will be trained in site and plant safety awareness. In general, visiting drivers who bring children to site in vehicle cabs will be required to ensure that they remain there throughout the vehicle's stay on site, unless escorted personally at all times.

Circumstances requiring special physical controls

- Transport on contaminated sites
- Transport of contaminated wastes

Selected references

- HSE Guidance Note: GS6 — Avoidance of danger from overhead electric lines
- HSE Booklets: HSG 144 — Safe use of vehicles on construction sites
  HSG 151 — Protecting the public — your next move

37 SKIPS

Significant hazards

- Trapping between skip and fixtures during raising and lowering
- Unintentional release of skip during raising and lowering
- Road traffic incidents, including falls of materials from skip

Regulatory requirements include

- Construction (Health, Safety and Welfare) Regulations 1996
- Provision and Use of Work Equipment Regulations 1998 and ACoP (L22)
- Highways Act 1980
- Builders' Skips (Marking) Regulations 1991
- Environmental Protection (Duty of Care) Regulations 1991

Pre-planning issues

Check the proposed use of skips for compliance with the requirements, and stipulate any special requirements in the safety plan. Liaison with the local authority is desirable, as it may reveal local specific restrictions on the use of skips. Contractual arrangements should be checked to ensure that responsibilities under the Duty of Care Regulations are met.

Basic physical control measures

- Skips must be marked with the name of the supplier, and must be provided with adequate lighting and signage if sited on a public highway. Level ground should be chosen where possible for skip placing, and a safe means of access will be required if tipping into a skip is necessary. Skips must not obstruct planned site traffic routes.
- When used in conjunction with a debris chute, measures such as covers are necessary to ensure that dropped materials do not bounce out or create unacceptable dust levels.
- Fires should not be permitted in skips, and are often forbidden by the supplier's contract terms.
- Special arrangements must be made when it is proposed to lift skips by crane or any lifting appliance other than the skip's delivery/collection vehicle. These may include the use of spreader bars, but in all cases the proposed arrangements should be cleared with the supplier of the skip.

Management tasks

- The role of site management is to monitor the arrival and use of skips, and to reject any which do not comply with the above.
- Sufficient numbers of skips should be made available to allow the separation of controlled wastes as required.
- Skip lugs and the lifting arrangements must be checked before lifting. Skips must not be loaded beyond the lifting capacity of appliances.
- Management on site must ensure that transfer notes are completed accurately, and this will involve monitoring the use being made of skips on a regular basis.

Training needs

- Site and office management should have an awareness of the requirements of the Environmental Protection Act's Duty of Care provisions.
- Correct use of skips and control of waste disposal should be included in operative induction training.

Circumstances requiring special physical controls

- The need for disposal of asbestos, lead, radioactive or special wastes
- Lifting operations involving skips

Selected reference

- Department of Transport Code of Practice — Safety in Streetworks and Roadworks

38   STEAM AND WATER CLEANERS

Significant hazards

- Electrocution
- Steam (burns and scalds)
- Contact with hazardous substances (cleaning agents and detergents)

Regulatory requirements include

- Provision and Use of Work Equipment Regulations 1998 and ACoP L22
- Supply of Machinery (Safety) Regulations 1992 and Amendment 1994
- Control of Substances Hazardous to Health Regulations 1999
- Electricity at Work Regulations 1989

Pre-planning issues

- COSHH assessments for detergents and cleaning agents must be available before this equipment is used.
- Steam/water pressure machines should be subject to a planned maintenance programme.
- Supplied/hired machines will be 110 volts supply or less, and be fitted with waterproof connections.

Basic physical control measures

- Operators must visually inspect machines, together with their leads and hoses, before they are used. The inspection should include a check for signs of physical damage or poor electrical safety.
- Supply leads must be positioned so as to avoid physical damage and ingress of water.
- Electrical supplies should be protected by 30 mA/30 ms residual current devices.
- PPE must be worn by operators as required by operating instructions and applicable COSHH assessments.

Management tasks

- Management must ensure that only trained operatives use these machines. Clear operating instructions must be provided and be readily available.

Training needs

- Operators must be trained specifically in the safe operation, and how to carry out the before-use inspection, of these machines. Only operatives who have experience of the work and are physically fit should be authorised to use these machines.

Circumstances requiring special physical controls

None

Selected reference

- HSE Guidance Note: PM29(Rev) — Electrical hazards from steam/water pressure cleaners

39 STORAGE OF MATERIALS ON SITE

Significant hazards

- Falling materials
- Unauthorised contacts by trespassers, especially children
- Environmental contamination

Regulatory requirements include

- Manual Handling Operations Regulations 1992
- Construction (Health, Safety and Welfare) Regulations 1996
- Control of Substances Hazardous to Health Regulations 1999

Pre-planning issues

- Check manufacturer's recommendations on temperature and humidity requirements, and stacking policy.
- Arrange for minimum quantities of materials to be delivered to site.
- For COSHH substances and compounds, COSHH assessments must be available on site before the material is delivered to site.
- Decide on storage areas and appropriate practicable security/means of preventing access.

Basic physical control measures

- Provide/arrange secure storage for all hazardous substances, to prevent access by unauthorised persons.
- Provide/arrange for trays and bunds where necessary beneath containers, to prevent ground contamination.
- Ensure that loads are lifted in the correct manner, avoiding manual handling where reasonably practicable and not employing makeshift systems.
- Limit material stacks in height to ensure stability: heights over 2 m must be specifically authorised by site management.
- Store compressed gas cylinders upright.
- Stabilise stacks of cylindrical objects such as pipes and drums, using chocks or wedges.
- Ensure that palletised loads are not stacked more than two pallets high.
- Mark drums and containers clearly to indicate their contents.
- Fit storage areas with edge protection where persons could fall more than 2 m.
- Where storage racking is constructed from scaffold tube and fittings, have the design loading calculated by a competent person, and the design verified.

Management tasks

- Stockpiles and storage areas must be inspected regularly to check on their stability and to ensure that the necessary basic control measures are in place.

Training needs

Verbal instructions on good housekeeping standards should be given at site induction meetings. Any specific stability and stacking instructions should be given to workers as needed.

Circumstances requiring special physical controls

- Storage of petrol
- Storage of highly flammable liquids
- Storage of LPG

Selected reference

- HSE Guidance Booklet: HSG 151 – Protecting the public – your next move

40   STORAGE AND USE OF LPG

Significant hazards

- Fire
- Explosion
- Asphyxia

Regulatory requirements include

- Highly Flammable Liquids and Liquefied Petroleum Gases Regulations 1972
- Pressure Systems Safety Regulations 2000

Pre-planning issues

- Prior to starting work with LPG, quantities must be estimated to ensure that adequate storage facilities will be available. Liaison between contractors, owners and clients must be maintained to ensure storage is adequate. Only the minimum quantity required should be held on site.

Basic physical control measures

- Cylinders must be stored upright in lockable open-mesh containers away from buildings, drains and excavations.
- Only cylinders connected to equipment are to remain in work areas.
- Cylinders must be kept away from flammable materials and heat sources.
- Adequate ventilation must be available and a dry powder fire extinguisher must be provided in areas where LPG is in use.
- Equipment using LPG should be subject to a planned maintenance programme.
- Empty cylinders must to be treated as full, except in designated storage areas where they should be segregated.
- Storage areas must be marked with appropriate safety signs and warnings. Direct heat must not be applied to cylinders.

Management tasks

- Management must ensure that storage facilities are adequate and are maintained to the specified standard. During inspections, they should check to ensure LPG equipment is being used properly, cylinders not in use are removed from the workplace, fire extinguishers are present, and storage areas are in good order. Hot work using LPG must be inspected at the end of work periods to ensure that the risk of fire is minimised.

Training needs

- Workers using LPG for any reason must be given training. This should include the approved method of leak detection, fire precautions and the use of fire extinguishers.

Circumstances requiring special physical controls

- Every occasion where LPG is carried in motor vehicles
- Storage on site of more than five full cylinders
- Use of LPG in confined spaces
- Storage of LPG inside buildings or structures

Selected references

- HSE Guidance Booklets: COP37 — Safety of pressure systems
  HSG 168 — Fire safety in construction work
- HSE Guidance leaflet: INDG 308 — Safe use of gas cylinders

41 STORAGE AND USE OF HFL

Significant hazards

- Fire
- Explosions

Regulatory requirements include

- Highly Flammable Liquids and Liquefied Petroleum Gases Regulations 1972

Pre-planning issues

- Prior to starting work with HFL, quantity requirements must be estimated to ensure that only the minimum quantities are ordered.
- A suitable fire-resistant store with warning signs displayed must be provided for storage of HFL.

Basic physical control measures

- HFL must be kept only in suitable containers. For example, petrol must be kept in marked plastic containers of 5 litres maximum capacity.
- Containers used to store HFL must be marked accordingly to show that HFL, with a flashpoint of less than 32°C, is present.
- Lids of containers must be secure.
- Where the presence of HFL vapour is foreseeable, all means of ignition must be excluded.
- Areas where HFL is stored or used must be kept clear of combustible material as far as practicable.
- Spillages must be mopped up immediately and materials removed to a safe place.
- HFL must not be used for unauthorised purposes, such as starting fires.

Management tasks

- Management must ensure that storage facilities are adequate and are maintained to the specified standard.
- During inspections, supervisors and others must check to ensure that HFL is being used properly, that spillages are cleared up promptly and that the correct fire precautions are being taken.

Training needs

- Operatives using HFL must be given training, which includes the use of fire extinguishers.

Circumstances requiring special physical controls

- Every occasion where HFL is carried in motor vehicles
- Storage on site of more than 20 litres of HFL
- Use of HFL which is known to be toxic or to cause asphyxia
- Storage of HFL inside buildings or structures
- Spraying of HFL

Selected references

HSE Guidance Booklets: HSG 168 — Fire safety in construction work
HSG 140 — Safe use and handling of flammable liquids

42 TEMPORARY ELECTRICAL SUPPLIES

Significant hazards

- Electrocution
- Fire
- Damage to equipment

Regulatory requirements include

- Provision and Use of Work Equipment Regulations 1998 and ACoP L22
- Electricity at Work Regulations 1989
- Low Voltage Electricity (Safety) Regulations 1989
- Electrical Equipment (Safety) Regulations 1994

Pre-planning issues

- Temporary electrical supplies to sites must be planned to take into account foreseen load requirements, environmental conditions, progress of work and the compatibility and maintenance requirements of the equipment. The competence of electrical subcontractors must be established before contracting, and before they begin work.
- The temporary installation should be certified before bringing it into use. Locked supply cabinets should form part of the system.
- Offices, stores, drying rooms and canteen will be regarded as permanent installations, and therefore the current IEE Wiring Regulations will apply to them.

Basic physical control measures

- Supply and distribution units must be lockable, and issue and return of the keys controlled. Signs warning of electrical hazards will be displayed on supply units, conforming to the Safety Signs and Signals Regulations. Fire extinguishers (carbon dioxide) will be available adjacent to distribution units. Rubber gloves and mats complying with current British Standards must be used for live work. All cables will be routed and pulled so as to prevent damage to the cable and avoid tripping hazards.

Management tasks

- When work on live systems is foreseen, a permit-to-work system or other suitable control system should be used. Operatives must not be allowed to work alone on live systems. Only competent electricians should be authorised to instal or modify temporary supplies; certification should be provided to cover the modifications made to the premises and the circuits.
- Generally, systems should be checked and re-checked at intervals for any physical damage. They should be checked and recertified in any event every 3 months.

Training needs

- Additional training or raised company and client standards will be given to the site management team, also electric trades handling installation should be supplied with copies of current guidance material.

Circumstances requiring special physical controls

- Work in damp or wet conditions, including sewers, tunnels and quarries
- Where there is an increase to the designed/planned load
- Work in flammable and potentially explosive atmospheres
- Following design changes to electrical installation and equipment

Selected references

- HSE Guidance: HSG 141 — Electrical safety on construction sites
- British Standards: BS7375:1996 — Code of Practice for distribution of electricity on construction sites

43 WORKING ALONE

Significant hazards

- Violence
- Fall from height
- Electric shock

Regulatory requirements include

- Construction (Health, Safety and Welfare) Regulations 1996
- Electricity at Work Regulations 1989

Pre-planning issues

- These will include assessment of the work to consider the likelihood of injury and the possible consequences to the lone worker. Where hazardous substances, live electrical work, work at height or near water may be involved, lone working should normally be avoided.

Basic physical control measures

- Lone worker alarms or other means of communication will be provided.
- First-aid facilities to treat minor injuries will be available during work periods and those working alone will be advised of this.
- Suitable means of access and egress will be provided, which can be handled safely by a single person.

Management tasks

- Peripatetic workers should be supervised on a regular basis, and management will ensure that their whereabouts is known at all times. Only experienced and trained operatives with no adverse medical history should be considered for working alone. The limits of that work which is permitted, and any limits on the initiative of the individual must be clearly specified before work is authorised.

Training needs

- Levels of training and experience for lone workers will include full understanding of the work, hazards, emergency procedures and the limits of the work which have been authorised and the limits on the use of their own initiative.

Circumstances requiring special physical controls

- Lone workers will not be allowed to enter confined spaces
- Lone workers holding or handling cash
- Use of hazardous substances having asphyxiant or toxic properties
- Electrical work involving live circuits

Selected references

- HSE Guidance Booklet: HSG 85 — Electricity at work — safe working practices
- HSE leaflet IND(G)73(L): Working alone in safety

44 WORK ON FRAGILE ROOFS

Significant hazards

- Falls of persons through roof materials

Regulatory requirements include

- Provision and Use of Work Equipment Regulations 1998 and AcoP L22
- Construction (Health, Safety and Welfare) Regulations 1996

Pre-planning issues

- Fragile materials must be identified before work begins. Where identified, a risk assessment must be carried out in order to provide a safe system of work which complies with the guidance in HSG 33, and which takes into account the nature of the work, access/egress and the need for protection of the area below.

Basic physical control measures

- Suitable means of access, such as roof ladders, crawling boards, scaffolding and staging must be provided.
- Where access is possible alongside fragile materials such as roof lights, covers will be needed or the fragile material will have to be fenced off.
- Barriers and signs must be provided so as to isolate the area below fragile materials while work is in progress.
- No person should be permitted to walk upon suspected fragile materials for any purpose, including access and surveying.

Management tasks

- The role of management is to define a safe work method prior to the commencement of work, and to arrange for provision of suitable access equipment and trained personnel as required by the safe system devised.
- Managers must check the method statements supplied by subcontractors and others, including the self-employed, to ensure that the proposed work method is safe.

Training needs

- All operatives must be given specific instructions on the system of work to be used in each case. Only those people who have experience of the work and who are physically fit should be selected.

Circumstances requiring special physical controls

None

Selected reference

HSE Guidance Booklet: HSG 33 — Safety in roofwork

45 WORK WITH FLAT GLASS

Significant hazards

- Falls from height
- Sharp edges on sheet glass and cullet

Regulatory requirements include

- Provision and Use of Work Equipment Regulations 1998 and ACoP L22
- Personal Protective Equipment at Work Regulations 1992
- Construction (Health, Safety and Welfare) Regulations 1996
- British Standard BS6262:1992 — Code of Practice for glazing of buildings
    BS6226: Part 4: 1994 — Safety related to human impact

Pre-planning issues

- Pre-contract planning must include the provision of a safe means of access for operatives and materials.
- The effect of wind pressure must be taken into account for glazing work at height or involving large sheets.
- Safe storage racks must be available to receive glass delivered to site.

Basic physical control measures

- Handling, drilling and cutting glass will require the use of PPE. This may include eye and foot protection, strong industrial gloves with wrist protection and aprons, and will be provided as required for the task. Provision of suitable PPE is subject to assessment.
- Suction pads, leather or plastic straps must be used for lifting cut sheets of glass. Glass must be stored in prepared racks at angles not greater than 3°. Newly-glazed surfaces, particularly full length sheets, should be marked with whiting or tape
- Where suction lifting pads are used, the load should be tested first near to the ground with the power turned off for at least 5 minutes to prove the integrity of the vacuum seals.

Management tasks

- Management will ensure that storage facilities are adequate and are maintained to the required standard. Broken glass or offcuts (cullet) will be cleared from work areas immediately on completion of work. Access equipment must be inspected by management on erection and before use, to ensure it is suitable and safe for operatives and their equipment. Weather conditions will be monitored to ensure safe systems of work can be maintained at all times.

Training needs

Operatives will be trained in the safe handling, storage and transport of flat glass, and in the safe use of access equipment provided. Supervisors will be trained in safe systems of work applied to glass handling, and in the inspection of access equipment provided.

Circumstances requiring special physical controls

- The use of suspended cradles as access for glazing

Selected references

None

46   WORKING AT HEIGHTS, GENERAL

Significant hazards

- Falls of persons
- Falls of material

Regulatory requirements include

- Provision and Use of Work Equipment Regulations 1998 and ACoP L22
- Construction (Health, Safety and Welfare) Regulations 1996
- Relevant British Standards include:
    BS 7883: 1997 — Code of Practice — application and use of anchor devices conforming to BS EN 795
    BS 8213 Part 1: 1991 — Code of Practice for cleaning of windows and doors
    BS EN 355, 358, 361—365 — Personal protective equipment against falls from a height

Pre-planning issues

- Work must be planned to ensure that a safe means of access to all work areas is provided. All equipment must be provided and maintained to the required legal and other standards.

Basic physical control measures

- Suitable signs and barriers must be positioned directly below works to warn of overhead operations.
- Edge protection must be erected at places of work where falls of more than 2 m could occur.
- Where edge protection is removed for access, or is not reasonably practicable, fall arrest or restraint equipment must be all used by operatives working at or near the edge.
- Where there is likely to be debris falling, fans, chutes or full enclosures must be used to protect third parties.
- All operatives working below overhead operations must wear safety helmets.

Management tasks

- All equipment used must be checked to ensure that it is in good order, to correct specification, and in date for inspection.
- Work must be monitored to ensure that additional precautions and equipment are taken into use if edge protection is removed.

Training needs

- Training and instruction must be provided to all operatives and supervisory staff involved in the use of fall arrest or restraint equipment such as lines and harnesses, including how to inspect and assess PPE of this type before use.

Circumstances requiring special physical controls

- Working over water
- Work from bosun's chairs or requiring abseiling

Selected references

HSE Guidance Booklets: HSG 33 — Safety in roofwork
                             HS(G)150 — Health and safety in construction

47 WORKING IN OCCUPIED PREMISES

Significant hazards

- Electrocution
- Fire
- Falls from heights
- Third party exposure to falls and falling objects

Regulatory requirements include

- Electricity at Work Regulations 1989 and Memorandum HS(R)25
- Construction (Health, Safety and Welfare) Regulations 1996
- Control of Substances Hazardous to Health Regulations 1999 (COSHH)
- Management of Health and Safety at Work Regulations 1999

Pre-planning issues

Will include exchange of information with owner/occupier to ensure full reciprocal knowledge of existing hazards, nature of the work, and work hazards, and demarcation of areas of responsibility. Access equipment and arrangements must ensure maximum safety of workers, occupants and third parties including the public. Details of existing services required before work begins. COSHH and noise assessments required for materials and tools to be used.

Basic physical control measures

- Physical barriers and notices will be required to isolate and give warning of work to occupants and members of the public. Fire exit routes must be kept free from obstruction, or alternative routes must be found and signed. Any hot work must be fully controlled, with extinguishers to hand.
- Where work at height is involved, debris netting, fans or other suitable measures to protect the public must be installed where assessment of the risk requires. Flammable and hazardous materials must be identified in advance and correctly controlled and stored.

Management tasks

- Monitoring will be required, to include initial checks to ensure safe systems of work are in place before work begins, and that areas are left safe and secured at the end of each work period. A fire check may be required, especially following hot work.
- Regular liaison must be maintained with occupants in order to co-ordinate work activities and eliminate hazards.
- Use of PPE and ventilation equipment must be monitored to ensure compliance with relevant COSHH assessments.

Training needs

Induction training at the site should include any hazards with residual risks uncontrolled, and the necessary precautions required for the workplace. Worker training should include the safe systems of work to be followed and the precautions designed to prevent injury to third parties. Those supervising the work should have been trained in safety management and supervision, with particular reference to working in co-operation with others likely to be affected by the work.

Circumstances requiring special physical controls

- Use of, or work with, highly flammable substances
- Working directly overhead building occupants
- Maintenance of live electrical systems

Selected references

- HSE Guidance: HSG 85 — Electricity at work — Safe working practices
- British Standard BS 5973: 1993 — Code of Practice, access and working scaffolds, etc.

48   WORK NEAR UNDERGROUND SERVICES

Significant hazards

- Contact with electricity or gas
- Flooding from water services
- Contact with sewage
- Explosion or asphyxia from gas leak

Regulatory requirements include

- Highways Act 1980
- New Roads and Streetworks Act 1991
- Traffic Signs Manual, Chapter 8
- DoT ACoP — Safety at Street Works and Roadworks
- Electricity at Work Regulations 1989
- Construction (Health, Safety and Welfare) Regulations 1996

Pre-planning issues

All work should be planned in advance, taking account of the above. Full details of underground services must be obtained in advance from the relevant authorities, including television relay companies, fibre optic cable owners and private property owners.

Basic physical control measures

Plans and cable location equipment must be available before work known to be close to buried services starts. Plans cannot be assumed to be accurate, and location devices should be used in addition. Trial holes should be dug, using hand digging to confirm locations, taking account of physical indications such as junction boxes and manholes. The lines of services should then be marked, using paint, wooden pegs, etc. All services must be assumed to be live until proven otherwise. Services crossing excavations must be supported. Services in concrete must be isolated before breaking operations begin.

Management tasks

Management must ensure that services are identified and marked before further work begins. Full consultation must be held with relevant authorities to agree precautions to be followed before work begins. All staff, machine operators and subcontractors must be fully briefed before they begin work. All temporary services will be properly marked.

Training needs

Supervisory staff should be trained in the operation of cable locating equipment, and contents of HSG 47. Operatives locating services must be similarly trained.

Circumstances requiring special physical controls

None

Selected references

- HSE Guidance Booklet: HSG 47 — Avoiding danger from underground services
- National Joint Utilities Group publications:
    No 3 — Cable locating devices
    No 4 — Identification of small buried mains and services

49 WORK UNDER POWER LINES

Significant hazards

- Arcing or contact by plant or vehicles
- Contact by long metal objects
- Contact with workers

Regulatory requirements include

- Electricity at Work Regulations 1989 (Regulation 14)
- British Standard BS7121: Code of Practice — Safe use of cranes

Pre-planning issues

Pre-contract liaison must be undertaken with the local electricity utility to agree diversions or safe clearance distances, and any other steps needed. Work which will require plant to be in the vicinity of lines must be identified, preferably within the pre-construction safety plan.

Basic physical control measures

Barriers and solid goalposts must be erected (detail in accordance with GS6 advice) and as agreed with the electricity utility. Appropriate signs must be positioned. Operations involving movement of long metal objects (such as ladders and scaffold tubing) in the vicinity of overhead lines must be subject to specific authorisation and supervision. Mobile plant which can penetrate the safe distance zone (including cranes and excavators) should have physical limiting equipment fitted.

Management tasks

Movements of visiting vehicles and plant must be controlled. Barriers and warning signs must be continuously monitored to ensure that they remain intact and in place. All crane operations and crane movements in the vicinity of overhead lines must be supervised continuously. A permit-to-work system may be required for some tasks beneath lines. Written instructions should be issued for briefing drivers of visiting vehicles on the hazards.

Training needs

Operatives and subcontractors must be briefed on the specific site hazards and the requirements of GS6. Drivers of visiting vehicles affected must be briefed on the hazards and the crossing points.

Circumstances requiring special physical controls

- Any work within the established safe zone
- Any passage of equipment which cannot fit beneath the goalposts or within the barriers
- Work on overhead power lines will require a detailed method statement, and compliance with local Electricity Supply standards

Selected reference

- HSE Guidance Note GS6(Rev) Avoidance of danger from overhead lines

50   WORK OVER WATER

Significant hazards

- Falls of site workers into water
- Drowning — workers, visitors, trespassers

Regulatory requirements include

- Provision and Use of Work Equipment Regulations 1998 and ACoP L22
- Construction (Health, Safety and Welfare) Regulations 1996
- Personal Protective Equipment at Work Regulations 1992

Pre-planning issues

Investigation must be made to establish whether any local regulations or bye-laws apply. In each case, an assessment of risk is required, taking into account the nature of the work, access/egress requirements and protection of the area beneath the work area.

Basic physical control measures

- Edge protection must be provided where practicable. Safety lines and harnesses will have to be worn where edge protection cannot be provided. Buoyancy/life jackets must be worn by persons working at the water's edge. Sufficient lifebuoys and rescue lines must be available and checked daily. A rescue boat or other means of prompt rescue must be available at immediate notice, appropriately staffed.
- Where there is fast-flowing water, consideration should be given to the provision of grablines downstream.
- Gangways and areas near water will be kept clear of obstructions. Suitable lighting will be provided at edges adjacent to water. Where trapping could occur, such as during construction of jetties, consideration should be given to employing swimmer rescuers during all working hours in addition to a rescue boat.

Management tasks

- The role of management is to define a safe work method prior to commencement of work, including for any transport over water. Method statements supplied by subcontractors and others, including the self-employed, must be checked to ensure that the proposed work method is safe.
- Rescue equipment must be checked daily. First-aid equipment and a trained first-aider must be made available.
- Supervisory staff must ensure that all persons wear buoyancy aids where necessary.

Training needs

- All workers must be given specific instructions on the system of work to be used in each case.
- Only those who have experience of the work and are physically fit should be selected.
- Anyone required to use harnesses and buoyancy aids must be trained in their proper use and maintenance.
- All personnel should be trained in the theory and local practice of rescue procedures.

Circumstances requiring special physical controls

- Work at docks
- Work within the jurisdiction of port or harbour authorities
- Work at or near water-filled mines or quarries

Selected references

None

Part 3
Legal Requirements

17 Construction Health and Safety Law

The law concerning health and safety matters is a mixture of **statute law** and **common law.** Breaches of health and safety (and environmental) statute laws are criminal offences for which financial penalties, and even prison sentences in special and relatively uncommon situations, can be imposed.

The common law has evolved over hundreds of years as a result of the decisions of courts and judges. When it relates to health and safety issues, the sector of common law known as the **tort of negligence** applies (a **tort** is defined as a type of civil wrong). Common principles or accepted standards fill the gaps where statute law has not supplied specific requirements, and the test applied is one of reasonable conduct in particular circumstances. The relationship between the employer and the employees is a special case of the application of common law principles, coming into play where the employer is alleged to have failed to carry out his duty of care towards an employee or towards another party to whom he owes a duty of care.

The essential difference between common law and statute law is that there is a penalty provided by the law for a breach of a statute law, regardless of whether an injury has occurred, whereas at common law an employer's actions will be called into question only if there is damage or loss to a person (such as an employee) who is owed a duty of care by the employer.

Decisions of higher courts are binding on those below — these are called precedents. Lower court decisions may be helpful in reaching decisions; these are called persuasive. Decisions of the House of Lords bind all lower courts, and can only be overridden by the House itself reversing its earlier decision, or by an Act of Parliament.

Statute law

Statute law is the written law of the land and consists of Acts of Parliament and the rules, regulations or orders made within the parameters of the Acts. Acts of Parliament usually set out a framework of principles or objectives and use specific regulations or orders to achieve these, sometimes after a delay written in to the Act. Most of the laws relating to health and safety at work are contained within a large body of regulations (some rules and orders are still operative) aimed at specific subject areas. For the construction industry, they are usually enforced by the Health and Safety Executive (HSE), and occasionally by the Environmental Health Departments of local authorities, using their delegated powers to do so.

Acts of Parliament are therefore the primary or principal legislation; regulations made under Acts are secondary, subordinate, delegated legislation. For example, the Health and Safety at Work etc. Act 1974 is a primary Act of Parliament, and regulations made under it have the same terms of reference and applicability — the Noise at Work Regulations 1989 and the Electricity at Work Regulations 1989 being only two of many.

Regulations are made by Ministers of the Crown where they are enabled to do so by an Act of Parliament. Regulations have to be laid before Parliament, but most do not need a vote and become law on the date specified in the regulations. They can still be vetoed by a vote within 40 days of being laid before Parliament. Regulations can apply to employment conditions generally, for example the Noise at Work Regulations; they can control specific hazards in specific industries (such as the Construction (Health, Safety and Welfare) Regulations); and they can have the most general application, as in the case of the Reporting of Injuries, Diseases and Dangerous Occurrences Regulations. Section 15 of the Health and Safety at Work etc. Act gives a lengthy list of the purposes for which regulations can be made under it.

Health and safety regulations are made by the Secretary of State (usually for the Environment), acting on proposals made by the Health and Safety Commission (HSC) or after consultation with the Commission. Normally, the HSC issues consultative documents to all interested parties before making such recommendations.

Approved Codes of Practice can supplement Acts and regulations, in order to give guidance on the general requirements which may be set out in the legislation (thus effectively enabling the legislation to be kept up to date by revising the Code of Practice rather than the law). The HSC has power to approve codes of practice of its own, or of others such as the British Standards Institution. Approval cannot be given without consultation, and consent of the Secretary of State. Failure to comply with an Approved Code of Practice is not an offence of itself, but failure is held to be proof of contravention of a requirement to which a Code applies unless a defendant can show that compliance was achieved in some other equally good way (Health and Safety at Work etc. Act, section 17).

Guidance notes are documents issued by the HSC or its executive arm, the Health and Safety Executive (HSE), as opinions on good practice. They have no legal force, but because of their origin and the experience employed in their production, they will be persuasive in practice to the lower courts and useful in civil cases to establish the reasonable standards prevailing in an industry. Approved Codes and guidance notes can therefore be said to have a 'quasi-legal' status, rather like the Highway Code.

Breaches of statute law can be used in civil claims to establish negligence, unless this is prohibited by the Act or regulation itself. The Health and Safety at Work etc. Act contains such a provision, so breaches of it cannot be used to support civil claims for compensation.

Types of statutory duty

There are three levels or types of duty imposed by statute, which allow different responses to hazards. These are: absolute duty, duty to do what is practicable, and duty to take steps that are reasonably practicable. Over the years, a body of case law has built up which gives guidance on the meaning of these duties in practice. Mostly, the cases have been decided in common law, in the course of actions for personal injuries which were based on alleged breach of statutory duty.

Recently the attention of the criminal courts has been directed to the interpretation and application of section 3

of the Health and Safety at Work etc. Act 1974 in relation to the responsibility of the occupier of premises to contractors and others.

(1) **Absolute duty** (this is also known as **strict liability**) — There are circumstances when the risk of injury is very high unless certain steps are taken, and in consequence acts and regulations have recognised these by placing an absolute duty on the employer to take specific steps to control the hazard. The words 'must' or 'shall' appear in the section or regulation to indicate this — there is no choice or evaluation of risk or feasibility to be made by the employer. The best-known example is probably the former section 12 of the Factories Act 1961, which required 'the fencing of every moving part of every prime mover'. The absolute nature of this duty was upheld by courts, even in circumstances where a machine had become practically or financially unusable because of the strict guarding requirement. However, an 'easement' is sometimes given where such absolute requirements are subject to a 'defence' clause where, in any proceedings for an offence consisting of a contravention of an absolute requirement, it shall be a defence for any person to prove that he took all reasonable steps and exercised all due diligence to avoid the commission of that offence. This is known as the 'due diligence' defence, and an example can be found in Regulation 29 of the Electricity at Work Regulations 1989.

(2) **'Practicable'** — Some regulations specify that steps must be taken 'so far as is practicable'; for example Regulation 11(2)(a) of PUWER 98 requires the provision of 'fixed guards enclosing any dangerous part or rotating stock-bar where and to the extent that it is practicable to do so'. 'Practicable' means something less than physically possible. To decide on whether the requirement can be achieved the employer would have to consider current technological knowledge and feasibility, not just difficulty of the task, its inconvenience or cost. Some degree of reason can be applied and current practice can be taken into account.

(3) **'Reasonably practicable'** — The phrase 'so far as is reasonably practicable' qualifies almost all the general duties imposed on employers by the Health and Safety at Work etc. Act. Its use allows the employer to balance the cost of taking action (in terms of time and inconvenience as well as money) against the risk being considered. If the risk is insignificant against the cost, then the steps need not be taken. This analysis of cost versus benefit must be done prior to any complaint by an enforcing authority. Employers

need to keep up to date with health and safety developments affecting their activities, as steps might become reasonably practicable over a time. Also, a single one-time analysis of precautions could be insufficient, in the sense that other less effective but cheaper alternatives might become available over time, in addition to the one being evaluated. 'Reasonably practicable' steps are, simply, those which are the best under the circumstances.

Development of health and safety law

Since the early part of the nineteenth century, when more and more people began to earn their living in factories and workshops instead of agricultural and handicraft work, Parliament has become correspondingly active in passing laws to promote health and safety at work. The first such law was passed in 1802.

In the Factories Act of 1937, construction work was first specifically included into safety law, which classed it as 'building operations and works of engineering construction' (building sites being classed as 'factories'). Further amendments and alterations were made to the 1937 Factories Act in 1948 and 1959, with a final consolidating measure in 1961 to produce the final version. In 1961 and 1966, four sets of Regulations dealing with the construction industry were introduced and lasted for more than 30 years.

Until 1974, health and safety law was aimed at types of premises (e.g. factories, construction sites, shops, and mines), and safety in defined types of work activity or processes — for example the grinding of metals or use of special types of machinery.

One of the difficulties with the Factories Act had been that it only applied to premises defined within it (notably factories, of course), and not to other workplaces such as hospitals and schools. This was one of the reasons behind the change of course begun in 1974 by the Health and Safety at Work etc. Act and later EU-driven regulations. These had their obligations based more on relationships between organisations and employees than on narrow definitions of types of premises. The Factories Act 1961 has now been repealed almost completely, and the few remaining sections are of little potential relevance to construction work.

The result of these changes has left the Health and Safety at Work etc. Act 1974 as the centrepiece of legislation for all industries, including construction. This provides the general principles for employers and others to follow. Other, subordinate, laws define in more detail the application of those general principles. In particular the Management of Health and Safety at Work Regulations 1992 (revised in 1999) introduced the key concept of risk assessment to all types of business, as well as several other detailed requirements.

For the construction industry, the two sets of regulations that are industry-specific are the Construction (Health, Safety and Welfare) Regulations 1996 and the Construction (Design and Management) Regulations 1994. It must be remembered that important topics relevant to the industry are covered in other, more general, regulations such as:

- Provision and Use of Work Equipment Regulations 1998
- Lifting Operations and Lifting Equipment Regulations 1998
- Control of Substances Hazardous to Health Regulations 1999, and
- Manual Handling Operations Regulations 1992.

Common law

The accumulation of common law cases has resulted in a system of precedents, or decisions in previous cases, which are binding on future similar cases unless over-ruled by a higher court or by statute. As a result of these cases, there is now a body of precedents which defines the duties of both employers and employees at common law.

The employer's duty of care is to take reasonable care to protect his employees from the risk of foreseeable injury, disease or death at work by the provision and maintenance of a safe place of work, a safe system of work, safe plant and equipment, adequate supervision and reasonably competent fellow employees. The formal basis of the employer's duty to employees derives from the existence of a contract of employment, but the duty to take reasonable care may also extend to matters which affect others but are within the employer's control.

To avoid common law claims (and to avoid the accidents which prompt them), the employer has to do what is reasonable under the particular circumstances. This will include providing a safe system of work, instruction on necessary precautions and indicating the risks if the precautions are not followed.

The employer can also be held liable for the actions of his/her employees causing injury, death or damage to others, providing the actions were committed in the course of employment. This form of liability of the employer is known as **vicarious liability**.

Employees have a general duty of care in common law towards themselves and to other people — so they can also be sued by anyone injured as a result of lack of reasonable care on their part. In practice this happens rarely, and the liability is assumed by the employer under the principle of vicarious liability.

The burden of proof in common law cases rests with the claimant — the injured party — who must show that a duty was owed by the party who is being sued, that there was a breach of the duty, and that as a result of that breach damage was suffered by the claimant. The claimant can be helped by procedural rules and rules of evidence. One of these saves the claimant having to show exactly how an accident happened if all the circumstances show that there would not have been an accident except because of lack of reasonable care on the part of the person being sued. This principle is known as *res ipsa loquitur* — let the facts speak for themselves.

Defences can be raised against common law claims: there was no negligence, there was no duty owed, the accident was the sole fault of the claimant, the accident did not result from the lack of care, contributory negligence (very rarely a complete defence), and *volenti non fit injuria* — the employee knowingly accepted the risk.

Civil actions have to be started within three years of the date when the accident happened, or the date when the injured person became aware that there was an injury suffered which could be the fault of the employer. The court has discretion to extend the three-year period in appropriate cases.

The Woolf reforms

Reforms to the civil claims procedures proposed by Lord Woolf were adopted on 26 April 1999. These were designed to ensure that courts deal with claims justly and quickly, and the resulting Civil Procedure Rules constitute probably the most far-reaching single set of changes in the administration of civil justice. At the time of writing it remains to be seen what effect the introduction of the Human Rights Act will have on the reforms and the system as a whole.

With the virtual disappearance of legal aid for injury claims, the courts deal with cases to ensure as far as practicable that the parties are on an equal footing without excessive cost, that each case is dealt with in proportion to its claim value, complexity and importance, and that each case is dealt with promptly and fairly. The larger the size of the claim, the more the courts will direct management of the proceedings and appoint a target trial date. Small claims up to £15 000 enter a fast-track system ensuring that trials will not last more than one day and will take place within about 33 weeks of the commencement of proceedings.

To assist in this, the Personal Injury Protocol gives a mechanism for early exchange of documentation so that each side can establish what their position will be. A detailed initial letter written by a claimant's lawyer must be acknowledged by the employer within 21 days. A maximum of 3 months is given for exchange of documents and investigation. If liability is denied, written reasons for that denial are to be provided to the claimant. Contributory negligence can be argued, but again the reasons and any documents to be used for relying on that argument must be given. Failure to respond, or otherwise not to work within the set time periods is likely to prejudice both claimant and defence in a variety of ways.

The system requires an active response from lawyers and employers, rather than allowing lengthy delays to develop and costs to mount up. A further small but important change is that the injured party is now known as the claimant, not as the plaintiff.

Damages awarded are difficult to quantify but they depend on loss of a faculty, permanent nature of the injury and its effect on the ability to earn a living, the expenses incurred (including estimated and agreed loss of earnings) and a range of other factors. There is a 'tariff' of awards used informally by judges, but this is often not followed. English courts do not use the jury system (except in libel cases), which decides the amount of damages in other jurisdictions. Awards may be reduced by a percentage if there is found to be a contribution to the injury caused by the claimant's own negligence, but the trend of the courts is to minimise this as a significant factor.

THE HEALTH AND SAFETY AT WORK ETC. ACT 1974

This Act is the major piece of health and safety legislation in Great Britain. It provides the legal framework to promote, stimulate and encourage high standards.

The Act introduced for the first time a comprehensive and integrated system dealing with workplace health and safety and the protection of the public from work activities. By placing duties of a general character upon employers, employees, the self-employed, manufacturers, designers and importers of work equipment and materials, the protection of the law, rights and responsibilities are available and given to all at work.

An 'enabling' Act, much of the text is devoted to the legal machinery for creating administrative bodies, combining others and detailing new powers of inspection and enforcement. The Health and Safety Commission (HSC) carries responsibility for policy-making and enforcement, answerable to the Secretary of State for the Environment.

One of the key features, echoed by European directives, is the principle of consultation at all levels in order to achieve consensus and combat apathy. This consultative process starts within the HSC, and continues to the workplace, where employers are required to consider the views of workers in the setting of health and safety standards.

The Act consists of four parts:

Part 1: contains provisions on:
- Health and safety of people at work
- Protection of others against health and safety risks from work activities
- Control of danger from articles and substances used at work
- Controlling certain atmospheric emissions

Part 2: establishes the Employment Medical Advisory Service

Part 3: amends previous laws relating to safety aspects of building regulations

Part 4: contains a number of general and miscellaneous provisions

General duties of employers

These are contained in sections 2, 3, 4 and 9. By section 40, the burden of proof is transferred from the prosecution to the defence in prosecutions where it is alleged that the accused person or employer failed to do what was practicable or reasonably practicable as required, in the particular circumstances. **Sections 36 and 37** provide for the personal prosecution of members of management in certain circumstances, notably in section 37 where they can be charged as well as, or instead of, the employer if the offence in question was due to their consent, connivance or neglect.

Employers must, as far as is reasonably practicable, safeguard the health, safety and welfare of employees (**section 2**). In particular, this extends to the provision and maintaining of:

- Safe plant and safe systems of work
- Safe handling, storage, maintenance and transport of (work) articles and substances
- Necessary information, instruction, training and supervision
- A safe place of work, with safe access and egress
- A safe working environment with adequate welfare facilities

There is an absolute duty on employers with five or more employees to prepare and revise as necessary a written statement of safety policy, which details the general policy and the particular organisation and arrangements for carrying it out. The policy must be brought to the notice of all the employees (**section 2(3)**).

Employers must consult with employees on health and safety matters. They must also set up a statutory health and safety committee on request of trade union-appointed safety representatives, with the main function of keeping under review the measures taken to ensure the health and safety at work of employees (**section 2 (4—7)**). Regulations made in 1977 expanded employers' duties to consult with trade union appointees. These are the Safety Representatives and Safety Committees Regulations, which describe the functions of safety representatives but do not impose any duties upon them.

The functions are: investigations of potential hazards and dangerous occurrences; examining the causes of accidents; investigation of complaints by employees; making representations to the employer on those matters; carrying out inspections of several kinds; representing employees in consultations with enforcing authorities; receiving information from inspectors and attending meetings of safety committees. A safety committee must be set up at the request of two or more safety representatives. Safety representatives have the right to training to carry out their functions; their entitlements are covered in an Approved Code of Practice and Guidance Notes.

In 1996 a wider entitlement was introduced which gave

formal consultation rights to all employees, regardless of trade union membership or representation. It offers employers the choice of consulting with the whole of the workforce or of allowing (and funding) elections of representatives with whom the employer must consult. These representatives have not been given the detailed functions of safety representatives, or their immunity from prosecution while carrying out their functions. The topic of employee consultation is covered in detail in Chapter 12, where both sets of Regulations are summarised.

The self-employed, other employees and the public must not be exposed to danger or risks to health and safety from work activities (**sections 3, 4**).

General duties of the self-employed

Similar duties to the above rest upon the self-employed, with the exception of the safety policy requirement (unless they in turn employ others) (**section 3(2)**). It should be noted that there have been many successful prosecutions and civil claims against people who believed they were self-employed, or who found the words a useful hiding-place. Although the exact circumstances can only be determined when a particular contract and situation is examined thoroughly, in general courts will not accept that a person is self-employed unless a number of conditions are satisfied. These include working for a number of different employers, providing the tools and equipment and not requiring detailed supervision. The tax status of an individual is not a reliable guide to whether a court will accept the person as self-employed. From an employer's perspective, it is usually best to assume that a person is employed for the purposes of health and safety law.

General duties of employees

By **section 7**, employees must take reasonable care of their own health and safety and that of others who may be affected by their acts or omissions. They must also co-operate with their employer so far as is necessary to enable the employer to comply with his duties under the Act. By **section 8**, it is an offence for anyone intentionally or recklessly to interfere with or misuse anything provided in the interests of health, safety or welfare. Members of management, who are also employees, are vulnerable to prosecution under section 7 if they fail to carry out their health and safety responsibilities (as defined in the safety policy statement), in addition to their liability under sections 36 and 37 as noted above.

General duties of manufacturers and suppliers

Section 6 of the Act focuses attention on the role of the producer of products used at work, and includes designers, importers and hirers-out of plant and equipment in the list of those who have duties under it. The section refers to articles and substances for use at work, and requires those mentioned to ensure so far as is reasonably practicable that articles and substances are safe when being 'used' in the widest sense. They must be tested for safety in use, or tests are to be arranged and done by a competent authority. Information about the use for which an article or substance was designed, including any necessary conditions of use to ensure health and safety, must be supplied with the article or substance.

Since the introduction of the Act, section 6 has been modified so as to apply to fairground equipment, to provide a more complete description of the kinds of use an article may be put to, and to ensure that necessary information is actually provided and not merely 'made available', as was originally required.

Hire purchase companies are not regarded as suppliers for the purposes of section 6. An interesting provision requires anyone installing an article for use at work to ensure, so far as is reasonably practicable, that nothing about the way in which it is installed or erected makes it unsafe or a risk to health when properly used. This obviously holds consequence for the construction industry, as the provision over-rides any contract provision or exclusion.

Charges

Section 9 forbids the employer to charge his employees for any measures which he is required by statute to provide in the interests of health and safety. In particular this refers to the provision of personal protective equipment, which is to be provided following the results of risk assessment in particular cases.

The Health and Safety Commission and Executive

These bodies were established by the Act, in **section 10**. Their chief functions are as follows:

The **HSC** takes responsibility for developing policies in health and safety away from government departments. Its nine members are appointed by the government from industry, trade unions, local authorities and other interest groups (currently consumers). It can also call on government departments to act on its behalf.

The **HSE** is the enforcement and advisory body to the HSC. For lower risk activities and premises such as offices, the powers of enforcement are delegated to local authorities' environmental health departments, whose officers have the same powers as HSE inspectors when carrying out these functions.

Powers of inspectors

These are contained in sections 20 to 22 and 25 of the Act. An appointed inspector can:

- Gain access without a warrant to a workplace at any time
- Employ the police to assist in the execution of the inspector's duty
- Take equipment or materials onto premises to assist the investigations
- Carry out examinations and investigations as seen fit
- Direct that locations remain undisturbed for as long as is seen fit
- Take measurements, photographs and samples
- Order the removal and testing of equipment
- Take articles or equipment away for examination or testing
- Take statements, records and documents
- Require the provision of facilities needed to assist the inspector with the enquiries
- Do anything else necessary to enable the inspector to carry out these duties

Enforcement

If an inspector discovers what is believed to be a contravention of any Act or Regulation, the inspector can:

- **Issue a prohibition notice**, with deferred or, more usually, immediate effect, which prohibits the work described in it if the inspector is of the opinion that the circumstances present a risk of serious personal injury. The notice is effective until any steps which may be specified in it have been taken to remedy the situation. Appeal can be made to an employment tribunal within 21 days, but the notice remains in effect until the appeal is heard.

- **Issue an improvement notice**, which specifies a time period for the rectification of the contravention of a statutory requirement. Appeal can be made to an employment tribunal within 21 days; this has the effect of postponing the notice until its terms have been confirmed or altered by the tribunal.
- **Prosecute** any person who contravenes a requirement or fails to comply with a notice as above.
- **Seize, render harmless or destroy** any article or substance which is considered to be the cause of imminent danger or serious personal injury.

The financial consequences of receiving prohibition and improvement notices are often severe, because of work delays and disruption. There are rights for businesses to make representations under the Health and Safety Commission's Enforcement Policy. Details of rights of appeal to tribunals are given in writing with notices as they are issued.

THE MANAGEMENT OF HEALTH AND SAFETY AT WORK REGULATIONS 1999

The Regulations implement most of the European Directive No 89/391/EEC of 29 May 1990, the 'Framework Directive', Council Directive 91/383/EEC dealing with the health and safety of those who are employed on a fixed term or other temporary basis, and also Council Directive 94/33/EC on the protection of young people. The Regulations are in addition to the requirements of the Health and Safety at Work etc. Act 1974, and they extend the employer's general safety obligations by requiring additional specific actions on the employer's part to enhance control measures. Originally one of the set of six issued at the end of 1992 and widely known as the 'six-pack', the Management of Health and Safety at Work Regulations (MHSWR) were enlarged and extended in 1994, and were renumbered, updated and reissued at the end of 1999.

Generally, a breach of a duty imposed by MHSWR does not confer a right of action in any civil proceedings. Major exceptions to this are the Regulations covering risks to young persons and to new and expectant mothers (Regulations 16(1) and 19).

There are some overlaps between these more general Regulations and specific Regulations such as the Manual Handling Operations Regulations 1992. Where the MHSWR general duties are similar to specific ones elsewhere, as in the case of general risk assessments and those required by other specific Regulations, the legal requirement is to comply with both the general and the specific duty. However, specific assessments required elsewhere will also satisfy the general requirement under MHSWR for risk assessment (as far as those operations alone are concerned). The employer will still have to apply the general duties in all work areas.

Because they provide the detail lacking in the Health and Safety at Work etc. Act 1974, a knowledge of the MHSWR is important for every employer. They contain specific requirements on risk assessments, training, and dealing with other employers — to name only three of the significant areas which require action by employers in the construction industry, and the development of arrangements which need to be written into the health and safety policy.

Summary of the Regulations relevant to the construction industry

Regulation 3 requires every employer and self-employed person to make a suitable and sufficient **assessment** of the health and safety risks to employees and others not in his employment to which his undertakings give rise, in order to put in place appropriate control measures. It also requires review of the assessments as appropriate, and for the significant findings to be recorded (electronically if desired) if five or more are employed. Details are also to be recorded of any group of employees identified by an assessment as being especially at risk.

In regard to young persons, an employer shall not employ a **young person** unless a risk assessment has been made or reviewed in relation to the risks to health and safety of young people. In making or reviewing them, anyone employing or about to employ a young person must take into account in particular:

- The inexperience, immaturity and lack of awareness of risks of young people
- The fitting-out and layout of the workstation and workplace
- The nature, degree and duration of exposure to physical, chemical and biological agents
- The form, range and use of work equipment and the way in which it is handled
- The organisation of processes and activities
- Risks from special processes which are listed

Regulation 5 requires employers to make appropriate **arrangements** given the nature and size of their operations, for effective planning, organising, controlling, monitoring and review of the preventive and protective measures they put in place. If there are five or more employees the arrangements are to be recorded. **Regulation 6** requires the provision by the employer of appropriate **health surveillance**, identified as being necessary by the assessments.

Adequate numbers of **'competent persons'** must be appointed by employers under **Regulation 7**. They are to assist the employer to comply with obligations under all the health and safety legislation, unless (in the case only of sole traders or partnerships) the employer already has sufficient competence to comply without assistance. This Regulation also requires the employer to make arrangements between competent persons to ensure adequate co-operation between them, to provide the facilities they need to carry out their functions, and provide specified health and safety information (on temporary workers, assessment results, risks notified by other employers under other Regulations within MHSWR). Regulation 7 repays a study because it sets out clearly what is required of the competent person and the appointment arrangements, including the new Regulation 7(8) 'preference' for internal appointment.

Regulation 8 requires employers to establish and give effect to procedures to be followed in the event of **serious and imminent danger** to persons working in their undertakings, to nominate competent persons to implement any evacuation procedures and restrict access to danger areas. Persons will be classed as competent when trained and able to implement these procedures. For example, one effect of this Regulation is to require fire evacuation procedures with nominated fire marshals in every premises, including construction sites.

Regulation 10 requires employers to give employees comprehensible and relevant **information** on health and safety risks identified by the assessment, the protective and preventive measures, any procedures under Regulation 8(1), the identities of competent persons appointed under that Regulation, and risks notified to the employer by others in shared facilities as required by Regulation 11. Regulation 10 also contains two paragraphs on the employment of children and 'young persons' — the difference between the definitions of 'child' and 'young person' is that the latter refers to anyone under the age of 18, and the former to a person who is not over compulsory school age.

The first requirement is that an employer must provide a parent of a child (before employing the child) comprehensible and relevant information on assessed health and safety risks and the preventive and protective measures, and also details of third-party risks notified to the employer under Regulation 11. The second new paragraph defines the word 'parent' in terms of people who have parental rights (in Scotland) and parental responsibility (in England and Wales).

Regulation 11 deals with the case of two or more **employers or self-employed persons sharing a workplace** temporarily or permanently. Each employer or self-employed person in this position is to co-operate and co-ordinate with any others as far as necessary to enable statutory duties to be complied with, reasonably co-ordinate his own measures with those of the others, and take all reasonable steps to inform the other employers of risks arising out of his undertaking. This Regulation applies whether the workplace is shared on a permanent or temporary basis, and therefore to many construction operations.

Regulation 12 covers **sharing of information on workplace risks**. It requires employers and the self-employed to provide employers of any employees, and every employee from outside undertakings, and every self-employed person, who is working in their undertakings with appropriate instructions and comprehensible information concerning any risks from the undertaking. Outside employers, their employees and the self-employed also have to be told how to identify any person nominated by the employer to implement evacuation procedures.

Regulation 13 requires employers to take capabilities as regards health and safety into account when giving tasks. Adequate **health and safety training** must be provided to all employees on recruitment, and on their exposure to new or increased risks because of: job or responsibility change, introduction of new work equipment or a change in use of existing work equipment, introduction of new technology, or introduction of a new system of work or a change to an existing one. The training must be repeated where appropriate, take account of new or changed risks to the employees concerned, and take place during working hours.

Regulation 14 introduces **employee duties**. They must use all machinery, equipment, dangerous substances, means of production, transport equipment and safety devices in accordance with any relevant training and instructions, and inform their employer or specified fellow employees of dangerous situations and shortcomings in the employer's health and safety arrangements.

Regulation 15 covers **temporary workers**. Employers and the self-employed are to provide any person they employ on a fixed-term contract or through an employment business (an agency) with information on any special skills required for safe working and any health surveillance required, before work starts. The employment business (agency) must be given information about special skills required for safe working, and specific features of jobs to be filled by agency workers where these are likely to affect their health and safety. The agency employer is required to provide that information to the employees concerned.

Regulation 19 is set in the most general terms, and contains a duty on employers to ensure that employed young persons are protected whilst at work from any risks to their health or safety which result from their lack of experience, of awareness of risks or any lack of maturity (paragraph 1). In particular (paragraph 2), young persons cannot be employed to do work beyond their physical or psychological capacity, or which involves:

- Harmful exposure to toxic, carcinogenic or other chronic agents of harm to human health
- Harmful exposure to radiation
- Risks of accidents which it can reasonably be assumed cannot be recognised or avoided by young persons because of lack of experience or training, or because of their insufficient attention to safety
- A risk to health from extreme heat or cold, noise or vibration.

In deciding whether the work will involve such harm or risks, employers must pay attention to the results of their risk assessments (which are required by Regulation 4).

Paragraph 3 does not prevent the employment of a young person who is not a child where this is necessary for training, or where the young person will be supervised by a competent person and where any risk will be reduced to the lowest level that is reasonably practicable. These provisions are without prejudice to any other provisions or restrictions on employment arising elsewhere.

Regulation 21 denies the employer a defence in criminal proceedings because of any act or default by an employee or a competent person appointed by him. Breach of the MHSWR cannot be used to give a right of civil action (**Regulation 22**), with the important exceptions of Regulation 19.

Readers should note that the foregoing only attempts to summarise a complex set of Regulations, and that they should consult the Regulations and the Approved Code of Practice for full details and an interpretation to ensure accuracy.

PROVISION AND USE OF WORK EQUIPMENT REGULATIONS 1998 AND MACHINERY SAFETY REQUIREMENTS

These Regulations (PUWER 98) replace an earlier 1992 set of the same name, which implemented the first EC directive on work equipment — the Use of Work Equipment Directive (known as UWED). The European Council of Ministers agreed a further directive — the Amending Directive to the Use of Work Equipment (AUWED) — in 1995, and PUWER 98 (SI 1998 No 2306) implements that Directive in the UK.

The main **changes** between the 1992 and the 1998 Regulations are that:

- There are new requirements for mobile work equipment
- They cover lifting equipment which is also subject to the Lifting Operations and Lifting Equipment Regulations 1998 (LOLER)
- Previous schemes for inspections, thorough examinations and tests are changed
- Regulations for power presses and woodworking machinery have been incorporated into PUWER 98 (although both topics have their own separate Approved Codes of Practice explaining in detail how the Regulations and their principles are to be applied)

The 1998 Regulations came into effect on 5 December 1998. The requirements in respect of mobile work equipment in Regulations 25 to 30 do not apply to such equipment provided for use in the establishment or undertaking before 5 December 1998 until 5 December 2002.

The Regulations revoke the Abrasive Wheels Regulations 1970, the Woodworking Machines Regulations 1974 and Regulation 27 of the Construction (Health, Safety and Welfare) Regulations 1996. They are accompanied by an

Approved Code of Practice, the study of which is essential to an understanding of the application of the Regulations.

The Regulations set objectives to be achieved, rather than establish prescriptive requirements. They simplify and clarify previous legislation using a clear set of requirements to ensure the provision of safe work equipment and its safe use, irrespective of its age or place of origin. As a result some equipment, notably that used for lifting, is subject to at least two sets of Regulations, and these should be read in conjunction to determine particular obligations.

Some significant definitions

Work equipment means any machinery, appliance, apparatus, tool or installation for use at work, whether exclusively or not. This definition covers almost any equipment used at work and any assembly of components which, in order to achieve a common end, are arranged and controlled so that they function as a whole. Examples in the construction industry include: 'tool-box tools' such as hand tools, including hammers, knives, handsaws; single machines including circular saws and dumpers; lifting equipment such as hoists, fork-lift trucks, elevating work platforms, and lifting slings (all of which are also covered by LOLER); other equipment including ladders and pressure cleaners; and installations such as scaffolding (except where the Construction (Health, Safety and Welfare) Regulations 1996 impose more detailed requirements). Not included in the definition are substances such as acids, alkalis, slurry, cement, water, and structural items such as walls, stairs, roofs and fences.

Inspection means such visual or more rigorous inspection by a competent person as is appropriate for the purpose.

Thorough examination means a thorough examination by a competent person, including testing, the nature and extent of which are appropriate for the purpose described in Regulation 32.

Use means any activity involving work equipment. This includes starting, stopping, programming, setting, transporting, repairing, modifying, maintaining, servicing and cleaning.

Danger zone means any zone in or around machinery in which a person is exposed to a risk to health and safety from contact with a dangerous part of machinery or a rotating stock-bar.

Summary of the Regulations

Almost all employment relationships and the places of work involved, where the Health and Safety at Work etc. Act 1974 applies, are within scope. This includes all industrial sectors including the construction industry. All activities involving work equipment are covered, as is any work equipment made available for use in non-domestic premises. All the requirements apply to employers, the self-employed in respect of personal work equipment, and persons holding obligations under section 4 of the Health and Safety at Work etc. Act 1974 (control of premises) in connection with the carrying on of a trade, business or other undertaking.

Regulation 4 requires employers and others to ensure that work equipment is constructed or adapted so as to be suitable for the purpose for which it is used or provided. In selecting it, employers must have regard to the working conditions and the risks to health and safety which exist where the equipment is used, as well as any additional risks posed by its use. The equipment is to be used only for the operations for which, and under conditions for which, it is 'suitable'. 'Suitable' is defined as being suitable in any respect which it is reasonably foreseeable will affect the health or safety of any person.

Regulation 5 requires the equipment to be maintained in an efficient state, efficient working order and in good repair. Where any machinery has a maintenance log, that log is to be kept up to date. There is no requirement to keep such a log, but the guidance to the Regulations recommends that one be kept in respect of high-risk equipment.

Regulation 6 covers the general inspection requirements. The employer must ensure that, where the safety of work equipment depends on the installation conditions, it is inspected by a competent person to ensure that it has been installed correctly and is safe to operate:

- After installation and before being put into service for the first time, and
- After assembly at a new location

Work equipment that is exposed to conditions causing deterioration liable to result in dangerous situations must be inspected to ensure that health and safety conditions are maintained, and that any deterioration can be detected and remedied in good time. This requirement affects most construction equipment exposed to varying weather conditions. Inspection has to be done at suitable intervals and each time that exceptional circumstances liable to jeopardise the safety of the work equipment have occurred.

The 'competent person' must have the necessary knowledge and experience. The level of competence will vary according to the type of equipment and how and where it is used. Employers must ensure that the results of inspections are recorded and kept until the next inspection is recorded.

No work equipment is to leave an employer's undertaking or, if it is obtained from another person, is to be used in his/her undertaking, unless it is accompanied by physical evidence that the last inspection which was required to be carried out has been carried out. This evidence could be a copy of the record of inspection, or for smaller items, a tag, colour code or label attached to the item.

The minimum inspection regime should be set by a competent person on behalf of the owner/supplier of the equipment, based on the manufacturer's information and other statutory obligations. The need, if any, for additional inspections will be identified by the user. Factors to be taken into account by the user include the work being carried out, any specific risks on site that may affect the condition of the equipment, and the intensity of use.

An inspection could include visual checks, a strip-down of the equipment and functional tests. Advice should be sought from the manufacturer's instructions and a competent person for guidance on what an inspection should consist of for each piece of equipment.

A number of parties have responsibilities for inspections. Hire companies must ensure that the equipment they hire out complies with PUWER 98. Those using hired equipment must agree with the plant hire company on who is to carry out the inspections and when they will be carried out. The employer or self-employed person has a duty to ensure that equipment they use or provide complies with PUWER 98, including ensuring that inspections are carried out by a competent person. Those using equipment provided by another employer have a duty themselves to ensure that the equipment is safe for use.

If the equipment is provided for common use on site, such as a scaffold, it must be established who is to take responsibility for ensuring that it complies with the Regulations. Each employer must still ensure that the equipment is suitable and safe for use by his employees.

Equipment that poses a significant risk will need to be considered by a person who is competent to determine a suitable regime. These inspections are in addition to the daily checks by the operator and must be carried out by a competent person. For the majority of equipment the formal inspection will be undertaken weekly (as was the case with certain items of construction plant under earlier regulations). Some equipment will need more frequent inspections, for example items used in confined spaces may require inspection before each shift.

There are some exceptions to these inspection requirements where other legislation details a particular inspection regime. Examples are the Construction (Health, Safety and Welfare) Regulations 1996 and the Lifting Operations and Lifting Equipment Regulations 1998 (LOLER).

Low-risk equipment used for low-risk activities will not require a formal inspection. A visual check may be required by the user before each use to ensure that it is in good condition, and the person carrying out these checks must be competent. There is no requirement for the operator to record the results of his or her own visual check, but this may be an appropriate practice in some circumstances. In other circumstances, where additional hazards exist, low-risk equipment may need a more detailed check.

Equipment of a higher risk, and that which has moving parts, should have a visual check before each use and may require a more formal check at specified intervals. A competent person should ascertain how often these formal checks should take place.

Specific risks are dealt with by **Regulation 7**, which restricts use of equipment likely to involve a specific risk to health or safety to those persons given the task of using it. Where maintenance or repairs have to be done, this work is restricted to those who have been specifically designated to do it, having received adequate training for the work.

Users and supervisors of work equipment must have available to them adequate health and safety information and, where appropriate, specific written instructions pertaining to the use of the equipment (**Regulation 8**). User and supervisor training, including work methods, risks and precautions, are covered in **Regulation 9**. Some specific training which is referred to in the Approved Code of Practice includes young persons, and operators of self-propelled work equipment and chainsaws.

Employers are required by **Regulation 10** to ensure that work equipment has been designed and constructed in compliance with any 'essential safety requirements'. These are those relating to its construction or design in any of the Statutory Instruments listed in Schedule 1 to the Regulations (probably the most significant is the Supply of Machinery (Safety) Regulations 1992 (see below), but there are other requirements for tractors, noise, industrial trucks and simple pressure vessels, for example). This requirement applies only to work equipment provided for use in the employer's premises or undertaking for the first time after 31 December 1992. In practice, this Regulation means that employers must check that operating instructions are present, and that the equipment has been checked for obvious faults. Products should also carry the CE Mark.

Regulation 11 deals with specific hazards associated with the use of work equipment. Mostly, this and the remainder of the Regulations are aimed at the provision of safe equipment, fitted where necessary with safety devices or protected against failure; in other cases the requirements affect the use of the equipment. The Regulation sets out requirements to reduce the risk to employees from dangerous parts of machinery. They include measures to prevent access to those dangerous parts and to stop the part's movement before someone (or part of them) enters the danger zone.

In selecting preventive measures, the Regulation sets out a hierarchy of them on four levels. These are:

- Fixed enclosing guards to the extent practicable to do so, but where not —
- Other guards or protection devices to the extent practicable, but where not —
- Protection appliances (jigs, push-sticks, etc.) to the extent practicable, but where not —
- Provision of information, instruction, training and supervision

Guards and protection devices are to be:

- Suitable for the purpose
- Of good construction, sound material and adequate strength
- Maintained in an efficient state, in good repair and efficient working order
- Not the source of additional risk to health or safety
- Not easily bypassed or disabled
- Situated at sufficient distance from the danger zone

- Not unduly restrictive of any necessary view of the machine, and
- Constructed or adapted that they allow maintenance or part replacement without removing them

Exposure of a person using work equipment to specified hazards must be prevented by the employer as far as is reasonably practicable, or adequately controlled where it is not (**Regulation 12**). The protection is to be by provision of personal protective equipment or of information, instruction, training and supervision so far as is reasonably practicable, and is to include measures to minimise the effects of the hazard as well as to reduce the risk. The specified hazards are:

- Ejected or falling objects
- Rupture or disintegration of parts of the work equipment
- Fire or overheating of the work equipment
- Unintended or premature discharge of any article or of any gas, dust, liquid, vapour or other substance produced, used or stored in the work equipment
- Unintended or premature explosion of the work equipment or any article or substance produced, used or stored in it

This Regulation does not apply where other specified regulations apply, including those on COSHH, ionising radiations, asbestos, noise, lead and head protection in construction activities.

Measures must be taken to ensure that people do not come into contact with work equipment, parts of work equipment or any article or substance produced, used or stored in it which are likely to burn, scald or sear (**Regulation 13**).

Specific requirements relate to the provision, location, use and identification of control systems and controls on work equipment (**Regulations 14–18**). They relate to controls for starting or making a significant change in operating conditions, stop controls, emergency stop controls, controls in general and control systems. They apply to construction work equipment in general.

All work equipment is to have a means to isolate it from all its sources of energy (**Regulation 19**). The means is to be clearly identifiable and readily accessible, and reconnection of the equipment to any energy source must not expose any person using the equipment to any risk to health or safety. Work equipment is to be stabilised by clamping or

otherwise where necessary, so as to avoid risks to health or safety (**Regulation 20**). Places where work equipment is used have to be suitably and sufficiently lit, taking account of the kind of work being done (**Regulation 21**).

So far as is reasonably practicable, maintenance operations are to be done while the work equipment is stopped or shut down. Otherwise, other protective measures are to be taken, unless maintenance people can do the work without exposure to a risk to health or safety (**Regulation 22**).

Employers must ensure that all work equipment has clearly visible markings where appropriate (**Regulation 23**) and any warnings or warning devices appropriate for health and safety (**Regulation 24**). Warnings will be inappropriate unless they are unambiguous, easily perceived and easily understood.

The following duties placed on the employer in respect of **mobile work equipment** do not apply to such equipment provided for use in the establishment or undertaking before 5 December 1998 until 5 December 2002. It must be noted that the requirements apply in full to all equipment hired out on or after 5 December 1998.

No employee can be carried on mobile work equipment unless it is suitable for carrying passengers and it has features for reducing to as low as is reasonably practicable risks to their safety, including risks from wheels or tracks (**Regulation 25**). Risks from moving parts other than wheels and tracks are covered by Regulation 11, while mounting and dismounting are covered by the Health and Safety at Work etc. Act 1974.

Regulation 26 deals with risks arising from roll-over while the equipment is travelling. Where there is a risk to an employee riding on mobile work equipment from its rolling over, it must be minimised by:

- Stabilising the work equipment
- Providing a structure which ensures that the work equipment does no more than fall on its side
- Providing a structure giving sufficient clearance to anyone being carried if it overturns further than on its side, or
- A device giving comparable protection

Where there is a risk of anyone being carried by mobile work equipment being crushed by its rolling over, the employer must ensure that it has a suitable restraining system for him.

These requirements do not apply to:

- A fork-lift truck having a structure which prevents it falling more than on its side and giving sufficient clearance if it overturns further than that (see below)
- Where such provisions would increase the overall risk
- Where it would not be reasonably practicable to operate the mobile equipment in consequence
- Where, in relation to an item of work equipment provided for use in the undertaking or establishment before 5 December 1998, it would not be reasonably practicable

Fork-lift trucks (which have vertical masts which give effective protection to seated operators, or which have full roll-over protection (ROPS) fitted) carrying an employee are to be adapted or equipped to reduce to as low as is reasonably practicable the risk to safety from over-turning (**Regulation 27**). This means that restraint systems must be provided, following a risk assessment and if they can be fitted, to fork-lift trucks with a mast or a ROPS.

Where self-propelled work equipment in motion could involve a risk to anyone's safety, by **Regulation 28** it must have:

- Facilities for preventing it being started by an unauthorised person
- Facilities for minimising the consequences of a collision (where it is rail-mounted)
- Physical means of braking and stopping
- Appropriate aids where necessary to assist the driver's field of vision
- Lighting for use in the dark
- Fire-fighting appliances, depending on the circumstances of use

Where remote-controlled, self-propelled equipment (such as a radio-controlled machine) present a risk to safety while in motion, it must stop automatically once it leaves its control range. Where the risk is of crushing or impact, it must incorporate features to guard against such risks unless other appropriate devices are able to do so (**Regulation 29**).

Drive shafts are covered by **Regulation 30**. These are devices conveying power from mobile work equipment to any work equipment connected to it — the agricultural tractor power take-off shaft is a common example. The basic requirement here is to safeguard against the effect of any seizure and the shaft itself.

Regulations 31—35 deal specifically with power presses and have no significance for the construction industry. In brief, compliance with the former Power Presses Regulations will achieve compliance with PUWER 98, although there are some differences in terminology and scope.

Supply of Machinery (Safety) Regulations 1992

These Regulations came into effect on 1 January 1993. They remain the principal law giving duties to suppliers, as against PUWER 98 which applies mostly to the user of machinery and work equipment generally. The objectives of the Regulations include harmonisation throughout the Community. 'Machinery' is not such a wide term as 'work equipment', and is usually defined as equipment with moving parts and a drive unit.

The basic requirements of the Regulations are that:

- Machinery supplied or imported into the United Kingdom must satisfy the 'essential health and safety requirements'
- The appropriate conformity assessment procedure has been carried out
- An EC declaration of conformity has been issued
- CE marking has been properly applied, and
- The machinery is in fact safe

Responsibility for showing that the machinery or safety component satisfies the essential safety requirements is given to the manufacturer or importer, who will usually have to put together a technical file. This process is known as 'attestation'. Following this, a self-declaration of conformity is issued, and CE marking is applied. Enforcement of the process is carried out by the Health and Safety Executive for construction equipment. The Regulations provide a defence of due diligence, and also for proceedings to be taken against someone else other than the principal offender if it is that person's fault.

If the machine is made or imported by an individual for use at work, that person has the responsibilities of the supplier. There are several exclusions, which exempt the following types of machinery (among others) from these Regulations:

- Unrefurbished second-hand machinery
- Storage tanks and pipelines for petrol, diesel fuel, flammable liquids and dangerous substances
- Passenger transport vehicles
- Construction site hoists
- Some lifts

The 'essential safety requirements' themselves are detailed and complex. Full details can be found at Annex B of the DTI guidance referenced below, and Annex 1 of the Machinery Directive (Council Directive 89/392/EEC), which is also reproduced as Schedule 3 to the Supply of Machinery (Safety) Regulations 1992.

References

The Supply of Machinery (Safety) Regulations 1992, SI 1992 No. 3073

The Supply of Machinery (Safety)(Amendment) Regulations 1994, SI 1994 No. 2063

INDG229	Using work equipment safely
INDG270	Supplying new machinery
INDG271	Buying new machinery
URN 95/650	Product standards — machinery. A guide on UK Regulations, HMSO/DTI 1995

THE LIFTING OPERATIONS AND LIFTING EQUIPMENT REGULATIONS 1998 (LOLER)

These Regulations (SI 1998 No 2307) apply to all lifting equipment, whether it is existing, second-hand, leased or new, and to all premises and work situations covered by the Health and Safety at Work etc. Act 1974. LOLER sets out a coherent strategy for the control of lifting operations and the equipment associated with them. LOLER implements the Amending Directive to the Use of Work Equipment Directive (AUWED, 95/63/EC). Other legislation, including the Construction (Lifting Operations) Regulations 1961, was replaced with effect from 5 December 1998. LOLER builds on the requirements of the Provision and Use of Work Equipment Regulations 1998 (PUWER 98) which came into force on the same day, and should be read in conjunction with them. Dutyholders 'providing' lifting equipment will need to comply with relevant parts of both PUWER 98 and LOLER.

The LOLER strategy is based on the requirement within the Management of Health and Safety at Work Regulations 1999 for a risk assessment to identify the nature and level of risks associated with a proposed lifting operation. The factors to be considered in the LOLER assessment process are the:

- Type of load being lifted — its nature, weight and shape
- Risk of the load falling or striking something, and its consequences
- Risk of the lifting equipment striking something or someone, and its consequences
- Risk of the lifting equipment failing or falling over while in use, and the consequences

LOLER applies to any item of equipment used for lifting or lowering loads and any operation concerned with the lifting or lowering of a load. The Regulations now include equipment not previously considered to be lifting equipment (described in former regulations as 'lifting appliances' or 'lifting machines'). The definition therefore includes cranes, goods and/or passenger lifts, mobile elevating work platforms (MEWPs), scissor lifts, vehicle hoists, lorry loaders (such as HIABs and tail-lifts), gin wheels, ropes used for access pulleys, and fork lift trucks and site handlers.

'Accessories for lifting' are also lifting equipment, and include chains, ropes, slings, and components kept for attaching loads to machinery for lifting, such as hooks, eyebolts, lifting beams or frames (described in former regulations as 'lifting gear' or 'lifting tackle').

Some significant definitions

- **Lifting equipment** — work equipment for lifting or lowering loads, including attachments used for anchoring, fixing or supporting them
- **Lifting accessory** — work equipment for attaching loads to machinery for lifting
- **Lifting operation** — an operation concerned with the lifting or lowering of a load
- **Load** — includes a person, material or animal as well as the lifting accessories and hook block
- **Examination scheme** — a suitable scheme drawn up by a competent person for such thorough examination of lifting equipment at such intervals as may be appropriate for the purpose of Regulation 9
- **Thorough examination** — means a thorough examination by a competent person and includes any appropriate testing
- **Work equipment** — includes any machinery, appliance, apparatus, tool or installation for use (whether exclusively or not) at work

Equipment outside the scope of LOLER includes conveyor belts moving on a horizontal level and winches used to haul a load on level ground. However, PUWER 98 requires a very similar level of safety, so the distinction is academic. Escalators are covered by more specific requirements in Regulation 19 of the Workplace Regulations 1992. LOLER's Approved Code of Practice is essential reading, partly because it devotes considerable space to commentary on the application to lifting equipment of the general principles of other regulations, especially PUWER 98, the Noise at Work Regulations 1989 and COSHH 99. It also discusses responsibilities between dutyholders such as owner/landlord and tenant, and hirers and users.

Summary of the Regulations

Lifting equipment selected for a lifting operation must be suitable for the activity it is to carry out (**Regulation 3**). Factors to be considered in the assessment of risk have been mentioned above. Every part of a load and any lifting attachment must be of adequate strength and stability (**Regulation 4**). Particular attention must be paid to any mounting and fixing points. These terms include the fixing of the equipment to a supporting structure and where two parts of the equipment are joined, such as with two sections of a crane. Where lifting equipment is used for lifting people, **Regulation 5** stipulates the operating conditions. The lifting carrier has to prevent crushing, trapping, being struck and falls from it so far as is reasonably practicable, and allow a trapped passenger to be freed. The lifting equipment must have suitable devices to prevent the carrier falling, except in the case of mines, where other enhanced safety coefficient suspension rope or chain can be used.

Regulation 6 applies to both fixed and mobile lifting equipment, requiring it to be positioned or installed to reduce the risk as low as is reasonably practicable of either the equipment or the load striking anyone, or of the load drifting, falling freely or being released unintentionally, and otherwise to be safe. There is also a general requirement to prevent falls down hoistways and shafts by means of gates.

Lifting equipment must be marked with the safe working load (SWL) for each configuration except where information clearly showing this is kept with the machinery. **Regulation 7** also requires lifting accessories to be marked to show their safe use characteristics; lifting equipment designed for lifting people is to be marked and that which is not so designed is clearly marked to that effect.

Organisation of lifting operations is covered in **Regulation 8**. The duty of the employer is to ensure that every lifting operation under LOLER has been properly planned by a competent person, is appropriately supervised and is carried out safely.

The previous regime of testing, thorough examination and inspection under the various industry-specific regulations, has been replaced by new thorough examination and inspection requirements contained in Regulations 9 to 11. Inspection and thorough examination requirements are detailed in **Regulation 9**. A thorough examination may include visual examination, a strip-down of the equipment and functional tests. Advice should be sought from the manufacturer's instructions and a competent person for guidance on what a thorough examination should include for each piece of lifting equipment. Any need for additional inspections will be identified by the user of the equipment. Factors that must be taken into account by the user include the work being carried out, any specific risks on site that may affect the condition of the equipment, and the intensity of use of the equipment.

All inspections and thorough examinations must be undertaken by competent persons. The level of competence will depend on the type of equipment and the level of thorough examination or inspection required. A thorough examination must be carried out as follows:

1. When the equipment is put into service for the first time unless:
   (a) it is new equipment that has not been used before and an EC declaration of conformity has been received which was made not more than 12 months before the lifting equipment is put into service; or
   (b) the equipment was obtained from another undertaking, and it is accompanied by a copy of the previous report of a thorough examination.
2. Where safety depends on the installation conditions to ensure that it has been installed correctly and is safe to operate:
   (a) after installation and before being put into service for the first time
   (b) after assembly and before being put into service at a new site or in a new location
   (*Note:* This applies to equipment erected or built on site such as tower cranes, hoists or gantry cranes, but not to equipment such as a mobile crane that is not 'installed'.)

3. Lifting equipment for lifting persons must be thoroughly examined at least once in every 6 months.
4. Accessories for lifting must be thoroughly examined at least once in every 6 months.
5. All other lifting equipment must be thoroughly examined at least once in every 12 months.
6. After exceptional circumstances liable to affect the safety of the lifting equipment. (This thorough examination is required even if an appropriate examination scheme has been developed.)

These intervals can be followed as laid down in Regulation 9, or there is an option for competent persons to develop an examination scheme for different pieces of plant. These schemes will normally specify a more frequent interval rather than a less frequent one. The requirement is therefore to possess **either** a written examination scheme **or** to follow the specified periods for examination and testing within the Regulations. The need for and the nature of a test will be a matter for the competent person to decide on, taking into account the manufacturer's information.

An inspection may also be required at suitable intervals for certain types of lifting equipment, following the identification of significant risk to the operator or other workers by a competent person. Unless indicated otherwise by the manufacturer, it will be appropriate to continue inspecting lifting equipment, such as hoists and excavators, at weekly intervals. For the majority of lifting equipment the driver should be competent to carry out the regular inspection.

The employer receives a notification forthwith if the competent person carrying out the thorough examination finds a defect which could in his opinion become a danger to anyone. The owner must then take immediate steps to ensure the equipment is not used before the defect has been rectified. Reports of thorough examinations are to be sent to the employer and any owner/hirer, and if there is found to be an imminent or existing risk of serious personal injury, the competent person is required to notify the enforcing authority. **Regulation 10** sets out inspection requirements, which include making the competent person inspecting the equipment notify the employer forthwith of any defects and to make a written record of the inspection.

Regulation 11 contains requirements for record-keeping. Any EC conformity declaration received must be kept for as long as the equipment is operated. The reports of thorough examination must be kept as follows:

- For equipment first put into use by the user — until he/she no longer uses the lifting equipment (except for an accessory for lifting)
- For accessories for lifting first put into use by the user — for 2 years after the report is made
- For equipment dependent on installation conditions — until the user no longer uses the lifting equipment at that place
- For all other thorough examinations — until the next report is made or for 2 years, whichever is the longer

If the intervals laid down in the Regulations are followed, all thorough examination reports will need to be kept for 2 years from the date that the last report was made. The initial declaration must be kept until the equipment is disposed of. Reports of inspections must be kept available until the next report is made. Reports of thorough examinations and inspections should be kept available for inspection at the place where the lifting equipment is being used. If this is not possible the information should be readily accessible. Reports must be readily available to HSE or local authority inspectors if required. No lifting equipment should leave any undertaking unless accompanied by physical evidence that the last thorough examination has been carried out.

CONSTRUCTION (HEALTH, SAFETY AND WELFARE) REGULATIONS 1996

These Regulations (CHSWR) took effect from 2 September 1996. They replaced the Construction (General Provisions) Regulations 1961, the Construction (Health and Welfare) Regulations 1966 and the Construction (Working Places) Regulations 1966, which were all revoked. They give effect to those parts of Annex 4 of the Construction Directive 92/57/EEC which deals with minimum health and safety requirements at construction sites. The CDM Regulations also came from the Directive.

CHSWR provides the 'construction equivalent' of the Workplace (Health, Safety and Welfare) Regulations 1992, which do not apply to construction sites (and are therefore outside the scope of this book). As a result CHSWR is drafted so as to be much less prescriptive than its predecessors, frequently using terms such as 'suitable', 'sufficient' and 'adequate' in relation to requirements. In general, the Regulations have less to do with physical features of construction work than the hazards and risks arising from them. For example, there is no longer a Regulation entitled 'Scaffolding' but there is one entitled 'Falls', and another is 'Prevention of drowning'. Detailed specifications are generally absent from the Regulations; if they are provided they are normally to be found in one of the nine Schedules to the Regulations rather than within the 35 Regulations themselves. There is no Approved Code of Practice to interpret the Regulations.

Another notable omission is dimensions — there are only four dimensions now contained within the Regulations (minimum height of guardrails, maximum gap beneath guardrails, minimum height of toeboards and minimum width of working platforms). Also, the use of Schedules to provide some of the details of standards of protection and provision to be achieved makes it harder to find the answer to any particular question. For ease of comprehension, in the summary which follows the Schedules are placed in context.

Some significant definitions

- A **construction site** (a term new to regulations) is a place where the principal work activity is construction work
- **Construction work** is a significant term which is repeated from the CDM Regulations and is set out in detail in the discussion of those Regulations below
- **Excavation** includes any earthwork, trench, well, shaft, tunnel or underground working
- **Fragile material** is any material that would be likely to fail if the weight of any person likely to pass across or work on it were to be applied (this includes the weight of anything being carried or supported by the person at the time)
- **Loading bay** is simply 'any facility for loading/unloading equipment or materials'
- **Personal suspension equipment** is another new term which includes abseiling equipment (but not a suspended scaffold or cradle), and boatswain's chairs
- **Vehicle** includes mobile plant and locomotives, and any vehicle towed by another vehicle
- **Working platform** includes any platform used as a place of work or for access

Summary of the Regulations

Regulation 3 restricts their application to persons at work carrying out construction work. CHSWR does not extend to workplaces on a construction site set aside for non-construction purposes, such as storage areas and site offices. In consequence, the Regulations covering traffic routes, emergency routes, exits and procedures, fire, cleanliness and site boundary markings only apply to construction work carried out on site.

General duties to comply with the Regulations are laid on employers and the self-employed as far as they affect him/her or persons under his/her control, or as far as they relate to matters under his/her control (**Regulation 4**). CHSWR assigns 'controllers' of construction work general duties to comply with the Regulations, by Regulation 4(2), regardless of their status as an employer of workmen. Employees must also comply with the Regulations, and all at work are to co-operate with duty-holders. When working under the control of another person, they must report any defect which may endanger anyone.

Under **Regulation 5**, the requirement for a safe place of work extends to 'every place of work', and 'every other place provided for the use of any person while at work'. It includes the duty to prevent risks to health and for there to be sufficient working space suitably arranged for any person working there. Where the safe place cannot be provided, access to a place which is not safe must be barred. All of this Regulation's requirements are to be observed so far as is reasonably practicable, except for the fourth part, which disapplies them to persons working on making the place of work safe, provided that all practicable steps are taken to ensure their safety.

The primary requirement of **Regulation 6** is for 'suitable and sufficient' steps to be taken so far as is reasonably practicable to prevent any person falling. Compliance with Schedules 1 and 2 is required for edge protection and working platforms used as those steps. The main points in these are:

Schedule 1 applies to all edge protection:

- Guardrails and toeboards, and their support structures, have to be of sufficient strength and rigidity for the purpose of use
- Guardrails are to be fixed at least 910 mm above the edge from which someone can fall
- The maximum unprotected gap is to be 470 mm (a reduction from the previous maximum gap of 765 mm). No intermediate guardrail is specified to produce this effect. Thus, the objective of closing the unprotected gap can also be achieved using extra toeboards or brickguards, subject to rigidity
- Guardrails, toeboards etc. are to be placed so far as is

practicable to prevent falls of person, objects or materials from any place of work

Schedule 2 applies to all working platforms:

- The term 'supporting structure' describes any structure used to support a working platform, including plant and equipment. A 'safe' resting surface is required for supporting structures, which includes being of suitable composition, stable and strong enough to support the platform and any load intended. The supporting structure must be strong enough, rigid and stable
- The working platform itself is to be rigid and stable, not able to be accidentally displaced so far as is reasonably practicable, and be dismantled in a way which prevents accidental displacement. Platform dimensions are limited to a minimum width of 600 mm and must further be sufficiently wide to permit free passage of people and safe use of equipment or materials
- Safety on working platforms is partly to be achieved by the principle of 'no gap in the surface giving rise to risk of injury'
- Slipping and tripping risks are identified as requiring prevention so far as is reasonably practicable, also the risk of being caught between the platform and any adjacent structure
- Schedule 2 also covers overloading of platforms and any supporting structure, which are not to be loaded so as to give rise to a danger of collapse or to any deformation which could affect their safe use

Regulation 6(3) continues with the mandatory requirement for edge protection where the height of fall is liable to be 2 m or more. Where work is of short duration (not defined) or because of the nature of the work, personal suspension equipment can be used. The specifics are to be found in Schedule 3, and, where these cannot be met either, then fall arrest equipment must be used in the manner specified in Schedule 4.

Schedule 3 states that the main requirements for personal suspension equipment are:

- Suitable and sufficient strength for the purpose and loads anticipated
- Securely attached to plant or a structure strong and stable enough for the circumstances
- Suitable and sufficient steps are required to prevent falls or slips from it

- Installed or attached to prevent uncontrolled movement

Schedule 4 gives the main requirements for the means of arresting falls as:

- Equipment is to be of suitable and sufficient strength to arrest falls safely
- Securely attached to plant or a structure strong and stable enough for the circumstances
- Suitable and sufficient steps are required to ensure the equipment itself does not injure the faller, as far as practicable

Regulation 6(4) allows the removal of the means of fall prevention or protection in use for the time and to the extent necessary for the movement of materials, but they must be replaced as soon as practicable.

Stipulations on the use of ladders in Regulation 6(5) and (6) require prior consideration of the nature of the work and its duration, and risks to the safety of any person. They may only be used as a place of work or a means of access or egress from a place of work, if it is reasonable to do so in the circumstances. Exemption from the Regulation 6(3) requirement for edge and other fall protection is provided when ladders are used in compliance with Schedule 5.

Schedule 5 states that the main requirements for ladders are as follows:

- Stability, strength and suitable composition for the resting surface
- Ladders are to be suitable and sufficiently strong for their purpose, and erected so as not to become displaced
- Where over 3 m high in use, ladders are to be secured as far as practicable or footed at all times
- Access ladders must be sufficiently secured to prevent slipping or falling
- There must be sufficient height at the top, if no other handhold is present
- Ladder runs rising a vertical distance of 9 m or more require rest platforms where practicable

Ladders and other equipment covered by Regulation 6 are to be properly maintained. Installation, erection and substantial alteration of all equipment covered by the Regulation is to be done only under supervision of a competent person. There is no requirement in this Regulation for erection to be done by competent workmen, although this is covered by implication in the training requirement contained in Regulation 28. Regulation 6(9) permits toeboards not to be fitted to stairways or rest platforms used only as a means of access/egress, provided that nothing is kept or stored on them.

The general duty under **Regulation 7** is to take suitable and sufficient steps to prevent any person falling through fragile material. The second part specifies platforms, coverings or other support to be provided and used, and guardrails and warning notices to be placed at the approach to the fragile material.

Regulation 8 deals with falling objects. The duty to take suitable and sufficient steps to prevent falls of material or objects is qualified by 'so far as is reasonably practicable'. The principle is to prevent an object falling; if this is done by using barriers or platforms then they must comply with Schedules 1 and 2, and where those requirements are not reasonably practicable, then suitable and sufficient steps are required to prevent people being struck by the falling object which is likely to cause injury. The Regulation also prohibits tipping or throwing down where anyone could be injured, and requires storage of materials and equipment in such a way as to prevent their collapse, overturning or unintentional movement.

The stability of structures is governed by **Regulation 9**. Practicable steps are to be taken where necessary to prevent the accidental collapse of a structure which might result in danger to any person, not only to construction workers. The Regulation prohibits loading any part of a structure so as to render it unsafe to any person, and any erection or removal of supportwork is restricted to being done only under supervision of a competent person.

Suitable and sufficient steps are required to ensure that any demolition or dismantling of structures is done so as to prevent danger, as far as practicable (**Regulation 10**), under the supervision of a competent person. Suitable and sufficient steps have to be taken to ensure no person is exposed to risk of injury from the use of explosives, including risk of injury from material ejected by the explosion (**Regulation 11**).

Excavations are covered at length in **Regulation 12**. There is no Schedule to this Regulation. 'All practicable' steps are required to prevent danger to any person, ensuring that accidental collapses do not occur. The taking of suitable

and sufficient steps to prevent burying or entrapment by material falling, so far as is reasonably practicable, is required by Regulation 12(2) and the duty extends to 'any person'. Support of excavations is to take place as early as practicable in the work, where it is necessary, using adequate support material. Support for excavations is to be installed or altered only under the supervision of a competent person. Suitable and sufficient steps are to be taken to prevent falls by anything or anyone into excavations. Nothing is to be placed or moved near an excavation where it could cause a collapse. A provision concerns the identification and prevention of risk from underground services — the identification steps are to be suitable and sufficient and they are to prevent so far as is reasonably practicable any risk of injury from an underground cable or service.

Cofferdams and caissons must be of sufficient capacity and suitable design, as well as of suitable strong and sound material and properly maintained (**Regulation 13**). Supervision by a competent person is required for all structural work on them.

The prevention of drowning is addressed by **Regulation 14**, which requires the taking of reasonably practicable steps to prevent falls into, and to minimise the risk of drowning in, water or any liquid where drowning could occur. Suitable rescue equipment is required. Transport by water to or from a place of work must be done safely, and any vessel used must be suitably built and maintained, not overcrowded or overloaded, and in the control of a competent person.

Specific provisions on the safety of traffic routes are contained in **Regulation 15**. The main burden of the Regulation is to detail what constitutes a 'suitable' traffic route, which is one where:

- Pedestrians and/or vehicles can use it without causing danger to persons near it
- Doors or gates leading onto a traffic route are sufficiently separated to allow users to see approaching traffic while still in a place of safety
- There is sufficient separation between vehicles and pedestrians, or where this cannot be achieved reasonably then there are other means of protection for pedestrians and effective warnings of approaching vehicles
- Any loading bay has at least one exit point for the exclusive use of pedestrians
- Any gate intended mainly for vehicle use has at least

one pedestrian gate close by, which is clearly marked and free from obstruction

Other details of Regulation 15 include a prohibition on driving vehicles on an obstructed route, and on a route with insufficient clearance, so far as is reasonably practicable. Where it is not, steps must be taken to warn the driver and any passengers of the obstructions or lack of clearance. This can be done by signs, which are also required to indicate the traffic route where necessary for reasons of health or safety.

Doors, gates and hatches receive attention in **Regulation 16**, which does not apply to those items on mobile plant and equipment. The basic requirement is for the fitting of suitable safety devices where necessary to prevent risk of injury to anyone. Whether the door, gate or hatch complies with the suitability of the devices is covered in the second part of the Regulation which lists four features which must be present. An example is a device to stop a sliding gate coming off its track during use.

Regulation 17 contains requirements on vehicles and their use. Unintended movement prevention by suitable and sufficient steps is required, and warning is to be given by the effective controller of the vehicle to anyone liable to be at risk from the movement. Safe operation and loading is specified. Riders on vehicles must be in a safe place provided for the purpose, and a safe place must be provided if anyone remains on a vehicle during loading and unloading of any loose material. Where vehicles are used for excavating or handling/tipping materials, they must be prevented from falling into an excavation or pit, or into water, or over-running the edge of an embankment or earthwork. Derailed vehicles (the definition of 'vehicle' includes 'locomotives') must only be moved or replaced on a track by use of suitable plant or equipment.

The contents of **Regulation 18** deal with prevention of risk from fire 'etc.' — the 'etc.' extends the reasonably practicable suitable and sufficient steps to be taken to cover the risk of injury from any substance liable to cause asphyxiation, as well risks from fire, explosion or flooding. Carbon monoxide is an example of an asphyxiant, so in order to meet the requirement of this Regulation the use of internal combustion powered equipment in confined spaces will have to be examined closely at the planning stages of projects.

Emergency routes and exits are to be provided where necessary in sufficient numbers to enable any person to reach a place of safety quickly in the event of danger (**Regulation 19**). These are to lead as directly as possible to an identified safe area, be kept clear and provided with emergency lighting where necessary. A list of factors to be taken into account when determining suitability of these exits and routes is provided in paragraph 4.

Regulation 20 covers emergency procedures, which are arrangements to deal with any foreseeable emergency. It is aimed at, but not restricted to, sites where CDM health and safety plans are in force, and amounts to a requirement for management to take these matters into account when controlling the work, and also informing all those to whom the arrangements are likely to extend of the details, and testing the procedures at suitable intervals.

Fire detection and fire-fighting equipment (which works, is maintained and examined at appropriate intervals and is suitably located) is to be provided on sites where necessary (**Regulation 21**). The list of factors to be taken into account is the same as given in Regulation 19(4). Paragraphs (5) and (6) require instruction in the use of fire-fighting equipment for all persons on site so far as is reasonably practicable. This can be done during induction training. Special instruction will be required for welders, roofers, felters and others carrying out hot work. Suitable signs are required to indicate the equipment.

Regulation 22 sets out the requirements for the provision of welfare facilities. The duty to comply with the Regulation rests on the person in control of the site, who must comply so far as is reasonably practicable. No numbers of toilets and washbasins are prescribed — 'suitable and sufficient' will be the numbers determined by the risk assessment made by the person in control.

The principles to be observed are listed in **Schedule 6**, to which all the paragraphs of the Regulation refer. The reader is referred to the Schedule for detailed study. The main points which it contains are:

- Toilets — no numbers are specified in relation to the facilities provided, but adequate ventilation and lighting are required and they must be in a clean state. Separate male and female toilets are not required if they are each in a separate room with a door which can be secured from the inside.
- Washing facilities — these are to include showers if required by the nature of the work or for health reasons. They are to be in 'readily accessible places'

and to be provided in the immediate vicinity of every toilet and every changing room, include a supply of clean hot and cold, or warm, water (which is to be running water so far as is reasonably practicable), a supply of soap or other suitable cleanser and towels or drying facility, be sufficiently ventilated and lit, and in a clean and orderly condition. The Schedule provides that washing facilities can be unisex where only one person at a time can use them and they are in a room where the door can be secured from the inside. Otherwise, facilities for males and females are to be separate, unless they are only used for washing hands, forearms and faces.

- Drinking water — a supply of wholesome drinking water is to be readily accessible at suitable places, and so far as is reasonably practicable, marked with a conspicuous sign, with a supply of cups or other drinking vessels where the drinking water is not from a jet.
- Accommodation for clothing — this has to be made available for normal clothing not worn at work, and for special clothing which is not taken home. This accommodation has to include a drying facility so far as is reasonably practicable, and provide changing facilities where special clothing is required for the work and for reasons of health or propriety persons cannot be expected to change elsewhere.
- Rest facilities — these must be suitable and sufficient, readily accessible and in compliance with Schedule 6(14) so far as is reasonably practicable. This states that there must be one or more rest rooms or areas, including those with suitable arrangements to protect non-smokers from discomfort caused by tobacco smoke and, where necessary, include suitable facilities for workers who are pregnant or nursing mothers to allow them to rest. The facilities must include suitable arrangements for preparation and eating of meals, including the means for boiling water.

There is no specific requirement for the means of heating food. Nothing in the Regulation or the Schedule prevents contractors using facilities reasonably adjacent to a site, whether public or private.

Under **Regulation 23**, sufficient fresh or purified air is required to ensure, so far as is reasonably practicable, that every workplace or approach thereto is safe and without risks to health. Any plant used to achieve this must be fitted with a device to give an audible or visual warning of any failure.

Regulation 24 requires that a reasonable temperature is to be ensured indoors, having regard to the purpose of the workplace. Outdoors, the place of work should provide protection from the weather, taken in relation to reasonable practicability and any PPE provided, as well as the purpose of use of the workplace.

Lighting generally on-site is covered by **Regulation 25**. Every place of work and approach to it, and traffic routes must be lit, and the lighting shall be by natural light so far as is reasonably practicable. Regard is to be had for the ability of artificial light to change perception of signs or signals, so the colour of the lighting is to be controlled with this in mind. Secondary lighting shall be provided to any place where there would be a health or safety risk to any person in the event of failure of primary artificial lighting (paragraph 3).

'Housekeeping' is known as 'good order' in **Regulation 26**; good order and a reasonable standard of cleanliness is required, so far as is reasonably practicable (paragraph 1). Where necessary for reasons of safety, the extent of the site must be readily identifiable by others, and this includes marking the perimeter with signs. Timber or other material with dangerous projecting nails shall not be used or be allowed to remain in a place where it may be dangerous (paragraph 3).

Regulation 27 was directed at plant and equipment, and has been revoked by PUWER 98. The training requirement in **Regulation 28** introduces no age limits and incorporates all of a number of previous specific training requirements. Wherever training, knowledge or experience is necessary, it shall be possessed or the person must be under an appropriate degree of supervision by someone who already has it.

Inspection requirements are contained in **Regulation 29**. The main Regulation refers the seeker after inspection details to Schedule 7. The Regulation itself requires that a place of work specified in the Schedule cannot be used to carry out construction work unless it has been inspected by a competent person as required by the Schedule and that person is satisfied that the work can be done safely (paragraph 1). In particular (paragraph 2), where the place of work is a part of a scaffold, excavation, cofferdam or caisson, anyone controlling the way the work is done must ensure the scaffold, etc., is stable, of sound construction and that the safeguards required by the Regulations are in place before his/her employees or persons under his/her control first use the workplace.

Paragraph 3 requires that, where the competent person is not satisfied that the work can be done safely from the workplace and the inspection was carried out on behalf of another person, he/she shall inform that person of anything he/she is not satisfied about and the place of work is not to be used until the defects are remedied. Inspections of a place of work must include plant, equipment and any materials which affect the safety of the place of work (paragraph 4).

As above, **Schedule 7** lists the places of work requiring inspection, and the times for the inspections. These are:

1. Any working platform or part thereof (not just scaffolding), and any personal suspension equipment, which are to be inspected:
   (a) before being taken into use for the first time
   (b) after substantial addition, dismantling or other alteration
   (c) after any event likely to have affected its strength or stability
   (d) at intervals not exceeding seven days since the last inspection.
   This applies to platforms from which falls of more than 2 m can occur.
2. Any excavation which is supported as required by paragraphs 1, 2 or 3 of Regulation 12:
   (a) before work at the start of every shift
   (b) after any event likely to have affected its strength or stability
   (c) after any accidental fall of rock, earth or other material.
3. Cofferdams and caissons, which are to be inspected:
   (a) before work at the start of every shift
   (b) after any event likely to have affected its strength or stability.

Inspections, though, are not required to be the subject of reports in every case. **Regulation 30** sets out those cases in which reports are required and the manner of the reporting. The required contents of the report of inspection are detailed in Schedule 8, which supplements this Regulation. Note that the old inspection register used in the construction industry (Form 91 Part 1) does not allow for the recording of all the particulars now required, and there is no statutory form listed to replace it.

Schedule 8 states that the prescribed contents to be included in reports are:

- Name and address of the person on whose behalf the inspection was carried out

- Location of the work inspected
- Description of the place of work inspected
- Date and time of the inspection
- Details of matters found that could give rise to a risk to health or safety of any person
- Details of any action taken as a result of finding those matters
- Details of any further action considered necessary
- Name and position of person making the report

The basics of the requirements on reports in Regulation 30 are:

- Paragraph 1 — where an inspection is required, the report, conforming to Schedule 8, must be prepared by the person who carried out the inspection before the end of the working period in which the inspection was made.
- Paragraph 2 — the person preparing the report must provide a copy of it to the person on whose behalf the inspection was carried out within 24 hours of completing the inspection.
- Paragraph 3 — the report or a copy of it must be kept at the place of work where the inspection was carried out, and after the work is completed it must be kept at the office of the person for whom it was prepared for three months after the date of completion.
- Paragraph 4 — reports are open to inspection by inspectors, and can be required to be sent to an inspector.
- Paragraph 5 — no report is required following inspection of a working platform or alternative means of support where falls of less than 2 m could occur.
- Paragraph 6 — no report is required for mobile towers, unless they remain erected in the same place for 7 days or more. (Note that if a tower has to be dismantled and then re-erected because it has been moved, then it is probably no longer 'in the same place'.)
- For additions, dismantling or other alteration, not more than one report every 24 hours is required.
- For excavations, cofferdams and caissons, an inspection is required at the start of every shift, but a report is needed only every 7 days.

A main consequence of this Regulation is that inspections and reports are required on scaffolding over 2 m high before first use by anyone.

The remaining Regulations (31–35) deal with the issue of exemption certificates, application of the Regulations outside Great Britain, enforcement of the Regulations on fire, and modifications and revocations.

CONSTRUCTION (DESIGN AND MANAGEMENT) REGULATIONS 1994

These Regulations, usually referred to as CDM, implement the design and management content of the Temporary or Mobile Construction Sites Directive, which was adopted on 24 June 1992. The Directive contains two parts: the first applies the Framework Directive provisions to construction sites; the second refers the Workplace Directive to sites. The Construction (Health, Safety and Welfare) Regulations 1996 (CHSWR) implement the second part of the Directive. They took effect on 31 March 1995.

Recognising that the construction 'system' in the UK has characteristics not anticipated by the Directive, the CDM Regulations simplify it by requiring the client to appoint central figures for the planning and the carrying out of construction work falling within the scope of the Regulations. These are the **Planning Supervisor** and the **Principal Contractor**. The link between them is the **Health and Safety Plan**, initiated by the Planning Supervisor and adjusted by the Principal Contractor to provide health and safety details relevant to the work — risks, timings, information — arrangements which together make up a comprehensive 'how to do it safely' document for the work to be done. There is also a requirement for a **Health and Safety File** to be kept and given to the eventual owner of the construction work — a 'how we built it' collection of relevant information which will be of use if future modification is required. The Regulations give detailed job descriptions for identified parties to the contract, in addition to duties placed on them under the Management of Health and Safety at Work Regulations 1992 (MHSWR).

Safety by **design** is also an important feature of CDM, as it places duties on designers and others to make assessments of the impact the proposed design may have on contractors, those who maintain a structure, and those who use it.

Generally, the Regulations apply to construction work which:

- Lasts for more than 30 days, or
- Will involve more than 500 person-days of work, or
- Includes any demolition work (regardless of duration or size), or
- Involves five or more workers being on-site at any one time

There are some exceptions where CDM does not apply, although CDM always applies to any design work for the construction process. The Regulations do not apply where the work to be done is minor in nature, or where it is carried out in premises normally inspected by the local authority. 'Minor work' is done by people who normally work on the premises. It is either not notifiable, or entirely internal, or carried out in an area which is not physically segregated, normal work is carrying on and the contractor has no authority to exclude people while it is being done. Maintaining or removing insulation on pipes or boilers, or other parts of heating and hot water systems, is not classed as 'minor work'.

Some significant definitions

- **Agent** means any person who acts as agent for a client in connection with the carrying on by the person of a trade, business or other undertaking, whether for profit or not.
- **Client** means any person for whom a project is carried out, whether the project is carried out in-house or by another person. One of a number of clients, or the agent of a client, can volunteer to accept the duties of Regulations 6 and 8–12. This acceptance has to be made by way of a declaration to the HSE when the project is notified.
- **Cleaning work** means the cleaning of any window or transparent/translucent wall, ceiling or roof in or on a structure where the cleaning involves a risk of falling more than 2 m.
- **Construction work** means the carrying out of any building, civil engineering or engineering construction work, and includes any of the following:
  - □ construction, alteration, conversion, fitting-out, commissioning, renovation, repair, upkeep, redecoration or other maintenance (including cleaning which involves the use of water or an abrasive at high pressure or the use of corrosive or toxic substances), decommissioning, demolition or dismantling of a structure
  - □ preparation for an intended structure including site clearance, exploration, investigation (but not site survey) and excavation, and laying or installing the foundations of the structure
  - □ assembly or disassembly of prefabricated elements of a structure

- □ removal of a structure or part of a structure, or of any product or waste resulting from demolition or dismantling of a structure or disassembly of pre-fabricated elements of a structure
- □ installation, commissioning, maintenance, repair or removal of mechanical, electrical, gas, compressed air, hydraulic, telecommunications, computer or similar services which are normally fixed within or to a structure – but does not include mineral resource exploration or extraction activities.

- **Contractor** means any person who carries on a trade or business or other undertaking, whether for profit or not, in connection with which he/she undertakes to or does manage construction work, or arranges for any person at work under his/her control (including any employee, where he/she is an employer) to carry out or manage construction work. This definition can therefore be applied to the self-employed.
- **Design** in relation to any structure includes drawing, design details, specification and bill of quantities (including specification of articles or substances) in relation to the structure.
- **Designer** means any person who carries on a trade, business or other undertaking in connection with which he/she prepares a design (Amendment Regulations in force from October 2000 have refined this definition for clarification, under impetus from a court ruling).
- **Health and safety file** means a file or other record in permanent form containing the information required by Regulation 14(d) – about the design, methods and materials used in the construction of a structure which may be necessary for appropriate third parties to know about for their health or safety.
- **Project** means a project which includes or is intended to include construction work.
- **Structure** means any building, steel or reinforced concrete structure (not being a building), railway line or siding, tramway line, dock, harbour, inland navigation, tunnel, shaft, bridge, viaduct, waterworks, reservoir, pipe or pipeline (regardless of intended or actual contents), cable, aqueduct, sewer, sewage works, gasholder, road, airfield, sea defence works, river works, drainage works, earthworks, lagoon, dam, wall, caisson, mast, tower, pylon, underground tank, earth retaining structure or structure designed to preserve or alter any natural feature, and any similar structure to these, and any formwork, falsework, scaffold or other structure designed or used to provide support or means of access during construction work,

and any fixed plant in respect of work which is installation, commissioning, de-commissioning or dismantling and where that work involves a risk of falling more than 2 m.

Summary of the Regulations

Regulation 4 covers circumstances where a number of clients (or their agents, which means people acting with the client's authority) are about to be involved in a project. One of the clients, or the agent of a client, can take on the responsibilities of the effective sole client. Clients appointing agents must be reasonably satisfied about their competence to carry out the duties given to clients by the Regulations. A declaration is sent to the HSE, which is required to acknowledge and date the receipt of it.

Regulation 5 applies CDM to the developer of land transferred to a domestic client and which will include premises intended to be occupied as a residence. Where this happens, the developer is subject to Regulations 6 and 8–12 as if he/she were the client.

Regulation 6 requires every client to appoint a Planning Supervisor and Principal Contractor in respect of each project. These appointments must be changed or renewed as necessary so that they remain filled at all times until the end of the construction phase. The Planning Supervisor has to be appointed as soon as is practicable after the client has enough information about the project and the construction work to enable him to comply with Regulations 8(1) and 9(1). The Principal Contractor, who must be a contractor, has to be appointed as soon as practicable after the client has enough information to enable him to comply with Regulations 8(3) and 9(3). The same person can be appointed as both Planning Supervisor and Principal Contractor provided he/she is competent to fulfil both roles, and the client can appoint himself/herself to either or both positions provided he/she is similarly competent.

Regulation 7 requires the Planning Supervisor to give written notice of notifiable projects to the HSE, as soon as practicable after his/her appointment as Planning Supervisor (Parts I and II) and as soon as practicable after the appointment of the Principal Contractor (Parts I and III), and in any event before the start of the construction work. Where work which is notifiable is done for a domestic client and a developer is not involved, then the contractor(s) doing the work have the responsibility of notifying the HSE, and one of these can notify on behalf of any others. Again, notification must be made before any work starts.

Regulation 8 covers the requirements for competence of Planning Supervisor, designers and contractors. The client must be reasonably satisfied that a potential Planning Supervisor is competent to perform the functions required by these Regulations, before any person is appointed to the role (Regulation 8(1)). No designer can be hired by any person to prepare a design unless that person is reasonably satisfied the designer is competent to do so, and similarly no contractor can be employed by anyone to carry out or manage construction work unless the person employing the contractor is reasonably satisfied as to the contractor's competence (Regulation 8(3)). Those under a duty to satisfy themselves about competence will only discharge that duty when they have taken such steps as it is reasonable for a person in their position to take — which include making reasonable enquiries or seeking advice where necessary — to satisfy themselves as to the competence. 'Competence' in this sense refers only to competence to carry out any requirement, and not to contravene any prohibition, placed on the person by any relevant regulation or provision.

'Provision' for health and safety in a wide sense is covered by **Regulation 9**. Clients must not appoint any person as Planning Supervisor unless they are reasonably satisfied that the potential appointee has allocated or will allocate adequate resources to enable the functions of Planning Supervisor to be carried out. Similarly, anyone arranging for a designer to prepare a design must be reasonably satisfied the potential designer has allocated or will allocate adequate resources to comply with Regulation 13. No one is allowed to appoint a contractor to carry out or manage construction work unless reasonably satisfied that the contractor has allocated or will allocate adequate resources to enable compliance with statutory requirements.

The client is required by **Regulation 10** to ensure so far as is reasonably practicable that a Health and Safety Plan which complies with Regulation 15(4) has been prepared in respect of the project before the construction phase starts. There is no duty on either the client or the Planning Supervisor to ensure that the plan continues to be in compliance with Regulation 15 once work has begun; this duty to keep it up to date is laid on the principal contractor by Regulation 15(4).

Regulation 11 obliges the client to ensure that the Planning Supervisor is provided with information relevant to his functions about the state or condition of any premises where relevant construction work will be carried out. This will be information which the client has, or which the client could obtain by making reasonable enquiries, and it has to be provided as soon as reasonably practicable but certainly before work starts to which the information relates.

The client is required by **Regulation 12** to take reasonable steps to ensure that the information in any Health and Safety File which is given to him/her is kept available for inspection by anyone who may need the information in the file to comply with any law. A client who sells the entire interest in the property of the structure can satisfy this requirement by handing over the file to the purchaser and ensures the purchaser is aware of the significance and contents of the file.

Designers' duties are given in **Regulation 13**. Except where the design is prepared in-house, employers cannot allow employees to prepare a design, and no self-employed person can prepare a design, unless the employer has taken reasonable steps to make the client for the project aware of the duties of the client within these Regulations and any practical HSC/E guidance on how to comply.

The designer is firstly to ensure that any design he/she prepares and which he/she knows will be used for construction work includes adequate regard to **'three needs'**. The designer has to:

1. Avoid foreseeable risks to health and safety of those constructing or cleaning the structure at any time and anyone who may be affected by that work
2. Ensure that risks to constructors or cleaners of the structure at any time, or to anyone who may be affected by that work, are combated at source
3. Ensure that priority is given to measures which will protect all such persons over measures which only protect each person at work

Secondly, the designer is to ensure the design includes adequate information about any aspect of the project, structure or materials to be used which might affect the health or safety of constructors, cleaners or anyone who may be affected by their work.

Thirdly, a (stronger) duty to co operate with the Planning Supervisor and any other designer is placed on a designer, so far as is necessary to enable each of them to comply with health and safety laws.

The duties are then qualified by Regulation 13(3), which allows the first two of them to include the required matters only to the extent that it is reasonable to expect the designer to deal with them at the time the design is prepared, and as far as is reasonably practicable.

Regulation 14 details the main duties of the planning supervisor. This dutyholder must:

1. Ensure as far as is reasonably practicable the design of any structure in the project complies with the needs specified in Regulation 13 and includes adequate information
2. Take reasonable steps to ensure co-operation between designers to enable each to comply with Regulation 13
3. Be in a position to advise any client and any contractor to enable them to comply with Regulations 8(2) and 9(2) — competence of designer — and to advise any client on compliance with Regulations 8(3), 9(3) and 10 (competence of contractor and readiness of the Health and Safety Plan)
4. Ensure that a Health and Safety File is prepared in respect of each structure in the project, reviewing and amending it over time, and finally delivering it to the client on completion of construction work on each structure

Regulation 15 sets out the specific requirements for the Health and Safety Plan. This is to be prepared by the appointed Planning Supervisor so as to contain the required information and be available for provision to any contractor before arrangements are made for the contractor to carry out or manage construction work on the project.

The information required to go into the Health and Safety Plan is:

- A general description of the construction work
- Details of the intended timescale for the project and any intermediate stages
- Details of any risks known to the Planning Supervisor or which are reasonably foreseeable risks to the health and safety of constructors
- Any other information which the Planning Supervisor has or could reasonably get which a contractor would need to show he/she has the necessary competence or has or will get the adequate resources required by Regulation 9

- Information which the Principal Contractor and other contractors could reasonably need to satisfy their own duties under the Regulations

The Principal Contractor must take reasonable steps to ensure the Health and Safety Plan contains until the end of the construction phase the required features, which are:
- Arrangements for the project which will ensure so far as is reasonably practicable the health and safety of all constructors and those who may be affected by the work, taking account of the risks involved in the work and any activity in the premises which could put any people at risk
- Sufficient information about welfare arrangements to enable any contractor to understand how he/she can comply with any duties placed on him in respect of welfare
- Arrangements which will include where necessary the method of managing the construction work and monitoring of compliance with health and safety regulations

Regulation 16 specifies the duties and powers of the Principal Contractor. These are firstly to take reasonable steps to ensure co-operation between contractors so far as is necessary to enable each to comply with requirements imposed. This includes, but is not limited to, those sharing the construction site for the purposes of Regulation 9 of MHSWR.

Secondly, the Principal Contractor must ensure so far is reasonably practicable that every contractor and every employee complies with any rules in the Health and Safety Plan. The Principal Contractor can make any reasonable written rules and include them in the plan, and give reasonable directions to any contractor.

Thirdly, the Principal Contractor must take reasonable steps to ensure that only authorised persons are allowed where construction work is being carried out.

Fourth, he/she must ensure that required particulars are displayed in any notice covered by Regulation 7, and are displayed in a readable condition where they can be read by any person at work on the project.

Finally under this Regulation, he/she must provide the Planning Supervisor promptly with any information he possesses or could reasonably find out from a contractor which the Planning Supervisor does not already possess and which could reasonably be believed necessary to include in the Health and Safety File.

Duties on the giving of information and training requirements are set out in **Regulation 17**. The Principal Contractor must (as far as is reasonably practicable) ensure that every contractor is provided with comprehensible information on the risks to himself/herself and any employees or persons under his/her control which are present as a result of the work. In the same terms, the Principal Contractor must ensure that every contractor who employs people on the work provides his/her employees with the information and training required by Regulations 8 and 11(2)(b) of MHSWR.

Provision is made by **Regulation 18** for receipt of advice from employees and the self-employed by the Principal Contractor, who must ensure that there is a suitable mechanism in place for discussing and conveying their advice on health and safety matters affecting their work. Arrangements for the co-ordination of these views of employees or their representatives are to be made having regard to the nature of the work and the size of the premises concerned.

Contractors' duties are covered by **Regulation 19(1)**, where in relation to a project they must co-operate with the Principal Contractor as necessary, provide the Principal Contractor with any relevant information which might affect anyone's health or safety while on the project or who could be affected by it. This information includes relevant risk assessments, and information which might prompt a review of the Health and Safety Plan for the project. Contractors must also comply with any directions of the Principal Contractor, and any applicable rules in the plan. The Regulation requires contractors to provide the Principal Contractor with any information which is notifiable by the contractor to the enforcing authority under RIDDOR — details of injuries, diseases and dangerous occurrences as defined which are related to the project. Other information to be supplied by the contractor to the Principal Contractor includes anything which he/she knows or could reasonably find out which the Principal Contractor does not know and would reasonably be expected to pass to the Planning Supervisor if he/she did know, in order to comply with Regulation 16(e) — to amend the Health and Safety File.

Regulation 19 contains in parts (2), (3) and (4) general requirements to be observed by employers with employees working on construction work, and the self-employed. No employer can allow his employees to work on construction work, and no self-employed person can work, unless the employer has been given specific pieces of information:

- The names of the Planning Supervisor and Principal Contractor
- The Health and Safety Plan or relevant parts of it

A defence is provided against prosecution in part (5), which allows that this duty can be satisfied by showing that all reasonable enquiries had been made and the employer or self-employed person reasonably believed either that the Regulations did not apply to the particular work being done or that he had in fact been given the information required.

The Regulations have the same coverage as the Health and Safety at Work etc. Act 1974 by **Regulation 20**, and, except for two of them, do not confer a right of civil action (**Regulation 21**). Breaches generally can only result in criminal prosecution. This provision is the same as that contained in MHSWR. The enforcing authority for the Regulations is exclusively the HSE (**Regulation 22**). More detailed discussion of the roles of dutyholders under CDM is contained elsewhere in this book.

THE CONFINED SPACES REGULATIONS 1997

The Confined Spaces Regulations (SI 1997 No 1713) came into force on 28 January 1998, and apply in all premises and circumstances where the Health and Safety at Work Act applies. To comply with the Confined Spaces Regulations it will be necessary to consider other Regulations including COSHH 99, PUWER 98, PPE and MHSWR. The Regulations specify requirements and prohibitions to protect the health and safety of persons working in confined spaces and also those who may be affected by the work.

Summary of the Regulations

The definition of 'confined space' is very wide and includes any place such as trenches, vats, silos, pits, chambers, sewers, wells or other similar spaces which because of their nature could give rise to a 'specified risk'. 'Specified risks' are defined in **Regulation 1**. They include injury from fire or explosion; loss of consciousness through a rise in body temperature or by asphyxiation; drowning; or from free-flowing solids causing asphyxiation or preventing escape from a space.

Confined space hazards arise because of the confined nature of the place of work and the presence of substances or conditions which, when taken together, increase risks to safety and health. It must be remembered that a hazard can be introduced into a space that would otherwise be safe.

A 'confined space' has two defining features:

1. Such a space is substantially (although not necessarily entirely) closed, and
2. There is a foreseeable risk from hazardous substances or conditions within the space or nearby.

Obvious construction industry examples include manholes, shafts, inspection pits, cofferdams and brewing vats. Less obvious are building voids, plant rooms, cellars, and the interiors of plant, machines or vehicles.

Likely **hazards** include:

- Flammable substances, from either the contents of the space or a nearby area
- Oxygen enrichment, for example from a leaking welding cylinder
- Ignition of airborne contaminants
- Fumes or sludge remaining from previous processes or contents. These may release toxic or flammable gases when disturbed
- Oxygen deficiency, which can result from inert gas purging; from natural biological processes such as rusting, decomposition or fermentation; from processes such as burning and welding; and as a result of workers breathing within the space
- Liquids entering the space from elsewhere, and solid materials which can flow into it
- Heat exhaustion caused by working in the confined space or from nearby processes

Regulation 3 imposes a general duty on employers to ensure that their employees comply with the Regulations. The self-employed must also comply with the Regulations, and both employers and the self-employed must ensure that those over whom they have control comply as far as their control permits. In many cases the employer or the self-employed person will need to liaise with other parties to ensure that their duties under the Regulations are fulfilled.

Regulation 4 requires that no one shall enter a confined space to carry out work for any purpose, unless it is not reasonably practicable to achieve the purpose without entering the space. Under the Health and Safety at Work etc. Act 1974 and the Construction (Design and Management) Regulations 1994, there are duties on designers to ensure that articles and buildings are designed to minimise foreseeable risks to health and safety, and engineers, designers and others should aim to eliminate or minimise the need to enter confined spaces. Employers have a duty to prevent employees and others under their control from entering or working inside a confined space when it is reasonably practicable to do the work from outside.

Employers should consider (and be advised by designers to consider) modifying the space to avoid the need for entry, or allow work to be done from outside. Working practices may need to be changed to make atmosphere testing, cleaning or inspections possible from outside the space using suitable equipment, sight glasses or CCTV.

To comply with Regulation 4, risk assessment will be necessary. The priority in carrying out a confined spaces risk assessment is to identify the measures necessary to avoid work within the space. Where the assessment shows that it is not reasonably practicable to do the work without entering the space, it can be used to identify the precautions to be included in a safe system of work. A competent person must carry out the risk assessment, and in large or complex situations, more than one person may need to be involved.

The factors to be assessed will include:

- The general situation and the risks that may be present
- The previous contents of the space, including any residues and the effect of disturbing them
- The risk of contamination from adjacent spaces and nearby plant or machinery. Where confined spaces are below ground, there is a particular risk of contamination through soil strata or from outside machinery. The risk of ingress of substances from other areas should be considered
- The likelihood of oxygen deficiency or oxygen enrichment
- The physical dimensions and layout of the space, the possibility of gas accumulation at different levels and constraints on access and rescue
- Any hazards from the work process itself, including chemicals and sources of ignition
- The requirements for rescue procedures

Regulation 4(2) requires that no person shall enter a confined space unless a safe system of work is in place. Before deciding what precautions are needed for entry, priority should be given to eliminating any sources of danger. The factors to be considered in designing a safe system, and which may form the basis for a permit to work, will depend on the risk assessment and could include:

- Necessary supervision levels, including the possible need to appoint a competent person to supervise the work
- Competence and suitability of workers, including training and experience and physical and mental attributes
- Provision of adequate communications arrangements to summon help in an emergency, and to allow those inside the space to communicate with each other and those outside
- How the atmosphere within the space will be tested, the choice of test equipment, the type of contaminants and level of oxygen.
- Whether the space will require purging with air or inert gas, and the amount and method of ventilation that will be required
- Whether it is necessary to clean the space or remove residues and how this will be done
- The way in which gases, liquids and flowing materials, and mechanical and electrical equipment can be isolated
- Selection and use of suitable equipment for lighting, work or rescue inside the space, including PPE and RPE. Working time may need to be limited where RPE is being worn or where heat or humidity are high
- Whether petrol, diesel and gas-fuelled machinery can be excluded from the space, control procedures for gas hoses and pipelines and measures to prevent static electricity build-up including earthing and bonding
- Arrangements for access, egress and emergency and rescue arrangements
- Fire prevention, material storage and smoking control arrangements inside and around the space

Regulation 5 prohibits entry into a space unless suitable rescue arrangements have been provided. The arrangements must reduce the risks to those involved in the rescue to the lowest reasonably practicable level, and should include the provision of resuscitation equipment where conditions require it. Regulation 5 also requires that where any emergency situation arises, the rescue arrangements must be put into operation immediately.

Regulation 7 provides a defence, whereby an employer can claim that rescue arrangements were not put into immediate effect through the fault of another person who was not in his employment, and that he took all reasonable precautions to prevent this happening. If the defence is successful, the other person may become guilty of the offence.

The precise rescue arrangements will depend on the risks identified. When assessing rescue arrangements, general accidents such as incapacitation after falls should be considered as well as those resulting specifically from work in confined spaces. Recovering an injured person by rope and harness can be difficult if the escape route is very narrow, even when a full safety harness is worn so that the body can be raised vertically rather than bent at the waist.

Rescue equipment provided as part of the arrangements should be suitable for the risk, and could include 'self rescue' equipment (where there will be time to react to an anticipated emergency situation), lifelines and lifting equipment, first aid equipment and breathing apparatus.

In the case of prolonged or complex work, and where the risks require it, it may be necessary to warn the emergency services and provide them with information before starting. Whether the emergency services are notified in advance or not, there should be a procedure in place to ensure that they can be alerted rapidly in the event of an accident.

Workers and rescuers should be trained in the communication, emergency and rescue procedures and the use of equipment. Training should include refresher training and rehearsals and drills, and cover the following areas:

- Likely causes of an emergency
- Use of equipment, including donning procedures, tests and function checks, malfunctions and defects, and maintenance
- Emergency procedures and methods of raising the alarm, including contact and liaison with the emergency services
- Methods of shutting down adjacent plant
- Operation of fire fighting equipment

THE CONTROL OF SUBSTANCES HAZARDOUS TO HEALTH REGULATIONS 1999

Construction workers are often exposed to substances which have the potential to damage their health. Many of these are present as a direct result of their use – they are brought onto site. Some are formed during the work itself, for example the generation of toxic gases by internal combustion engines. Others are used in peripheral activities, such as maintenance cleaning. Some hazardous substances occur naturally, such as microbiological agents. These can cause diseases like leptospirosis (leptospiral jaundice, also known as Weil's disease) that affect construction and maintenance workers.

The move to control all such substances came as a result of EC Directive 80/1107/EEC on 'the protection of workers from the risk related to exposure to chemical, physical and biological agents at work'. The Directive resulted in the introduction of the Control of Substances Hazardous to Health Regulations, widely known as COSHH. These Regulations cover all people at work, including construction work, and they apply to virtually all substances hazardous to health. The original Regulations were introduced in 1988, and were revised and enlarged to incorporate provisions required by the Biological Agents Directive 90/679/EEC, and to prohibit importation into the UK of specified substances and articles from outside the European Economic Area. The latest revision of the Regulations – COSHH 99 – came into force on 25 March 1999.

The changes made in the 1999 version streamlined what had become an unwieldy set of controls, requiring constant changes to the Regulations. As a result, the 1999 Regulations no longer contain the schedule of substances assigned maximum exposure limits (MELs), which is now to be found inside the annually-revised HSE publication, EH40. Other 1999 changes included the expansion of the definition of 'substances hazardous to health' to include trigger limits for 'total inhalable dust' or 'respirable dust', refining the definition of 'suitable' PPE, giving members of the armed forces an appeal against suspension from work on medical grounds, and making minor changes to some technical definitions.

What is a 'substance hazardous to health'?

This term includes any material, mixture or compound used at work or arising from work activities, which is harmful to people's health in the form in which it occurs in the work activity. Categories specifically covered are:

1. Substances, or combinations of them, classed as dangerous (toxic, very toxic, harmful, corrosive and irritant) under other statutory provisions. These can be identified by the warning labels used on their containers. These will usually be in the form of orange squares with pictograms, and they will be labelled toxic or very toxic (pictograms of a skull and crossbones), harmful or irritant (a black cross) or corrosive (a dripping test tube). Any substance in this group must be properly labelled by the supplier and have a data sheet provided for the user.
2. Substances assigned an occupational exposure standard (OES) or MEL
3. Harmful biological agents, such as bacteria and other micro-organisms
4. Dust, other than a substance within 1. or 2. above, with an airborne concentration of more than $10\,mg/m^3$ (8-hour time-weighted average) of total inhalable dust, and more than $4\,mg/m^3$ similarly of respirable dust, where no lower value is given for the substance. 'Respirable' dust has a mean particle size that permits it to enter the deep lung, with potentially greater danger
5. Any substance creating a comparable hazard, not already covered in the previous categories. Note that asbestos and lead have their own specific regulations (see below).

Summary of the Regulations

Duties are imposed upon employers for the protection of employees who may be exposed to substances hazardous to health at work, and of other persons who may be affected by such work. Specific duties are also imposed upon the self-employed and employees. An employer must not carry on any work which is liable to expose any employee to a substance hazardous to health unless a suitable and sufficient assessment has been made of the risks to health created by the substance at work and about the measures necessary to control exposure to it (**Regulation 6**).

This is the key requirement; the assessment is a systematic review of the use of the substance present, its form and quantity, possible harmful effects, how it is stored, handled, used and transported as appropriate, the people who may be affected and for how long, and the control steps which are appropriate. The assessment should take all of these matters into account, taking care to describe actual conditions when assessing risk. It is important to realise that collection of suppliers' material safety data sheets does not constitute making a risk assessment, but rather the gathering of data to assist in making it. The assessment should be written down for ease of communication. All risk assessments should be reviewed every 5 years, in addition to the requirement to review them if they are suspected of being invalid.

The employer must ensure that the exposure of employees to substances hazardous to health is either prevented, or where this is not reasonably practicable, is adequately controlled (**Regulation 7**). This applies whether the substance is hazardous through inhalation, ingestion, absorption through the skin or contact with the skin. Prevention of exposure must be attempted in the first place by means other than the provision and use of personal protective equipment. Regulation 7(2) only permits use of personal protective equipment (PPE) where other control is not reasonably practicable. Any respiratory protective equipment supplied must comply with current requirements as to suitability (Regulation 7(5)). Prevention of exposure can be achieved by removing the hazardous substance, substituting it with a less hazardous one or using it in a less hazardous form, enclosing the process, isolating the worker(s), using partial enclosure and extraction equipment, general ventilation, and the use of safe systems of work.

Occupational exposure limits have been set (and are revised and updated yearly) for a wide range of substances. Where an MEL has been assigned, the level of exposure must be reduced as far as is reasonably practicable (Regulation 7(6)). The MEL should not be exceeded. For substances assigned an OES, it is sufficient to reduce the level of exposure to that standard (Regulation 7 (7)). Special provisions (Regulation 7(9)) are made to cover situations where control measures fail resulting in potential release of carcinogens into the workplace.

Employers providing control measures must ensure they are properly used, and every employee must make full and proper use of what is provided, reporting any defects to the employer and doing their best to return it after use to any accommodation provided for it (**Regulation 8**).

Regulation 9 requires those control measures which are provided to be maintained in efficient working order and good repair. It should be noted that the term 'local exhaust ventilation equipment' — one type of control specifically mentioned as requiring maintenance and inspection — can be applied to any device which captures, controls or contains airborne releases of pollutants at or near to the point of their emission by means of ventilation, and conveys the pollutant to a point where it can be safely collected or released. Construction examples include fixed extraction systems in joineries and on some hand tools.

If the controls are engineering controls, they must be examined and tested at prescribed intervals. For local exhaust ventilation equipment, this is defined as at least once in every 14 months generally. The Regulation also requires examination and testing of respiratory protective equipment at suitable intervals, where the equipment is provided under Regulation 7. Records of these tests and examinations have to be kept, and a summary of them has to be kept for at least 5 years. PPE is required to be kept in a clean condition.

Monitoring of exposure must be carried out (**Regulation 10**) where it is necessary to ensure that exposure is adequately controlled. For listed processes and substances, the frequency is specified. Records of this monitoring have to be kept for at least 5 years, or 40 years where employees can be personally identified. Monitoring would be required where:

- There could be serious risks to health if control measures should deteriorate
- It cannot be guaranteed without measurement that exposure limits are not being exceeded, or control measures are working properly

Regulation 11 requires health surveillance where appropriate, which would be where there is significant exposure to a substance listed in Schedule 5 to the Regulations and the employee is working at a process specified in the Schedule; or:

- Where there is an identifiable disease or adverse health effect which may be related to the exposure; and
- There is a reasonable likelihood that this may occur under the particular work conditions; and
- There are valid techniques for detecting indications of it

The surveillance can be done by doctors, nurses, or by trained supervisors making simple checks. Where the surveillance is carried out, records or copies of them must be kept for 40 years from the date of the last entry.

If any of his employees are or may be exposed to substances hazardous to health, the employer must provide them with suitable and sufficient information, instruction and training for them to know the health risks created by the exposure and the precautions which should be taken (**Regulation 12**). The information must include the results of any monitoring and the collective (non-personalised) results of health surveillance.

Provision is made in **Regulation 13** for the control of certain fumigation operations, generally requiring notices to be posted beforehand.

The Ministry of Defence is exempt from certain of the requirements of the Regulations, and there are a number of exceptions to them which are mostly covered in Regulation 3. Notably, health surveillance requirements are not extended to non-employees, but the training, information and monitoring requirements of Regulations 10, 12(1) and 12(2) are extended to them if they are on the employer's premises. Regulations 10 and 11 do not apply to the self-employed, but all the others do.

Regulations 6 to 12 do not have effect in some circumstances, which are detailed in **Regulation 5**. For the construction industry in general, these are where:

- The Control of Lead at Work Regulations 1998 apply
- The Control of Asbestos at Work Regulations 1987 apply
- The hazardous substance is hazardous solely because of its radioactivity, flammability, explosive properties, high or low temperature or high pressure, and
- The health risk to a person derives because the substance is administered in the course of medical treatment

THE PERSONAL PROTECTIVE EQUIPMENT AT WORK REGULATIONS 1992

The Personal Protective Equipment at Work Regulations replaced much of the detailed, prescriptive requirements of legislation covering a wide range of industries and particular tasks. They place duties on both employers and the self-employed. The Regulations provide a framework for the provision of personal protective equipment (PPE) in circumstances where assessment has shown a continuing need for personal protection. Generally, a breach of these Regulations can give rise to a civil claim against an employer.

PPE used for protection while travelling on a road, such as cycle helmets, crash helmets and motorbike leathers worn by employees on the highway, is not covered by the Regulations. However, the requirements do apply in circumstances where this type of equipment is worn elsewhere at work (other than on the highway). Self-defence equipment (personal sirens/alarms) or portable devices for detecting and signalling risks and nuisances (such as personal gas detection equipment or radiation dosimeters) are not within the scope of the Regulations.

The requirements of the Regulations do not apply to most respiratory protective equipment, ear protectors and some other types of PPE, because these are already covered by existing Regulations such as COSHH, the Ionising Radiations Regulations, the Control of Lead and Asbestos at Work Regulations and the Construction (Head Protection) Regulations. However, this legislation is amended in varying degrees by Schedule 1 of the Regulations, which introduces the concept of risk assessment where previously missing. The Regulations also amend earlier provisions by requiring equipment to comply with European conformity standards and inserting a requirement to return it to 'suitable storage' after use.

Personal protective equipment means all equipment designed to be worn or held by a person at work to protect against one or more risks, and any addition or accessory designed to meet this objective. Both protective clothing and equipment are within the scope of this definition, and therefore such items as diverse as safety footwear, waterproof clothing, safety helmets, gloves, high-visibility clothing, eye protection, respirators, underwater breathing apparatus and safety harnesses are covered by the Regulations. Ordinary working clothes or clothing provided which is not specifically designed to protect the health and safety of the wearer is not within the definition — such as clothes provided with the primary aim of presenting a corporate image.

Summary of the Regulations

The Regulations do not apply to the crews of ships while operational (**Regulation 3**). Shore-based contractors working on ships will be within scope of the Regulations while inside territorial waters. The requirements apply to PPE used on ships operating on inland waterways and aircraft while on the ground and operating in UK airspace.

Regulation 4 requires every employer to provide 'suitable' PPE to each of his employees who may be exposed to any risk while at work, except where any such risk has been adequately controlled by other means which are equally or more effective. A similar provision applies to the self-employed.

'Suitable' is defined as:

- Appropriate for the risks involved and the conditions
- Taking account of ergonomic requirements and the state of health of the person wearing it
- Capable of fitting the wearer correctly after adjustments
- Effective to prevent or adequately control the risk without leading to any increased risk (particularly where several types of PPE are to be worn) so far as is practicable
- Complying with national and European conformity standards in existence at the particular time. (This requirement does not apply to PPE obtained before the harmonising Regulations came into force, which can continue to be used without European Conformity marks, but the employer must ensure that the equipment he/she provides remains 'suitable' for the purpose to which it is put)

PPE is to be used as a last resort. Steps should first be taken to prevent or control the risk at source by making machinery or processes safer and by using engineering controls and systems of work. Risk assessments made under the Management of Health and Safety at Work Regulations 1999 will help determine the most appropriate controls.

Regulation 5 requires the employer and others to ensure the compatibility of PPE in circumstances where more than one item of equipment is required to control the various risks. Before choosing any PPE the employer or the self-employed person must make an assessment to determine whether the proposed PPE is suitable (**Regulation 6**). This assessment of risk is necessary because no PPE provides 100% protection for 100% of the time (it may fail in operation, it may not be worn correctly or at all, for example) and as with other assessments, judgement of types of hazard and degree of risk will have to be made to ensure correct equipment selection. The assessment must include:

1. An assessment of any risks which have not been avoided by any other means
2. The definition of the characteristics which PPE must have in order to be effective against the risks, including any risks which the PPE itself may create
3. A comparison of the characteristics of the PPE available with those referred to in (2.)

For PPE used in high-risk situations or for complicated pieces of PPE (such as some diving equipment) the assessment should be in writing and be kept available. Review of it will be required if it is suspected to be no longer valid, or if the work has changed significantly.

Regulation 7 requires all PPE to be maintained, replaced or cleaned as appropriate. Appropriate 'accommodation' must be provided by the employer or self-employed when the PPE is not being used (**Regulation 8**). 'Accommodation' includes pegs for helmets and clothing, carrying cases for safety spectacles and containers for PPE carried in vehicles. In all cases, storage must protect from contamination, loss or damage (particularly from harmful substances, damp or sunlight).

Regulation 9 requires an employer who provides PPE to give adequate and appropriate information, instruction and training to enable those required to use it — to know the risks the PPE will avoid or limit, and the purpose, manner of use and action required by the employee to ensure the PPE remains in a fit state, working order, good repair and hygienic condition. Information and instruction provided must be comprehensible.

By **Regulation 10**, every employer providing PPE must take all reasonable steps to ensure it is properly used. Further, Regulation 10 requires every employee provided with PPE to use it in accordance with the training and instruction given by the employer. The self-employed must make full and proper use of any PPE. All reasonable steps must be taken to ensure the equipment is returned to the accommodation provided for it after use.

Employees provided with PPE are required under **Regulation 11** to report any loss or defect of that equipment to their employer forthwith. It should be noted that section 9 of the Health and Safety at Work etc. Act 1974 requires that no charges can be made to workers for the provision of PPE which is used only at work. In practice, employers can and do make supplementary charges for providing PPE which is more attractive than the basic item, where these are requested by employees.

THE ELECTRICITY AT WORK REGULATIONS 1989

The dangers of the use of electricity are, or should be, well known. It is a curious yet disturbing fact that the majority of injuries due to electricity occur to electricians, who would appear to be those most aware of the dangers. Electrical injuries form a relatively small proportion of all construction lost-time injuries, but those suffering them have an enhanced risk of not surviving the experience as compared with the risk of death from exposure to other construction hazards. According to the Chinese authorities, deaths from electrical contact account for as many as 20% of the total, so there is some limited evidence that control measures in the UK are relatively effective.

Legislative control of electrical matters in the past has been concerned with not just the fundamental principles of electrical safety but also specific and detailed requirements relating to particular plant and activities. For many years, the Electricity (Factories Act) Special Regulations 1908 and 1944 controlled the use of electricity at work. These Regulations were enforced under the Factories Act 1961, and continued under the Health and Safety at Work etc. Act 1974. Some of the requirements of those Regulations had become outdated as technology advanced and working and engineering practices changed to accommodate it.

All the provisions of the Provision and Use of Work Equipment Regulations 1998 are relevant to the present Regulations, which came into force on 1 April 1990. They removed the above drawbacks and provide a coherent and practical code that applies to construction work as well as to all other work areas and to all workers.

Summary of the Regulations

The Regulations generally consist of requirements which have regard to principles of use and practice, rather than identifying particular circumstances and conditions. Action is required to prevent danger and injury from electricity in all its forms. Employers, self-employed people and employees all have duties of compliance with the Regulations so far as they relate to matters within their control; these are all known as dutyholders. Additionally, employees are required to co-operate with their employer so far is necessary for the employer to comply with the Regulations (**Regulation 3**).

All electrical systems must be constructed and maintained at all times to prevent danger, so far as is reasonably practicable (**Regulation 4(1) and 4(2)**). An 'electrical system' is a system in which all electrical equipment is, or may be, electrically connected to a common source of electrical energy, and includes the source and the equipment. 'Electrical equipment' includes anything used, intended to be used or installed for use to generate, provide, transmit,

transform, rectify/convert, conduct, distribute, control, store, measure or use electrical energy — a very comprehensive definition. 'Danger' means the risk of injury — where 'injury' in this context means death or personal injury from electric shock, burn, explosion or arcing, or from a fire or explosion initiated by electrical energy, where any such death or injury is associated with electrical equipment.

Every work activity (including operation, use and maintenance of and work near electrical systems) shall be carried out so as not to give rise to danger, so far as is reasonably practicable (**Regulation 4(3)** — see also the further requirements of Regulations 12, 13, 14 and 16).

Equipment provided for the purpose of protecting persons at work near electrical equipment must be suitable, properly maintained and used (**Regulation 4(4)**). The term 'protective equipment' as used here has a wide application, but typically it includes special tools, protective clothing or insulating screening equipment, for example, which may be necessary to work safely on live electrical equipment.

No electrical equipment shall be put into use where its strength and capability may be exceeded, giving rise to danger (**Regulation 5**). This requires that before equipment is energised the characteristics of the system to which it is connected must be taken into account, including those characteristics under normal, transient and fault conditions. The effects to be considered include voltage stress and the heating and electromagnetic effects of current.

Electrical equipment must be protected and constructed against adverse or hazardous environments (such as: mechanical damage; weather; temperature or pressure; natural hazards; wet, dirty or corrosive conditions; flammable or explosive atmospheres) (**Regulation 6**).

All conductors in a system which could give rise to danger must be insulated, protected or placed so as not to cause danger (**Regulation 7**). A 'conductor' means a conductor of electrical energy. The danger to be protected against generally arises from differences in electrical potential (voltage) between circuit conductors and others in the system, such as conductors at earth potential. The conventional approach is to insulate them, or place them so that people cannot receive electric shocks or burns.

Precautions shall be taken, by earthing or by other suitable means, to prevent danger from a conductor (other than a circuit conductor) which may become charged, either as a result of the use of the system or of a fault in the system (**Regulation 8**). A 'circuit conductor' means any conductor in a system which is intended to carry current in normal conditions, or to be energised in normal conditions. This definition does not include a conductor provided solely to perform a protective function as, for example, an earth connection. The scope of Regulation 8 is therefore such that it includes a substantial number of types of conductor as requiring to be earthed, including combined neutral and earth conductors, and others which may become charged under fault conditions such as metal conduit and trunking, metal water pipes and building structures.

Restrictions are made on the placing of anything which might give rise to danger (fuses, for example) in any circuit conductor connected to earth (**Regulation 9**). Connections used in the joining of electrical systems must be both mechanically and electrically suitable (**Regulation 10**). This means that all connections in circuits and protective conductors, including connections to terminals, plugs and sockets and any other means of joining or connecting conductors should be suitable for the purposes for which they are used. Applying equally to both temporary and permanent connections, this Regulation covers supplies to construction site conditions.

Systems must be protected from any dangers arising from excess current (**Regulation 11**). The means of protection is likely to take the form of fuses or circuit breakers controlled by relays. Other means are also capable of achieving compliance. There must be suitable means provided for cutting off energy supply to and the isolation of electrical equipment. These means must not be a source of electrical energy themselves (**Regulation 12(1) and (2)**). In circumstances where switching off and isolation is impracticable, as in large capacitors, for example, precautions must be taken to prevent danger so far as is reasonably practicable (**Regulation 12(3)**). Switching off can be achieved by direct manual operation or by indirect operation by stop buttons. 'Effective isolation' includes ensuring that the supply remains switched off and that inadvertent reconnection is prevented. This essential difference between 'switching off' and 'isolation' is crucial to understanding and complying with **Regulation 13**. This requires that adequate precautions are taken to prevent electrical equipment made dead to be worked on from becoming electrically charged, for example, by adequate isolation of the equipment.

No person shall be engaged in any work near a live conductor (unless insulated so as to prevent danger) unless: (a) it is unreasonable in all circumstances for it to be dead; (b) it is reasonable in all circumstances for persons to be at work on or near it while it is live; and (c) suitable precautions are taken to prevent injury (**Regulation 14**).

Adequate working space, means of access and lighting must be provided at all electrical equipment on which or near which work is being done which could give rise to danger (**Regulation 15**). Restrictions are placed on who can work where technical knowledge or experience is necessary to prevent danger or injury (**Regulation 16**). They must possess the necessary knowledge or experience or be under appropriate supervision having regard to the nature of the work.

Regulation 29 provides a means of defence in legal proceedings, where a person charged can show that all reasonable steps were taken and all due diligence was exercised to avoid committing the offence.

THE REPORTING OF INJURIES, DISEASES AND DANGEROUS OCCURRENCES REGULATIONS 1995

As discussed in Part 1 of this book, reliable information about types of accidents, incidents and the ways in which they have happened can be a very useful tool to work with in the prevention of future events of a similar kind. The information gained can be used to indicate how and where problems occur, and demonstrates trends over time. It is important to distinguish between accidents, incidents and injuries — they are not the same. Injury can occur as a result of an incident; the injury and the incident together amount to an accident (the common term). The reader is referred to Chapter 1 for a discussion of this point.

In 1985, when the first version of these Regulations replaced the Notification of Accidents and Dangerous Occurrences Regulations (NADO), the law properly recognised in the title of the Regulations that it requires the collection of specified information about incidents which result in specified types of injury; in some cases it also requires information about specified incidents with the potential to cause serious physical injury, whether or not they produced such injury. The enforcing authorities are interested in assembling such information because it gives them knowledge of trends and performance (failure) statistics. It also highlights areas for research, enforcement or future legislation. Following consultation on the functioning of its predecessor, these Regulations were implemented on 1 April 1996.

The main purpose of the Regulations is to provide enforcing authorities with information on specific injuries, diseases and dangerous occurrences arising from work activities covered by the Health and Safety at Work etc. Act. The authorities are able to investigate only a proportion of the total, so the Regulations aim to bring the most serious injuries to their attention quickly.

Summary of the Regulations

The Regulations cover employees, self-employed people and those who receive training for employment (as defined by the Health and Safety at Work etc. Act), and also members of the public, pupils and students, temporary workers and other people who die or suffer injuries or conditions specified, as a result of work activity.

Where anyone dies or suffers any of the injuries or conditions defined in Schedule 1, or where there is a 'dangerous occurrence' as defined in Schedule 2, as a result of work activities, the 'responsible person' must notify the relevant enforcing authority. This must be done by the quickest practicable means (usually by telephone) and a report must be sent to them within 10 days (**Regulation 3(1)**). If the personal injury results in an absence of more than 3 calendar days, but does not fall into the categories specified as 'major', the written report alone is required (**Regulation 3(2)**). The day of the accident is not counted when calculating absence, but any days which would not have been working days are counted.

Where the injured person is a member of the public, and where the person is either killed or the injury is sufficient to warrant the injured person being taken directly from the scene to hospital, the injury is reportable by the occupier of the premises to the relevant enforcing authority (see below).

Regulation 2 contains interpretation of all significant terms used within the Regulations. These include:

Accident — the term is still relevant as a generic descriptor of the event causing the injury. The definition includes 'a non-consensual act of physical violence done to a person at work', and 'an act of suicide which occurs on or in the course of operation of a relevant transport system'.

Enforcing authority — the body responsible for the enforcement of health and safety legislation relating to the premises where the injury or disease occurred. Usually, this will be the HSE or the local authority's environmental health department. In case of doubt, reports should be made to the HSE, which will forward the information to the correct enforcing authority if this is not the HSE.

Major injury — the following injuries are classified as major:

- Any fracture, except to fingers, thumbs or toes
- Any amputation
- Dislocation of shoulder, hip, knee or spine
- Loss of sight (temporary or permanent)
- Eye injury from chemical or hot metal burn, and any penetrating eye injury
- Injury from electric shock or electric burn leading to unconsciousness, or requiring resuscitation or admittance to hospital for more than 24 hours
- Loss of consciousness caused by asphyxia or by exposure to a harmful substance or biological agent
- Acute illness requiring medical treatment, or loss of consciousness arising from absorption of any substance by inhalation, ingestion or through the skin
- Acute illness requiring medical treatment where there is reason to believe it has resulted from exposure to a biological agent or its toxins, or infected materials

'Responsible person' — this may be the employer of the person injured, a self-employed person, someone in control of premises where work is being carried out or someone who provides training for employment. The 'responsible person' for reporting any particular injury or dangerous occurrence is determined by the circumstances, and the employment or other relationship of the person who is killed or suffers the injury or condition.

Where death results within one year of a notifiable work accident or condition, the person's employer must notify the relevant enforcing authority in writing (**Regulation 4**). There is no prescribed form for this purpose.

When reporting injuries and dangerous occurrences, the approved form must be used (F2508). The reporting of diseases (on form F2508A) which are specified in Schedule 3 is required only when the employer receives a written statement or other confirmation from a registered medical practitioner that the affected person is not only suffering from a listed disease but also that it has arisen in the manner also specified in the Schedule (**Regulation 5**).

Incidents involving death or major injury arising from the supply of flammable gas must be notified to the HSE forthwith and a written report must be sent on the approved form (F2508G) within 14 days (**Regulation 6**). Gas fittings that are used by consumers and defined by the Gas Safety (Installation and Use) Regulations 1994 and which are found on examination to be dangerous, must also be reported.

Records must be kept by employers and others of those injuries, diseases and dangerous occurrences which require reporting (**Regulation 7**). Records can be kept in the form of entries made in the accident book (Form BI 510) with reportable injuries and occurrences clearly highlighted, by keeping photocopies of reports sent to the enforcing authorities, or on computer provided that they can be retrieved and printed out. Keeping of computer records of this type will require registration under the Data Protection Act if individuals can be identified.

Records should be kept at the place of work or business, for at least three years from the date they were made. The enforcing authorities may request copies of such records which must then be provided (**Regulation 7(4)**). Further information may be requested by the enforcing authorities about any injury, disease or dangerous occurrence — in some situations analysis of the initial reports might highlight patterns or categories indicating a need for further legislation or action, and a deeper study requiring more detailed information may then be carried out by the HSE with the approval of the HSC.

Regulation 11 contains what amounts to a 'due diligence' defence provision, whereby a person can escape conviction for an offence under the Regulations if he can prove that he was not aware of the event which should have been reported and that he had taken all reasonable steps to have such events brought to his notice. (The presence of a suitable reporting procedure detailed as an arrangement within an organisation's health and safety policy would, of course, constitute such steps.)

The Regulations do not extend to cover: hospital patients who die or are injured while undergoing treatment in hospital, dental or medical surgeries; members of the armed services killed or injured while on duty; people killed or injured on the road (except where the injury or condition results out of exposure to a substance conveyed by road, unloading or loading vehicles, or by maintenance and construction activities on public roads) or during train travel (**Regulation 10**). They do, however, cover dangerous occurrences on public and private roads

Eight Schedules are attached to the Regulations. An extended and extensive list of dangerous occurrences can be found in Schedule 2, which also contains special parts relating specifically to mines and quarries, offshore workplaces and relevant transport systems. Schedule 3 contains the list of occupational diseases and their corresponding work activities which make them reportable.

Future refinements of RIDDOR may result following current consultation on the need for Regulations specifically requiring employers to investigate the circumstances of all accidents, and following the outcome of trials held in Scotland on the reporting of injury details by telephone.

18 Penalties

The ultimate penalty to the individual is personal injury and the consequences of accidents to a business in terms of disruption and loss of reputation extend far beyond the ability of a court to impose financial and other penalties on those who break the law. The use of Prohibition and Improvement Notices also contributes practical penalties in terms of delays to the work and involvement of senior management. Mostly, though, 'penalties' are seen as the outcome of court proceedings brought by the enforcement authorities against employers and employees.

Breaches of health and safety law are criminal offences, dealt with initially in Magistrates' Courts, and in most cases proceeding no further. Local magistrates adjudicate on about 97% of all criminal prosecutions, but can decline to deal with a case if they decide their powers of punishment are not sufficient. In that event, the case is heard in the Crown Court before a judge and jury. Magistrates' sentencing powers are limited at present to a fine of up to £20 000 per offence against Sections 2—9 of the Health and Safety at Work etc. Act 1974, and up to £5 000 for offences against other health and safety laws. At the time of writing, revised sentencing limits for magistrates are being proposed, and the maximum fine is likely to rise to £20 000 for all offences. In the Crown Court there is no maximum financial penalty, and imprisonment for up to two years is possible for a limited range of offences including disobeying a Prohibition Notice and contravening a licensing requirement (such as the unlicensed removal of asbestos).

Recently, general **sentencing policy** of the courts has become more consistent. Fines against construction companies following major incidents can easily run into six figures, especially when the costs of an investigation are added to a fine. Balfour Beatty Civil Engineering Ltd was fined £1 200 000 plus costs of £200 000 in 1999 following the collapse of tunnels beneath London Heathrow Airport.

Enforcement action is most often associated with physical injury or loss, as public interest demands punitive action. However, the breach lies in failure to comply with the law and not in the result of that failure. If an incident does occur, however, establishing a defence to the general requirements of (for example) the Health and Safety at Work etc. Act 1974 can be difficult when the results are plain for all to see and the standard required is one of reasonable practicability.

Since November 1998, guidelines based on the case of *R v Howe & Son (Engineers) Ltd*. (Court of Appeal case 97/10/101/Y3) have been used by courts deciding on financial penalty and costs. This company had appealed against a fine and costs imposed by Bristol Crown Court, on the grounds that the fine alone was 50% more than the company's profit at the time of the accident which was the subject of the prosecution.

The Court of Appeal reduced the fine but upheld the Crown Court's earlier view on costs. In giving judgement, it was said that the required standard of care does not depend on the size of the employer's business or on its financial position. The Court of Appeal also commented that the low level of risk involved in the task which gave rise to a fatal electrocution was not a mitigating factor. It accepted that the company's means and previous good record could be taken into account. The degree of risk, the extent and duration of the breach of the law, and the fact that active steps were taken to improve safety following the incident were also taken into account.

The key questions in sentencing are:

- How far short of the appropriate standard did the defendant fall?
- What actually happened?
- Was there a deliberate breach of the law (perhaps with a view to profit)?
- What attention had been paid to any previous warnings?

If reliance is to be placed on mitigation based on the limited means of a person or business, this will have to be mentioned at the earliest opportunity and backed with written proof.

Courts are now required to give a reduction of up to a third in a financial penalty where a 'guilty' plea has been entered at the earliest opportunity. It is necessary to be aware of this as a factor when deciding whether or not to plead guilty!

Moves towards the introduction of the new offence of **corporate killing** are considered likely to succeed in the future, following a major review by the Law Commission. At present, where directors and managers knew, or ought to have known, that a serious and obvious risk of death existed, a charge of manslaughter can be brought. The maximum penalty for this is life imprisonment. The present law requires that 'reckless' boardroom behaviour must be proved, and be attributable to an individual.

The current governing principle in English law on the criminal liability of companies is that those who control or manage the affairs of the company are seen as embodying the company itself. Before a company can be convicted of manslaughter, an individual who can be 'identified as the embodiment of the company itself' must first be shown to have been guilty of manslaughter. Only if that person is found guilty can the company be convicted. Where there is insufficient evidence, any prosecution of the company must fail. Introduction of the corporate killing offence is aimed to change that position.

A suspended one-year sentence was received by a manager in 1989 for a breach of the Health and Safety at Work etc. Act 1974. The company concerned, David Holt Plastics Ltd., and two of its directors, were fined a total of £48 000 after an incident, having ignored a Prohibition notice.

The **Company Directors Disqualification Act 1986** can be applied to health and safety breaches, as a Disqualification Order can be made against anyone convicted of 'an indictable offence connected with ... the running of a company'.

Index

Learning Resources Centre